Electromagnetic Cascade and Chemistry of Exotic Atoms

ETTORE MAJORANA
INTERNATIONAL SCIENCE SERIES
Series Editor:
Antonino Zichichi
European Physical Society
Geneva, Switzerland

(PHYSICAL SCIENCES)

Recent volumes in the series:

Volume 44 **HEAVY FLAVOURS AND HIGH-ENERGY COLLISIONS IN THE 1–100 TeV RANGE**
Edited by A. Ali and L. Cifarelli

Volume 45 **FRACTALS' PHYSICAL ORIGIN AND PROPERTIES**
Edited by Luciano Pietronero

Volume 46 **DISORDERED SOLIDS: Structures and Processes**
Edited by Baldassare Di Bartolo

Volume 47 **ANTIPROTON–NUCLEON AND ANTIPROTON–NUCLEUS INTERACTIONS**
Edited by F. Bradamante, J.-M. Richard, and R. Klapisch

Volume 48 **SAFETY, ENVIRONMENTAL IMPACT, AND ECONOMIC PROSPECTS OF NUCLEAR FUSION**
Edited by Bruno Brunelli and Heinz Knoepfel

Volume 49 **NONLINEAR OPTICS AND OPTICAL COMPUTING**
Edited by S. Martellucci and A. N. Chester

Volume 50 **HIGGS PARTICLE(S): Physics Issues and Experimental Searches in High-Energy Collisions**
Edited by A. Ali

Volume 51 **BIOELECTROCHEMISTRY III: Charge Separation Across Biomembranes**
Edited by G. Milazzo and M. Blank

Volume 52 **ELECTROMAGNETIC CASCADE AND CHEMISTRY OF EXOTIC ATOMS**
Edited by Leopold M. Simons, Dezsö Horváth, and Gabriele Torelli

Volume 53 **NEW TECHNIQUES FOR FUTURE ACCELERATORS III: High-Intensity Storage Rings—Status and Prospects for Superconducting Magnets**
Edited by Gabriele Torelli

Electromagnetic Cascade and Chemistry of Exotic Atoms

Edited by

Leopold M. Simons

Paul Scherrer Institute
Villigen, Switzerland

Dezsö Horváth

Central Research Institute for Physics
Budapest, Hungary

and

Gabriele Torelli

National Institute for Nuclear Physics
Pisa, Italy

Springer Science+Business Media, LLC

Library of Congress Cataloging-in-Publication Data

International School of Physics of Exotic Atoms (5th : 1989 : Erice, Italy)
Electromagnetic cascade and chemistry of exotic atoms / edited by Leopold M. Simons, Dezső Horváth, and Gabriele Torelli.
p. cm. -- (Ettore Majorana international science series. Physical sciences ; v. 52)
"Proceedings of the fifth course of the International School of Physics of Exotic Atoms ... held May 14-20, 1989, in Erice, Italy"--T.p. verso.
Includes bibliographical references and index.
ISBN 0-306-43686-8
1. Particles (Nuclear physics)--Multiplicity--Congresses. 2. Cascade shower--Congresses. 3. Electromagnetic interactions--Congresses. 4. Exotic atoms--Congresses. 5. Muons--Congresses. 6. Nuclear chemistry--Congresses. I. Simons, Leopold M. II. Horváth, Dezső. III. Torelli, Gabriele. IV. Title. V. Series.
QC794.6.M85I58 1989
539.7--dc20 90-48317
CIP

ISBN 978-1-4899-3703-2 ISBN 978-1-4899-3701-8 (eBook)
DOI 10.1007/978-1-4899-3701-8
Softcover reprint of the hardcover 1st edition 1990

Proceedings of the Fifth Course of the International School of Physics of Exotic Atoms on Electromagnetic Cascade and Chemistry of Exotic Atoms, held May 14–20, 1989, in Erice, Italy

Originally published by Plenum Press, New York in 1990.

PREFACE

This Workshop was organized to bring once more together the scientists of the rather heterogeneous field of exotic atoms. At present the main topic of the field seems to be the study of the atomic cascade. There are some who study it intentionally - let us call them *cascadeurs* - and others who think they investigate other features of the exotic atoms (like Coulomb capture, particle transfer, muon catalyzed fusion, chemical effects, fundamental properties, etc.) - *users* - while in fact they study some special consequences of the same atomic cascade.

We decided to get *cascadeurs* and *users* discuss the problems of exotic atoms at wonderful Erice, at the **5th Course of the International School of Physics of Exotic Atoms**. Our Workshop was quite successful, we have heard excellent talks from participants from a dozen countries and most of them have prepared written contributions for this volume.

The Organizers express their gratitude to all participants for their contributions, especially to David Measday for his concluding remarks (not printed here) and to James Cohen for jumping in for Leonid Ponomarev who had to leave unexpectedly in the middle of the meeting. We greatly appreciate the enthusiastic help of Marianne Signer in every stage of the organization work. And, of course, the Workshop could not happen at all without the incredibly efficient organization by the **Ettore Majorana Centre of Scientific Culture.**

Leopold M. Simons
Dezsö Horváth
Gabriele Torelli

CONTENTS

OPENING ADDRESS xi

I. DECELERATION AND COULOMB CAPTURE, MESIC CHEMISTRY

Formation of Exotic Hydrogen Atoms 1
J.S. Cohen

Slowing-Down and Atomic Capture of Exotic Particles in Elements with Z>2 23
F.J. Hartmann

Recent Results of Experiments with Slow Muons 41
W. Schott, H. Daniel, F.J. Hartmann, and W. Neumann

Recent Progress in Antiproton - Atom Collisions 47
L.H. Andersen, K. Elsener, P. Hvelplund, H. Knudsen, S.P. Moller, E. Morenzoni, J.O.P. Pedersen, and E. Uggerhoj (Presented by K. Elsener)

The Formation and Reactions of Pionic Hydrogen Atoms 53
D.F. Measday

Collision Processes in the Formation of Light Exotic Atoms 67
D. Horváth

II. CASCADE STUDIES IN EXOTIC HYDROGEN ATOMS

Cascade Processes in Exotic Atoms with Z=1 73
V. Markushin

Protonium: The Mainz Cascade Model 97
G. Reifenröther and E. Klempt

Determination of K X-Ray Yields in Pionic Hydrogen as a Function of Gas Pressure 105
A.J. Rusi El Hassani, W. Beer, M. Bogdan, J.F. Gilot, P.A.F. Goudsmit, H.J. Leisi, St. Thomann, W. Volken, D. Bovet, E. Bovet, J.P. Egger, G. Fiorucci, J.L. Vuilleumier, K. Gabathuler, and L.M. Simons

Measurement of the Kinetic Energy Distribution of Pionic Hydrogen Atoms in Liquid Hydrogen 111
J.F. Crawford, M. Daum, R. Frosch, B. Jost, P.-R. Kettle, R.M. Marshall, B.K. Wright, and K.O.H. Ziock (Presented by R. Frosch)

Results of X-Ray Measurements in Antiprotonic Hydrogen (PS175) 117
R. Bacher, R. Badertscher, P. Blüm, J. Eades, J.P. Egger, K. Elsener, D. Gotta, K. Heitlinger, and L.M. Simons (Presented by K. Heitlinger)

III. EXOTIC ATOMS AT HIGHER Z

Exotic-Atom Cascade Processes in $Z>2$ 127
F.J. Hartmann

Interplay of Exotic Atoms and Hypernuclei - Cascade Trapping and Deeply-Bound Exotic Atoms 141
R.S. Hayano

An Experimental Study of the $(\mu^- He)^+$ Lifetime in the Metastable 2S-State Using NaI-Crystals 155
H. Orth

Lifetime of Metastable Muonic Helium 163
L. Bracci and E. Zavattini

A Study of the Strong Interaction Effects of Deeply Bound Pionic Levels in Pionic Atoms 167
J. Konijn, C.T.A.M. de Laat, A. Taal, P. David, H. Hänscheid, F. Risse, Ch.F.G. Rösel, W. Schrieder, W. Lourens, C. Petitjean, and A. van der Schaaf

An Improved Parametrization of the Optical Potential for Pionic Atoms 179
J. Konijn, C.T.A.M. de Laat, A. Taal, and J.H. Koch

Adiabatic Invariant of Radiative Transitions in the Coulomb Field 197
S.P. Alliluev and S.S. Gershtein (Presented by S.S. Gershtein)

IV. MUON TRANSFER

Muon Capture and Transfer in Gas Mixtures with Hydrogen 205
H. Schneuwly

Muon Transfer in $H_2 + SO_2$ Gas Mixtures 217
F. Mulhauser, R. Jacot-Guillarmod, C. Piller, L.A. Schaller, L. Schellenberg, and H. Schneuwly

Muon Transfer to Low Z Elements 223
R. Jacot-Guillarmod, F. Mulhauser, C. Piller, L.A. Schaller, L. Schellenberg, and H. Schneuwly

V. MUON CATALYZED FUSION

Muon-Catalyzed Fusion - A Selection of Old and New Experiments and Open Problems 231
C. Petitjean

Preliminary Results on Muon Catalyzed pt Fusion 245
P. Baumann, H. Daniel, T. von Egidy, S. Grunewald, F.J. Hartmann, R. Lipowsky, E. Moser, W. Schott, P. Ackerbauer, W.H. Breunlich, M. Fuchs, M. Jeitler, P. Kammel, J. Marton, N. Nägele, J. Werner, J. Zmeskal, C. Petitjean, K. Lou, H. Bossy, K. Crowe, and R.H. Sherman

Collisions of Muonic Atoms 253
L. Bracci, C. Chiccoli, G. Fiorentini, V.S. Melezhik, P. Pasini, and J. Wozniak

More on the Cold Fusion Family 269
J.D. Davies and J.S. Cohen

VI. BEAMS AND METHODS

The ISIS Facility at RAL - The Pulsed Muon Beam 279
J.D. Davies and G.H. Eaton

Formation of Atomic Pionium 289
H. Orth

High Resolution Spectroscopy of X-Rays from Antiprotonic Atoms Using a Low-Energy Crystal Spectrometer 295
G.L. Borchert, D. Gotta, O.W.B. Schult, L.M. Simons, K. Elsener, K. Rashid, J.J. Reidy, R.D. Deslattes, E.G. Kessler, and T. Mooney (Presented by D. Gotta)

CONCLUSION (instead of) 301

PARTICIPANTS 303

INDEX 307

OPENING ADDRESS

Ladies and Gentlemen!

I am very happy to open the Fifth Course of the International School of Physics of Exotic Atoms. This School was founded in the year of 1977 when Nino Zichichi convinced me that even this rather narrow field was worth covering by a special school. And indeed, the idea was accepted by the community with much enthusiasm as it successfully brought together most of the researchers working in this field.

The friendly atmosphere of the first course will certainly be remembered with pleasure by all the participants. The second course only two years later reflected the success of the quark model in that time. Hence a major topic of that exotic atoms course was the discussion of even more exotic matter as quark atoms and molecules. The renaissance of muon catalyzed fusion has affected that course quite remarkably, too.

With the commissioning of LEAR in 1982 the third course was naturally devoted to antiproton physics and helped to stimulate and develop ideas which are still being pursued. The main objective of the fourth course in 1985 had been the tests of conservation laws as, e.g., for the case μ - e conversion where exotic atoms are used to reach the initial stage of the process. At the time of the course the joint experimental and theoretical efforts in the study of muon catalyzed fusion led to such a tremendeous surge in activity that did not allow to include it in a common course any longer.

Now, in 1989 we are back to the original theme of the School. We are motivated by the fact that many of the experiments and theoretical considerations have reached a precision where effects of the formation and deexcitation mechanisms of exotic atoms can no longer be neglected but need to be studied in detail again. Thus it is time again to interchange experimental and theoretical results and to discuss different ideas. This way we should be able to assess the present and future possibilities of exotic atoms in order to exploit this beautiful field as efficiently as possible.

In this spirit I hope the Fifth Course will be a success, too.

Gabriele Torelli

I

DECELERATION AND COULOMB CAPTURE, MESIC CHEMISTRY

The White Rabbitt put on his spectacles.
"Where shall I begin, please your Majesty?", he asked.
"Begin at the beginning," the King said gravely,
"and go on till you come to the end: then stop."

Lewis Carroll: Alice in Wonderland

FORMATION OF EXOTIC HYDROGEN ATOMS

James S. Cohen

Theoretical Division
Los Alamos National Laboratory
Los Alamos, New Mexico 87545 USA

1. INTRODUCTION

Formation of exotic hydrogen atoms has been studied theoretically for 40 years.[1] However, most of the work prior to 1980 utilized perturbative or two-state approximations that are now known to be inadequate because the slow ionizing collisions, which lead to muon capture, entail strong coupling of many intermediate states. Ironically the very first study, by Wightman[2] in 1950, has the most in common with modern treatments. His method, known as adiabatic ionization (AI), followed from the observation of Fermi and Teller[3] that there exists a critical strength of the dipole, formed by the projectile negative muon and target proton in μ^-+H collisions, for binding the electron. In collisions where the μ^- adiabatically approaches closer than this distance, the electron escapes and, if the muon is left with negative energy, the $p\mu^-$ atom is formed. The most serious failures of this simple model stem from trajectory deflection and nonadiabatic lag in the motion of the electron. The deviation from a straight–line trajectory is easily taken into account. Nonadiabatic effects have been included by a sequence of methods of increasing sophistication: diabatic states (DS),[4] classical–trajectory Monte Carlo (CTMC),[5] time–dependent Hartree Fock (TDHF),[6] and classical–quantal coupling (CQC).[7] The DS and CTMC methods have also been used to calculate the slowing–down cross sections. Consistent calculation of the slowing–down and capture cross sections is crucial for correct determination of the capture distributions in a medium.[8]

Results of all the above methods are discussed in Sec. 2 for μ^-+H collisions. Other exotic hydrogen atoms, in particular $p\pi^-$ and $p\bar{p}$, are also of interest and can be treated in a similar fashion as discussed in Sec. 3. Results for $\bar{p}$+H collisions obtained by the AI, DS, and CTMC methods are compared with the corresponding μ^-+H calculations. All exotic–atom experiments done to date actually use H_2 molecular targets. The similarities and differences of H and H_2 in the capture process are discussed. A clever new way to

Electromagnetic Cascade and Chemistry of Exotic Atoms
Edited by L. M. Simons *et al.*, Plenum Press, New York, 1990

form $p\bar{p}$ in collisions of $\bar{p}$ with the atomic hydrogen negative ion H^- has been suggested.[9] Theoretical calculations are presented that show capture by the negative ion is significantly different from capture by the neutral atom.

2. THEORETICAL METHODS FOR MUON CAPTURE BY THE HYDROGEN ATOM

2.1. *Adiabatic Ionization (AI)*

Interaction of an electron with a fixed dipole has the peculiar property of having either an infinite number of bound states or none at all; the critical dipole strength for unit charges (e.g., p and μ^-) occurs at a distance R_c= 0.639 a_0.[3] The adiabatic potential energy is shown in Fig. 1. Assuming a straight–line trajectory, Wightman[2] estimated the ionization cross section to be

$$\sigma_{AI}^{SLT} = \pi R_c^2 = 0.408\ \pi a_0^2 \tag{1}$$

independent of the collision energy as shown in Fig. 2 (chain–dashed curve). This model assumes that the electron escapes adiabatically with zero kinetic energy asymptotically, so capture occurs only if the collision energy is less than the ionization energy I_a. The

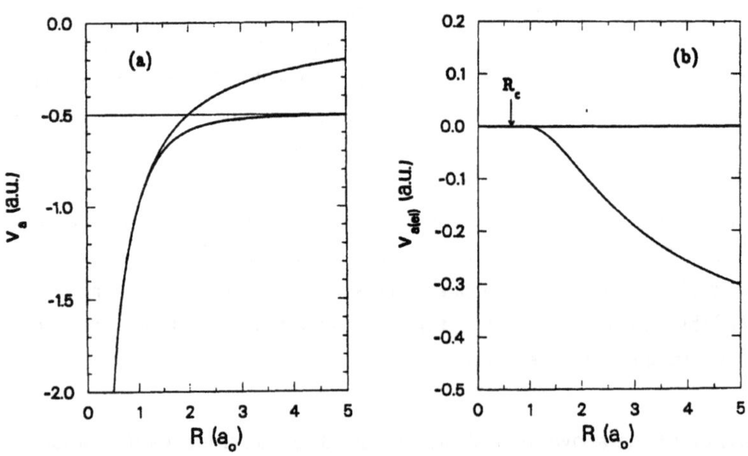

Fig. 1. Adiabatic electronic potential energy curves for μ^- + H and μ^- + H^+. In (a) the complete potential curve is shown, and in (b) the electronic energy only is shown –i.e., the -1/R Coulomb potential between the μ^- and p has been subtracted out. The potential curve is independent of the particle masses so also applies to other isotopes of hydrogen and other negative particles. The continuum is reached at R_c = 0.639 a_0.

corresponding wave function in Fig. 3 shows the electron first polarizing away from the approaching μ^-, then becoming more diffuse as the μ^- comes still closer, and finally passing into the continuum as the μ^- reaches the distance R_C.

Actually the deflection of the μ^- is quite significant at $E_{c.m.} \lesssim I_a$ and can be simply taken into account.[5] The adiabatic potential energy between μ^- and H is

$$V_a(R) = -\frac{1}{R} + V_{a(el)}(R) \; ; \tag{2}$$

the energy $V_{a(el)}$ of the electron for μ^- and p fixed at distance R has been exactly calculated.[10-11] The effective potential including the centrifugal repulsion can be written

$$V_{a(eff)}(R,b) = V_a(R) + \frac{b^2 E_{c.m.}}{R^2} \tag{3}$$

where $E_{c.m.}$ is the collision energy. If $V_{a(eff)}$ has no local maximum at $R > R_C$, then the

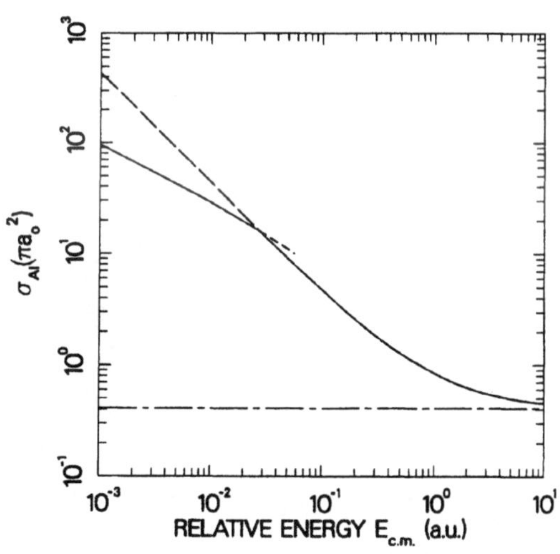

Fig. 2. Adiabatic ionization cross section with straight-line trajectory (chain-dashed curve) and curved trajectory (solid curve). In the latter case, the cross section is the lesser of Eq. (4) (long-dashed curve) and the cross section for orbiting collisions (short-dashed curve).

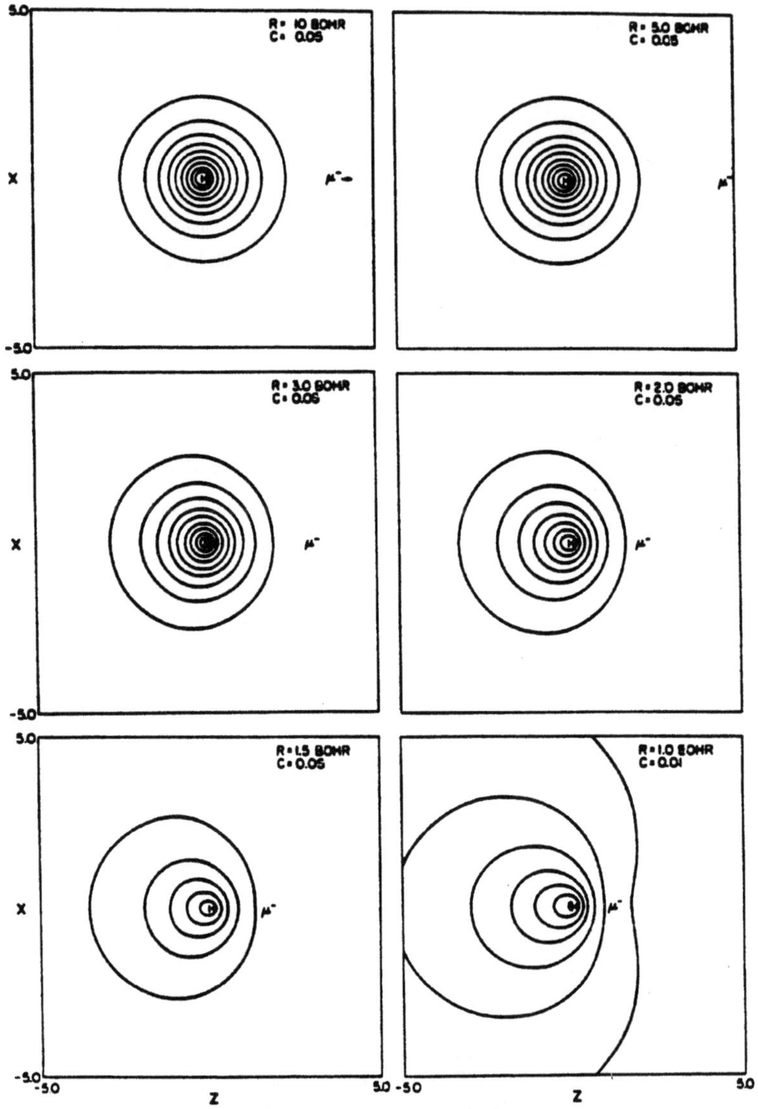

Fig. 3. Contours of the normalized adiabatic electronic wave function for various separations R between the μ^- and p. Contour values are C, 2C, 3C,... .

cross section at energy $E_{c.m.}$ is just

$$\sigma_{AI(c)}(E_{c.m.}) = \pi b_c^2 = \frac{\pi R_c^2}{E_{c.m.}}\left[E_{c.m.} + \frac{1}{R_c} - 0.5\right] \tag{4}$$

(long–dashed curve in Fig. 2) where b_c satisfies

$$V_{a(eff)}(R_c,b_c) = E_{c.m.} \tag{5}$$

For $E_{c.m.} < 0.06$ a.u.,* Eq. (3) does have a maximum at some $R_0 > R_c$ satisfying $V_{a(eff)}(R_0,b_0) = E_{c.m.}$ so there are three classical turning points. The value πb_0^2 is also shown in Fig. 2 (short–dashed curve). However, for $0.03 < E_{c.m.} < 0.06$ a.u. even the innermost classical turning point is outside R_c so it is only at $E_{c.m.} < 0.03$ a.u. that the "hump" in the effective potential actually limits the cross section. Tunneling through the barrier is not expected to be important. Note that Figs. 1 and 2 are independent of the mass of the negative particle so they apply to $\bar{p}$ as well as μ^-. Only at extremely low energies (~10^{-6} a.u. for μ^-) where the scattering is dominated by s–waves would adiabatic ionization depend on the mass other than through the energy in the c.m. system (at 10^{-3} a.u., 6 partial waves still contribute to μ^- capture). Of course, the adiabatic ionization model itself is more valid for the heavy antiproton and breaks down completely for electron collisions.

2.2. *Diabatic States (DS)*

A conspicuous feature of Fig. 1 is that the electronic energy is extremely close to the continuum at $R \approx 1\ a_0$ even though it doesn't actually reach the continuum until $R_c = 0.639\ a_0$. The diabatic–states treatment[4] takes into account the nonadiabatic behavior that allows the electron to be ionized at distances larger than R_c and also to carry off some kinetic energy. The nonadiabatic behavior is strengthened by the Coulomb attraction between the p and μ^- at distances $R \lesssim 1\ a_0$; here the relative velocity is high enough that the electron no longer has time to adjust adiabatically.

In the DS method the μ^-–H interaction is described in the Born–Oppenheimer framework (i.e., the μ^- is treated as a heavy particle), but diabatic rather than adiabatic electronic states are used. The diabatic potential curve $V_d(R)$, unlike the adiabatic potential curve $V_a(R)$, crosses into the electronic continuum (this is true even for cases like μ^- + He or μ^- + H_2 where adiabatic ionization cannot occur). In the extreme diabatic case, i.e. with the electronic orbital frozen as the 1s orbital of the H atom, this crossing

---

*In atomic units, $\hbar = e = m_e = 1$, so 1 a.u.(energy) = 27.21 eV, 1 a.u.(velocity) = 2.19×10^8 cm/s, 1 a.u.(time) = 2.42×10^{-17} s, and 1 a_0(distance) = 0.529 Å.

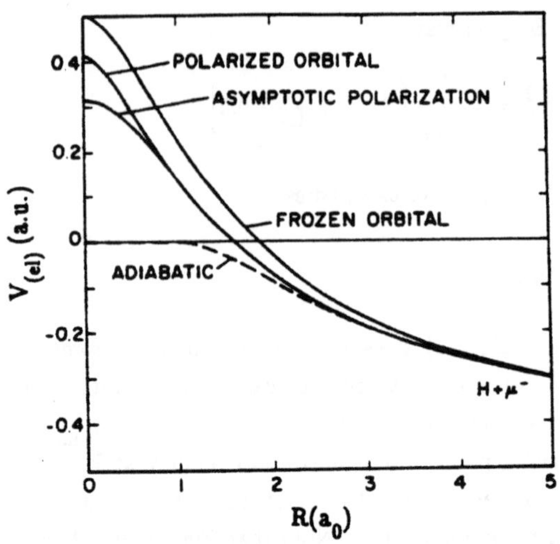

Fig. 4. Electronic potential energy curves for μ^- + H. The μ^--p Coulomb potential has been subtracted out. The dashed curve is the adiabatic energy and the solid curves are three different diabatic approximations. The DS cross sections in later figures are calculated with the polarized-orbital (po) potential.

occurs at a distance $R_x = 1.86\ a_0$ as shown in Fig. 4. This effect has been taken into account by introducing a polarizing orbital of p character. The energy is relatively insensitive to how this orbital is chosen; Fig. 4 shows the different results of optimizing the orbital (by minimizing the energy) at large distances and at the distance R_x. For definiteness the latter will be used henceforth. The effect of polarization moves the continuum crossing point in to $R_x' = 1.59\ a_0$. At $R > 2.5\ a_0$ the potential including polarization essentially coincides with the adiabatic potential.† At distances smaller than the crossing into the continuum, the diabatic state is embedded in the electronic continuum and hence has a finite autoionization width, which is calculated by "Fermi's golden rule". We avoid direct evaluation of continuum integrals by discretizing the continuum and utilizing Stieltjes moment theory; the width Γ is shown in Fig. 5.‡

†At the time Ref. 4 was published it was thought that the difference between the frozen and polarized orbital treatments was within the uncertainty of the model so cross sections were presented only with the frozen orbital. Now there seems to be empirical evidence that the method may be more accurate than expected. Where necessary, the two will be distinguished by DS–fo and DS–po.

‡This calculation was performed with the frozen–orbital wave function—the result with the polarized–orbital wave function is not expected to be too different. The nonzero value of Γ at $R > R_x$ reflects coupling to Rydberg states but since such coupling does not generally result in μ^- capture it is set to zero in the subsequent calculations.

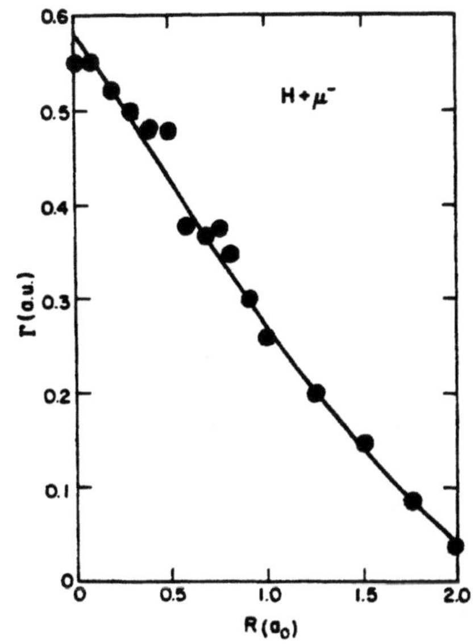

Fig. 5. Diabatic ionization width for μ^- + H. Calculations were done at the points shown.

The interaction is then formulated by the complex potential

$$W(R) = V_d(R) - i\Gamma(R)/2\,, \tag{6}$$

and the scattering in this potential is treated by the impact-parameter method with quasiclassical trajectories. The energy carried off by the electron is just the difference between the neutral and ionic potential curves, which is just $V_d(R) + 1/R$ for μ^-+H since the final state is devoid of electrons.†* For very slow collisions ($v \lesssim 0.1$ a.u.) most ionization occurs soon after crossing into the continuum so the electrons carry away little kinetic energy, but for faster collisions ($v \gtrsim 1.0$ a.u.) greater penetration is achieved and more energetic electrons are ejected (up to ~ 0.9 a.u.). Hence the two major objections to the AI model are eliminated in the DS model: (1) ionization can occur at a distance R_x' about 2.5 times larger than R_c so the cross section can be up to $(R_x'/R_c)^2 \approx 6.2$ times as big and (2) the electrons carry away kinetic energy so the stopping power is greater and the initial state of the resulting muonic hydrogen atom is determined.

†*In Ref. 4 it was speculated that the electron energy distribution might be *broadened* according to the width Γ by dynamic effects missing in the DS model. Subsequent CTMC calculations (see next section) indicate that such broadening is unimportant and this complication can be dispensed with.

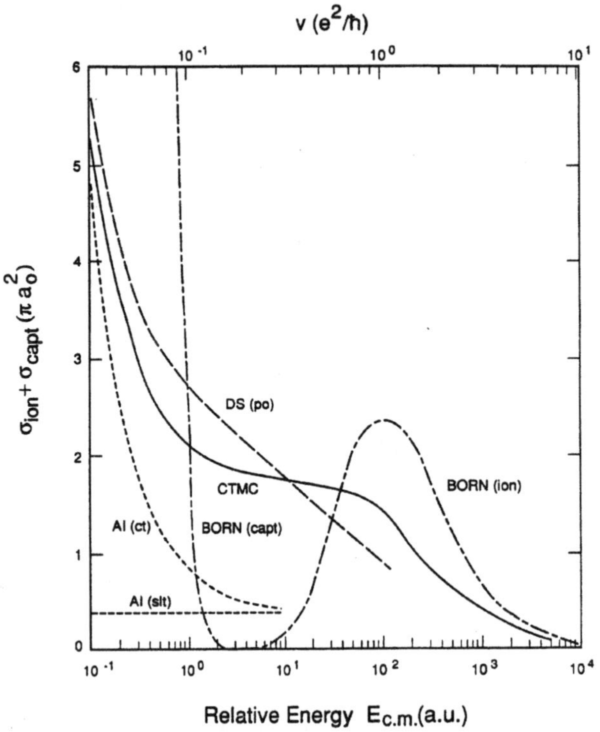

Fig. 6. Comparison of different total (ionization plus capture) cross sections for μ^- + H collisions: adiabatic ionization with straight-line trajectories and curved trajectories (short-dashed curves), diabatic states with polarized orbital (long-dashed curve), classical-trajectory Monte Carlo (solid curve), and Born approximation (chain-dashed curve—capture from Ref. 12a and ionization from Ref. 12b).

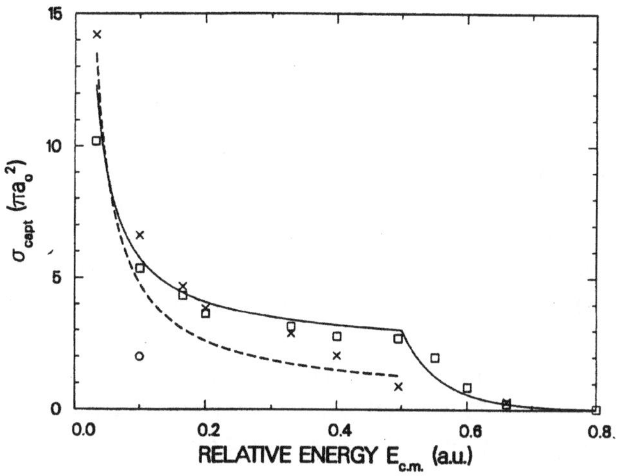

Fig. 7. Comparison of different capture cross sections for μ^- + H collisions: adiabatic ionization with curved trajectories (dashed curve), diabatic states with polarized orbital (solid curve), classical-trajectory Monte Carlo (□), time-dependent Hartree Fock (○), and classical-quantal coupling (×).

The total (ionization plus capture) cross section is shown in Fig. 6 and the capture cross section alone is shown in Fig. 7. In these two figures comparison can be made with the adiabatic ionization results as well as with the results of other methods discussed in subsequent sections. The DS cross section does not actually achieve the factor of $(R_x'/R_c)^2$ enhancement over the AI cross section but only a factor of 4.1 at $E_{c.m.} \approx 5$ a.u. At higher energies the DS cross section falls off due to the decreased interaction time and eventually behaves as $E_{c.m.}^{-\frac{1}{2}}$, whereas the AI cross section approaches a constant. However, even the DS cross section is not reliable at $v \gtrsim 1$ a.u. since it does not take into account the binary encounter mechanism of ionization shown in the next section to be dominant at high energies.

In the AI model, capture occurs only in collisions at $E_{c.m.} \lesssim 0.5$ a.u. In the DS model (including polarization), capture can occur at collision energies up to ~ 0.9 a.u., but the calculations show that the capture cross section decreases rapidly above 0.5 a.u. At such low energies, the width Γ is large enough to saturate the cross section; hence for $E_{c.m.} < 0.5$ a.u., the DS capture cross section could simply be calculated by Eq. (4) with R_c replaced by R_x'. The resulting formula yields a cross section larger than σ_{AI} at $E_{c.m.} > 0.05$ a.u. and slightly smaller at $E_{c.m.} < 0.05$ a.u. This crossing is visible in Fig. 7. However, σ_{DS} does not actually fall as far below σ_{AI} as indicated by Eq. (4) because at $E_{c.m.} < 0.016$ a.u. the DS and AI cross sections are limited by the same centrifugal barrier and hence become identical.

2.3. *Classical–Trajectory Monte Carlo (CTMC)*

The CTMC method[5] has a completely different character from the AI and DS methods but has served to verify the main conclusions of those models. It has also provided quantitatively accurate cross sections and has suggested a link to more rigorous quantum-mechanical methods. In the CTMC method no approximation is made except for a classical ensemble for the initial atomic coordinates and classical dynamics for the 3–particle motion. Numerous calculations have proved that the CTMC method is very good for treating collisions of stripped ions with the H atom. The method should be even better for μ^-+H collisions since the capture quantum numbers are large. Accurate quantum mechanics is very difficult at intermediate energies (i.e. for $v \sim 1$), especially when continuum states are involved. Usually different methods have to be used in different energy regimes; the CTMC method has the important advantages of being valid over a wide energy range and treating slowing down and capture consistently.

The CTMC procedure consists of three steps: (1) Monte Carlo selection of the initial conditions for H(1s) from the microcanonical distribution, $\delta(E_0+0.5$ a.u.), (2) integration of Hamilton's equations of motion, and (3) identification of the final state. These steps are repeated a sufficient number of times to get good statistics for the process "R" (e.g., capture in a particular state) of interest and the corresponding cross section is given by

$$\sigma_R = \frac{N_R}{N_{tot}} \pi b_{max}^2 \,. \tag{7}$$

The CTMC cross sections for μ^-+H are shown in Figs. 6 and 7. Note that the total (capture plus ionization) cross section is a smooth function of energy, in contrast to the Born approximation.[12] This characteristic was also true of the AI and DS methods but might be considered to be "built into" those models. At high energies the Born approximation is valid and is to be preferred at very high energies where the classical cross section falls off as $1/E_{c.m.}$ whereas the correct quantal behavior is $(\log E_{c.m.})/E_{c.m.}$.

Examination of the CTMC trajectories shows that the ionization mechanisms are quite different for high- and low-energy collisions. At high energies the mechanism is rather like that for high-energy collisions with positive particles (examination in greater detail exhibits the well–known Barkas effect[13]). The collision is sudden and can be viewed approximately as a binary encounter between the muon and a free electron which implies an escape energy limited by

$$E_e^{final} \lesssim 4 \frac{m_e}{m_\mu} E_\mu^{initial}$$

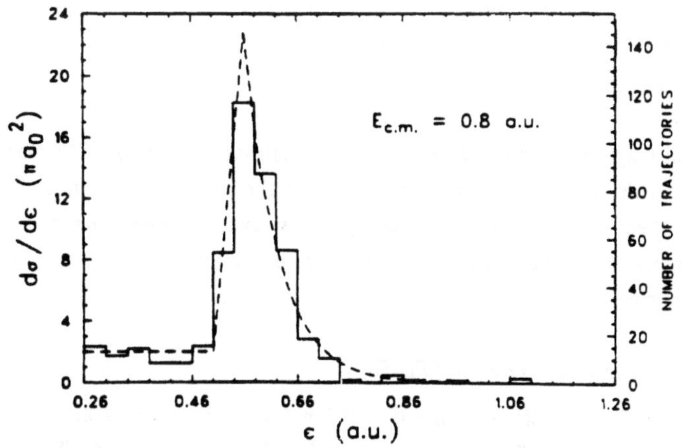

Fig. 8. Differential energy-loss cross section for μ^- + H collisions at relative energy 0.8 a.u. The dashed curve is a fit of the CTMC results (histogram).

(this limit is found to hold for $E_{c.m.} \gtrsim 1000$ a.u.). Note that this dynamic mechanism is unavailable in the electronic-structure-orientated AI and DS models.

For slow collisions (v << 1 a.u.), ionization occurs much as suggested by the DS method—the negative muon shields the electron from the atomic nucleus and the μ^- escapes with a velocity between 0 and its normal orbital velocity of 1 a.u. Hence capture,

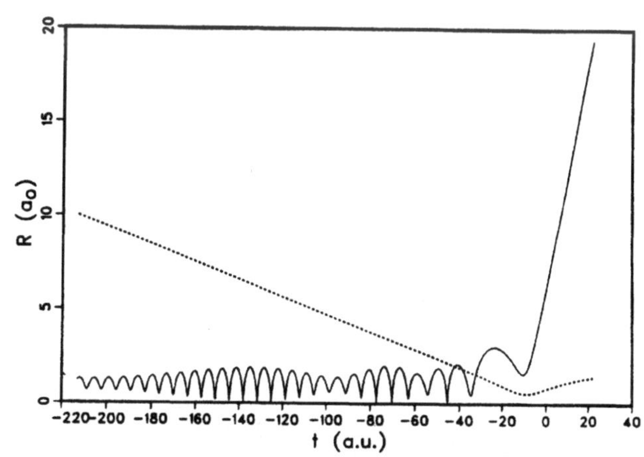

Fig. 9. Typical trajectory resulting in μ^- capture for μ^- + H collisions at relative energy 0.2 a.u.. The solid curve is the p-e^- distance and the dashed curve is the p-μ^- distance. The capture orbital has n = 14 and $\ell = 12$.

which requires the total energy loss, $I_a + E_e^{final}$, exceed the initial muon energy, only occurs in low-energy collisions. The energy of the ejected electron falls off approximately exponentially as shown in Fig. 8.

A typical trajectory resulting in capture is illustrated in Fig. 9. The electron orbital can be seen to exhibit a breathing-type motion as the μ^- approaches from a quite large distance and then escapes quickly as the μ^- intervenes between the e^- and p. This behavior demonstrates that angular-momentum exchange can occur at much larger distances than energy exchange.

2.4. *Time-Dependent Hartree Fock (TDHF)*

In the TDHF calculations for μ^-+H collisions[6] both the μ^- and e are treated quantum mechanically; the proton is fixed in space but reduced masses are used. This use of TDHF theory differs from the usual application in that one of the quantal particles, the μ^-, is initially free. This feature makes further use of one of the main advantages of the TDHF method for ionizing collisions, namely the ability to describe continuum orbitals. (The other advantage—the elimination of separate electron translational factors—is unimportant at the very low energies considered here.) Most time-independent methods employ L^2 basis sets for the electron orbitals and have great difficulty in treating continuum processes like ionizing collisions. This problem is exacerbated for μ^-+H since the wave function undergoes large distortion (so perturbative methods won't work) but is still nonadiabatic enough that a large number of excited and continuum electronic states are involved. This process presents no difficulty for the TDHF method which calculates the orbitals numerically.

Following the TDHF procedure the wave function is written in product form

$$\Psi(\vec{r}_\mu,\vec{r}_e;t) = \phi(\vec{r}_\mu,t)\Phi(\vec{r}_e,t)\,. \tag{8}$$

The single-particle orbitals ϕ and Φ are then solutions of the equations

$$\begin{aligned} i\hbar\frac{\partial\phi(\vec{r}_\mu,t)}{\partial t} &= \left[-\frac{\hbar^2}{2m_\mu}\nabla^2_\mu - \frac{e^2}{r_\mu} + \int d^3r_e\, e^2\frac{|\Phi(\vec{r}_e,t)|^2}{|\vec{r}_e-\vec{r}_\mu|}\right]\phi(\vec{r}_\mu,t) \\ i\hbar\frac{\partial\Phi(\vec{r}_e,t)}{\partial t} &= \left[-\frac{\hbar^2}{2m_e}\nabla^2_e - \frac{e^2}{r_e} + \int d^3r_\mu\, e^2\frac{|\phi(\vec{r}_\mu,t)|^2}{|\vec{r}_e-\vec{r}_\mu|}\right]\Phi(\vec{r}_e,t) \end{aligned} \tag{9}$$

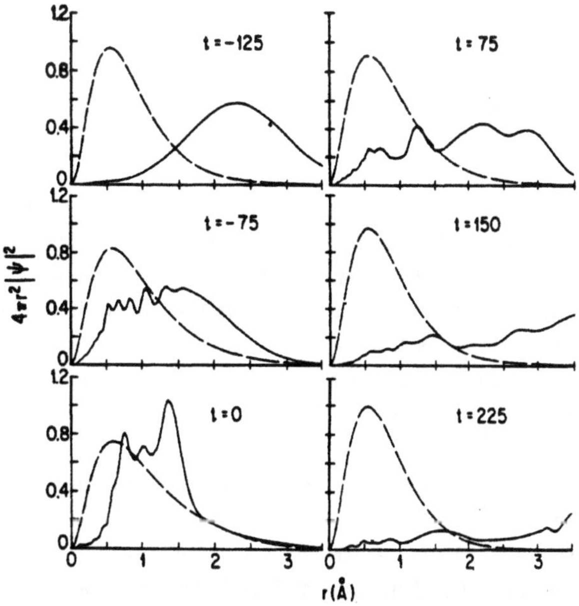

Fig. 10. TDHF calculation of the radial densities for the μ^- (solid curve) and the e^- (dashed curve) at various times (in units of 100 Å/c) for a 2.7 eV collision. The μ^- wave can be seen to move in from large radii, then recede as time increases.

where the coordinate origin is at the fixed position of the proton. Each equation is coupled to the other by the *mean* field due to the other particle. These nonlinear equations are solved on a cylindrical grid with boundary condition $\Psi = 0$ using the Peaceman–Rachford alternating direction implicit method.[14] The wave function is assumed to have cylindrical symmetry, i.e. m coupling is neglected. Typically a cylinder diameter of 8 Å and length of 12 Å was used with a grid spacing and time step chosen to ensure accurate solution.

The initial free muon wave function is chosen to be a Gaussian wave packet of width α, typically 1.5 Å in all dimensions. The transition probability is determined by projection at large times onto a final state Φ_f,

$$P_f = \lim_{t \to \infty} |\langle \Phi_f | \Psi(t) \rangle|^2 \, . \tag{10}$$

The transition probability P_f depends on α, and the physically meaningful cross section is given by

$$\sigma_f = \lim_{\alpha \to \infty} \pi\alpha^2 P_f \ . \tag{11}$$

In practice, a formula for σ_f involving the T–matrix is found to converge faster in $1/\alpha$ than do Eqs. (10)–(11).[6]

The radial density plots for the muon and electron are shown at various times in Fig. 10 for a collision at $E_{c.m.} = 2.7$ eV. The time t = 0 is the time at which the center of the *free* wave packet would have passed the center of the atom. The muon wave can be seen to move in from large radii, develop rapid spatial oscillations as it interacts with the polarized atom, and then recede as time increases leaving behind a small fraction. The electron density can be seen to undergo a breathing–type motion, somewhat like the variation in amplitude exhibited in Fig. 9 by classical dynamics.

The muon capture cross section for 2.7 eV collisions is shown in Fig. 7. The TDHF cross section is almost a factor of 3 smaller than the CTMC result (the latter agrees with the DS and AI calculations). At first it was unclear which result was more accurate, but now the balance seems to be tilted in favor of CTMC for two reasons: (1) the CTMC value is corroborated by the calculations discussed in the next section and (2) a fundamental difficulty seems to be inherent in this TDHF application. This difficulty stems from the conflict that arises from the definition of cross section for a wave packet, Eq. (11), and the neglect of correlation of the electron and muon motions. The broad wave packet interacts weakly with the *mean* field of the electron and, without correlation, has difficulty in keeping the initially captured part of the muon to asymptotically large times. This problem would appear to plague any mean field theory.

2.5. *Classical–Quantal Coupling (CQC)*

The successes and limitations of the CTMC and TDHF methods warrant a more accurate investigation. The TDHF method treats the electron dynamics quantum mechanically, but the neglect of its correlation with the muon motion has been identified as a serious problem. On the other hand, the CTMC method fully includes correlation, but the classical treatment of the electron needs to be tested. For the reasons mentioned in Sec. 2.3, the classical treatment of the muon is well justified. These observations suggest a method in which the muon and proton are treated classically but the electron is quantum mechanical. This mixing of classical and quantal degrees of freedom must be done with care to avoid inconsistencies.

Such a method, termed classical–quantal coupling (CQC) has been proposed by Kwong.[15] The equations are derived from the action variational principle and take the form of coupled Schrödinger and Newton equations for μ^-+H collisions that might have

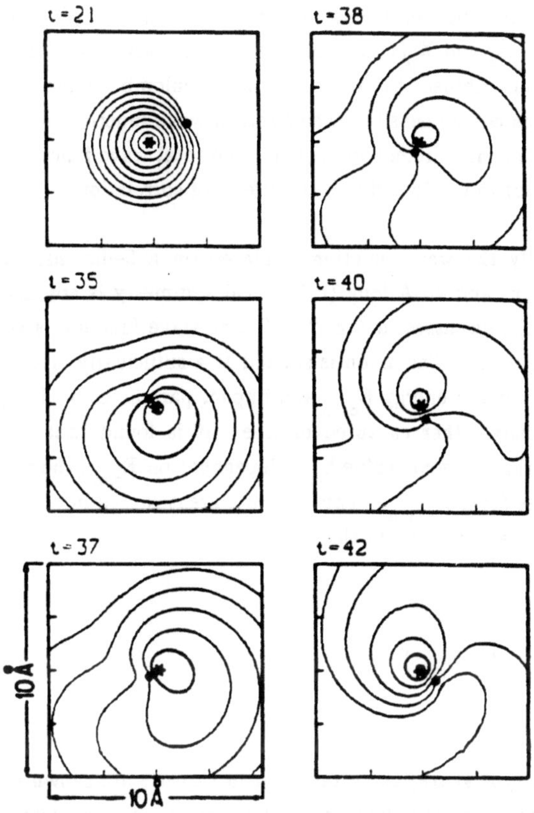

Fig. 11. Time frames showing the positions of the classical muon (•) and proton (⋆), and the density distribution of the quantal electron for a collision at relative energy 0.165 a.u. and impact parameter 0.5 Å. At t (in a.u.) = 0, the muon was at a distance of 5 Å from the proton.

been guessed,

$$
\begin{aligned}
i\hbar\, \dot{\psi}(\vec{x},t) &= \left[-\frac{\hbar^2}{2m_e} \nabla_x^2 + \frac{e^2}{|\vec{x}-\vec{X}_\mu(t)|} - \frac{e^2}{|\vec{x}-\vec{X}_p(t)|} \right] \psi(x,t) \\
m_i \ddot{\vec{X}}_i &= \vec{\nabla}_{X_i} \left[\frac{e^2}{|\vec{X}_\mu - \vec{X}_p|} \right] - \langle \psi(\vec{x},t) |\, \vec{\nabla}_{X_i} \frac{e^2}{|\vec{x}-\vec{X}_i|} \,| \psi(\vec{x},t) \rangle
\end{aligned}
\qquad (12)
$$

where i = μ or p. The time-dependent electron wave function $\psi(\vec{x},t)$ and the muon and proton trajectories $\vec{X}_\mu(t)$ and $\vec{X}_p(t)$ are solved for selfconsistently. The muon and proton follow trajectories determined by their mutual Coulomb attraction as well as the instantaneous electron charge density. The electron wave function evolves *nonadiabatically* in response to the moving muon and proton. The total wave function is a superposition of different trajectories and fully includes the effects of correlation.

Computationally the wave function is placed on a 3-dimensional cylindrical grid typically 8 Å in radius and 40 Å long (cylindrical symmetry is not assumed as it was in the TDHF calculations discussed in Sec. 2.4).[7] The wave function is taken to be zero on the boundary for all times. The muon is started at a sufficiently large distance (typically 5 Å) with specified incident energy $E_{c.m.}$ and impact parameter b. For given $E_{c.m.}$ and b the muon is eventually either captured or free. It is found that capture has a sharp dependence on b so that for some value b_{crit}, dependent on $E_{c.m.}$, the muon is captured for $b < b_{crit}$ and escapes for $b > b_{crit}$. This allows a great saving in computer time; i.e., the value of b_{crit} can be searched for and then

$$\sigma_{capt}(E_{c.m.}) = \pi\, b^2_{crit}(E_{c.m.}) \quad . \tag{13}$$

Snapshots of a typical trajectory resulting in capture are shown in Fig. 11. As the muon approaches the atom, the electron can be seen to polarize away. In the adiabatic molecular picture (e.g., the AI model), this polarized electron density would stay symmetric about the μ-p axis. This symmetry is clearly broken as the muon spirals into orbit about the proton and the electronic "wake" it leaves behind escapes into the continuum.

The capture cross sections calculated by the CQC method are shown in Fig. 7. The overall agreement between the CQC and CTMC methods is very satisfactory. The CQC cross sections seem to be somewhat larger at very low energies, but, more interesting, is the difference at $E_{c.m.} \approx 0.5$ a.u. While the CTMC result is rather flat just below 0.5 a.u. and decreases rapidly above 0.5 a.u. like the DS result, the CQC result behaves more smoothly through this region. In principle, the CQC calculation should be the better; however, we note that capture of a 0.5 a.u. muon usually occurs in a very large orbital and hence the integration box needs to be quite large. The CTMC calculations show that the electron usually carries off a kinetic energy of $\lesssim$ 0.1 a.u. in the capture process. For a collision at $E_{c.m.}$ = 0.5 a.u., this implies an average distance of ~ 10 a_0 for the captured muon.

3. OTHER PROJECTILES AND TARGETS FOR EXOTIC HYDROGEN FORMATION

3.1. *Capture of other Negative Particles by the H Atom*

Capture of π^-, K^-, or $\bar{p}$ differ from μ^- only by the reduced mass. The subsequent cascades are quite different because of the nuclear strong interactions of the particles other than μ^-, but the strong interaction is quite negligible in the diffuse, nonzero-angular-momentum states where capture usually occurs.

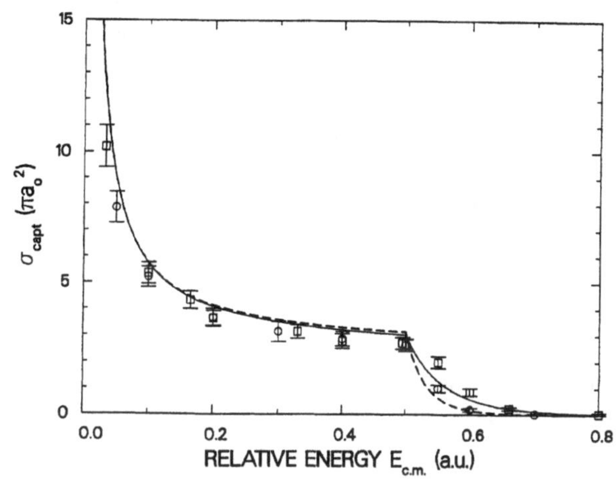

Fig. 12. Capture cross sections for $\bar{p}$+H (dashed curve = DS calculation, o = CTMC calculation) and μ^- + H (solid curve = DS calculation, □ = CTMC calculation). The error bars are 2 sigma.

In the AI model the cross section *as a function of relative energy* doesn't depend on the mass. This is true even when trajectory deflection is taken into account since the slower velocity of the heavier particle allows time for just the same deflection as in the case of the muon. In more accurate treatments there is a mass dependence. We will examine the further considerations now.

Figure 12 compares the capture cross sections for μ^-+H and $\bar{p}$+H obtained in DS and

CTMC calculations. At collision energies less than the target ionization potential of 0.5 a.u., the cross sections are almost the same. Here the DS calculations suggest that capture of $\bar{p}$ is slightly more probable because of the longer interaction time of the heavier particle,but the difference is within the error bars of the corresponding CTMC calculations.[16] At $E_{c.m.} > 0.5$ a.u. the difference is much more striking, and again the DS model offers a simple explanation. At these higher energies, some energy has to be carried away by the electron. As is clear from Fig. 4 the closer the penetration before ionization, the greater the kinetic energy given to the electron in a vertical transition to the continuum. Hence the higher velocity of the light μ^-, though a detriment to the total (ionization plus capture) cross section, enhances the capture component above 0.5 a.u.

The agreement between the DS and CTMC results is quite satisfactory in light of the simplicity of the DS model. The intuitive interpretation of the DS model is also suggestive of simple approaches to the more common experimental problem, namely μ^- capture by H_2 molecules.

3.2. *Capture by the H_2 Molecule*

Historically in the absence of specific information, the H_2 molecule has been represented by two H atoms. While this treatment is adequate for high-energy collisions, e.g. in μ^- stopping by H_2 at v >> 1 a.u., it is totally unjustified at the low energies where μ^- capture occurs. In fact, there are two sensible arguments for comparing capture by H and H_2 that lead to opposite conclusions about the relative magnitudes of their cross sections for capturing μ^- ![14b] On the one hand, H_2 cannot be *adiabatically* ionized since the electron can migrate to the opposite end of the diatomic molecule to form a stable atomic negative ion (note that dissociation is improbable since even an asymptotically very slow muon will be moving much faster than the internal molecular vibration when it is at distances $\lesssim 1\ a_0$). This view suggests that the cross section will be smaller for H_2 than for H. On the other hand, H_2 has two electrons, a higher ionization potential, and a larger geometric (core) size, suggesting H_2 may have a larger capture cross section.

No rigorous calculation has yet been done for μ^- capture by H_2, but a recent calculation by Korenman and Popov[17] suggests that both the above arguments are relevant. Their model is similar to our diabatic-states model with $\Gamma \to \infty$. It takes into account the anisotropy of the molecule but ignores molecular vibration and rotation. The results are very interesting. They find that the H atom cross section is larger than the H_2 molecular cross section at $E_{c.m.} < 0.3$ a.u., but smaller at higher collision energies. This relation is exactly that suggested by the qualitative arguments in the above paragraph. At energies where capture occurs, the atomic and molecular cross sections differ by $\lesssim$ 20%.

3.3. *Capture by the* H^- *Negative Ion*

Experiments have been proposed for LEAR in which protonium ($p\bar{p}$) will be formed in corotating beams of H^- and $\bar{p}$.[9] In the c.m. frame the relative energy of H^- and $\bar{p}$ can be made low enough (< 1 a.u.) for $\bar{p}$ to be captured by H^-. Since the electron affinity of H is small (0.754 eV) one might expect[18] the capture cross section for H^- to be similar to that of H, but we will see that such is not the case.

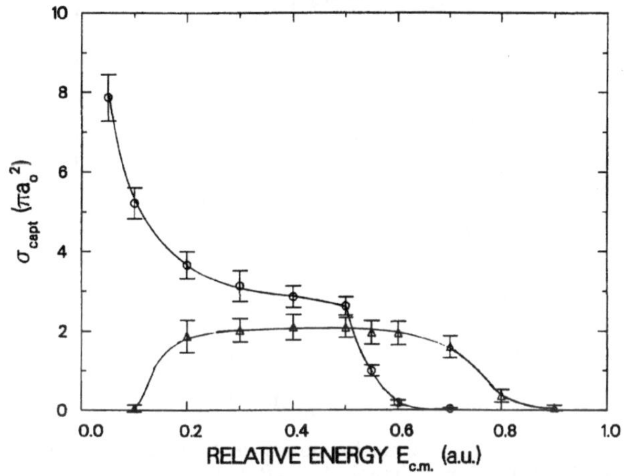

Fig. 13. Cross sections for $\bar{p}$ capture by H^- (Δ symbols) and H (o symbols) calculated by the CTMC method. The error bars are 2 sigma and the curves are just visual fits.

The $p\bar{p}$ formation cross section for $\bar{p}+H^-$ collisions has been determined in a 4–body CTMC calculation.[16] In Fig. 13 the result is compared with the $p\bar{p}$ formation cross section for $\bar{p}+H$ collisions determined in a 3–body CTMC calculation.[16] They are quite different from each other in both the low and high ranges of the energy, their magnitudes crossing just above 0.5 a.u. The difference at low energy is due to the Coulomb repulsion between $\bar{p}$ and H^-. The difference at $E_{c.m.} > 0.5$ a.u. might be considered more surprising in view of the small difference in the total binding energies of H and H^-. While the cross section for H falls steeply at $E_{c.m.} > 0.5$ a.u., a similar drop in the H^- cross section does not occur until $E_{c.m.} > 0.7$ a.u. The difference resides in the kinetic energies of the ionized electrons. It is found that both electrons are invariably removed before capture occurs.

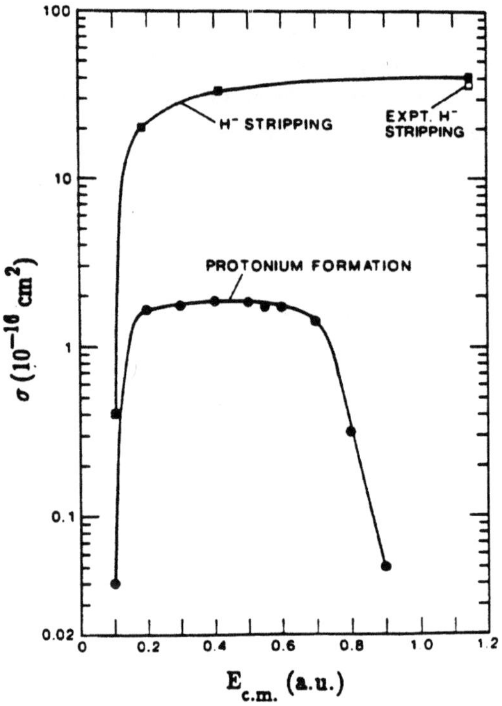

Fig. 14. Stripping and protonium-formation cross sections for $\bar{p}$ + H^- collisions. The experimental point is from Chanel *et al.* (Ref. 19).

Table I. Average n and ℓ quantum numbers and average time for annihilation (ATA) after $p\bar{p}$ formation in a $\bar{p}$ + H^- collision at energy $E_{c.m.}$.

$E_{c.m.}$(a.u.)	$\bar{n}$	$\bar{\ell}$	ATA (s)
0.20	30.2	18.6	4.51×10^{-6}
0.30	33.3	22.0	8.40×10^{-6}
0.40	37.7	25.6	1.77×10^{-5}
0.50	45.2	28.5	4.33×10^{-5}
0.55	51.8	30.2	7.36×10^{-5}
0.60	60.5	31.1	1.12×10^{-4}
0.70	131	34.8	1.46×10^{-3}

As in the case of μ^-+H, most of the energy loss to the strongly bound electron is due to its binding (*potential* energy)—it leaves with little kinetic energy (usually < 0.1 a.u.). The opposite situation prevails for the weakly bound electron. It is bound by only 0.0277 a.u., but leaves *first* in the repulsive Coulomb field of the remnant $p\mu^-e^-$ thereby gaining considerable *kinetic* energy (> 0.1 a.u.). The other electron then escapes in a weak field due to the $p\mu^-$ dipole. The balance of these two effects is an energy dependence of the $p\bar{p}$ formation cross section for $\bar{p}+H^-$ that displays a broad plateau in contrast to the energetically monotonic $\bar{p}$+H cross section.

The usual adiabatic ionization model ignores the kinetic energy of the ionized electrons and hence fails for $\bar{p}+H^-$ collisions. As an aside, I point out that it might be possible to modify the adiabatic ionization model to get a qualitatively reasonable description of $\bar{p}+H^-$. Recall that the AI cross section (neglecting trajectory curvature to simplify the discussion) for $\bar{p}$+H is given by πR_c^2, where R_c is the distance at which the adiabatic potential energy of $\bar{p}$+H becomes equal to the potential energy of $\bar{p}+H^+$. Capture occurs only at $E_{c.m.} < I_a$ in this model. For $\bar{p}+H^-$, two ionization stages need be considered. At some larger distance R_c' the adiabatic potential energy of $\bar{p}+H^-$ will become equal to the adiabatic potential energy of $\bar{p}$+H. Now since both electrons must be ionized for capture to occur, the magnitude of the cross section will be unchanged, but the cutoff will be extended to $I_a + 1/R_c'$ taking into account the kinetic energy removed by the weakly bound electron. Of course, the very low–energy cross section is also modified by the Coulomb repulsion between $\bar{p}$ and H^-, and comparison of the results of Ref. 18 with the more accurate CTMC results[16] suggests that simple electrostatics does not suffice for the calculation. It remains to be seen if the description outlined in this paragraph is viable.

Let us now return to the problem of $p\bar{p}$ formation in corotating beams. The protonium formation in collisions with H^- has to compete with the stripping reaction

$$H^- + \bar{p} \rightarrow H + e^- + \bar{p} \tag{14}$$

as well as the similar reaction of H^-+H^-. The cross section for (14) has been calculated in both 3–body[11] and 4–body[16] calculations, which are in good agreement and also agree well with an experimental measurement.[19] The two competing cross sections are shown in Fig. 14. The ratio of stripping to formation is ~ 20 in the relative energy range 5–20 eV where $p\bar{p}$ formation is possible. The $p\bar{p}$ atoms are generally formed in states with large n and ℓ quantum numbers. Hence the subsequent cascade, essentially in vacuum under beam conditions, occurs primarily via circular orbitals. This path tends to maximize the elapsed time before annihilation. As can be seen in Table I, the average values of the initial n and ℓ and consequently the average time before annihilation depend significantly on the relative energy of the $\bar{p}$ and H^- when capture occurs. The time was determined by Monte Carlo simulation[20] of the $p\bar{p}$ cascade with the initial conditions sampled from the CTMC-determined n and ℓ distributions. To be experimentally useful, the neutral $p\bar{p}$ atom must be observable for this time. If such a long observation time is difficult, then it will be beneficial to cool the beams to the lower end of the acceptable energy range where an observation time of a few μs might be adequate.

ACKNOWLEDGMENTS

I gratefully acknowledge the collaboration of a number of colleagues in this work over the last decade and especially the recent advances made possible by J. D. Garcia and Nai–Hang Kwong. I am indebted to Melvin Leon for numerous discussions of this subject. This work was performed under the auspices of the U.S. Department of Energy.

REFERENCES

1 S. S. Gershtein and L.I. Ponomarev, in *Muon Physics*, edited by V. W. Hughes and C. S. Wu (Academic, New York, 1975), Vol. III, p. 141.
2 A. S. Wightman, Phys. Rev. 77, 521 (1950).
3 E. Fermi and E. Teller, Phys. Rev. 72, 406 (1947).
4 J. S. Cohen, R. L. Martin, and W. R. Wadt, Phys. Rev. A (a) 24, 33 (1981); (b) 27, 1821 (1983).
5 J. S. Cohen, Phys. Rev. A 27, 167 (1983).
6 J. D. Garcia, N. H. Kwong, and J. S. Cohen, Phys. Rev. A 35, 4068 (1987).
7 N. H. Kwong, J. D. Garcia, and J. S. Cohen, in *Muon Catalyzed Fusion (Sanibel Island, FL 1988)*, American Institute of Physics Conference Proceedings 181, edited by S. E. Jones, J. Rafelski, and H. J. Monkhorst (AIP,New York, 1989), p. 236.
8 M. Leon, Phys. Rev. A 17, 2112 (1978).
9 U. Gastaldi and R. Klapisch, in *From Nuclei to Particles*, proceedings of the International School of Physics "Enrico Fermi," Course LXXIX, edited by A. Molinari (North-Holland, Amsterdam, 1981),p. 462.
10 T. J. Baird, Ph.D. thesis, Rensselaer Polytechnic Institute, 1976 (unpublished) and Los Alamos National Laboratory Report No. LA-6619-T (unpublished).
11 J. S. Cohen and G. Fiorentini, Phys. Rev. A 33, 1590 (1986).
12 (a) G. Ya. Korenman and S. I. Rogovaya, Yad. Fiz. 22, 754 (1975) [Sov. J. Nucl. Phys. 22, 389 (1975)]; J. Phys. B 13, 641 (1980); (b) D. R. Bates and G. Griffing, Proc. Phys. Soc. A66, 961 (1953).
13 W. H. Barkas, N. J. Dyer, and H. H. Heckman, Phys. Rev. Lett. 11, 26 (1963).
14 K. C. Kulander, K. R. S. Devi, and S. E. Koonin, Phys. Rev. A 25, 2968 (1982).
15 N. H. Kwong, J. Phys. B 20, L647 (1987).
16 J. S. Cohen, Phys. Rev. A 36, 2024 (1987).
17 G. Ya. Korenman and V. P. Popov, *op. cit.* Ref. 7, p. 145.
18 L. Bracci, G. Fiorentini, and O. Pitzurra, Phys. Lett. 85B, 280 (1979).
19 M. Chanel *et al.*, CERN Report No. PS 87-12 (LEA), 1987 (unpublished).
20 J. S. Cohen and N. T. Padial, to be published.

SLOWING-DOWN AND ATOMIC CAPTURE OF EXOTIC PARTICLES IN ELEMENTS WITH Z>2

F.J. Hartmann

Physik-Department, E18, Technische Universität München

Garching, F.R.G.

1 Introduction

Investigations on slowing-down and atomic capture of exotic particles started shortly after the pion as the first of these particles had been discovered [1]. Since then such a large number of publications has been written that the references alone fill a book [2]. What makes the study of the slowing-down and capture processes so interesting is the combination of atomic, nuclear and even elementary particle physics. Furthermore, Coulomb capture has to be understood whenever one wants to interprete processes involving stopped exotic particles like muons, pions, kaons and antiprotons. Let me give you two examples: To study the feasibility of parity violation experiments with muonic atoms it is essential to know the population of the 2s level in a suitably chosen muonic atom [3], which depends on the angular momentum distribution after capture. Another example: When thinking about using muons in the determination of the composition of ancient samples [4] it was essential to have an idea about the Coulomb capture process to derive atomic capture probabilities for the elements measured.

2 Slowing-down of exotic particles

When an exotic particle, e.g. a muon generated by pion decay at one of the modern meson factories, enters matter with nonrelativistic energies it is slowed down by independent collisions with the electrons of the medium; interaction with nuclei plays only a minor role for stopping (channeling as a collective process shall not be considered here). The stopping power S, i.e. energy loss per unit path length, can be described by the formula

$$S = -dE/dx = C * \frac{Z * z^2}{A * \beta^2} * (L_0 + L_1 * z + L_2 * z^2). \tag{1}$$

Here C is a constant, Z and A are the atomic number and mass of the target, z the projectile charge and β its velocity (normalized to the velocity of light, β= v/c). L_0 is the Bethe term [5]

Electromagnetic Cascade and Chemistry of Exotic Atoms
Edited by L. M. Simons *et al.*, Plenum Press, New York, 1990

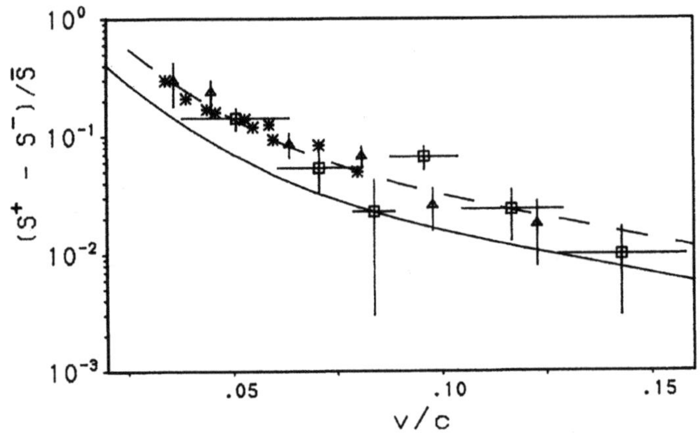

Figure 1. The Barkas effect at low velocities. Rectangles: Pions in emulsion [9]. Triangles: Muons in Ag [10]. Stars: Antiprotons in Si [11]. Solid line: Calculation [8]. Dashed line: Revised calculation [12].

$$L_0 = ln[\frac{2 * m_e c^2 * \beta^2}{I * (1 - \beta^2)}] - \beta^2 - ShC/Z, \tag{2}$$

with I the ionization potential and ShC/Z the shell correction term. L_2 is the Bloch term [6].

L_1 describes an effect first observed by Barkas and coworkers [7] when comparing stopping powers for positive and negative particles. The effect is attributed (see e.g. [8]) to a displacement of the electrons from their basic position by the field of the passing projectile. Fig. 1 shows these differences as a function of velocity. The stopping power for positive particles becomes up to 30 % larger at low velocities. The solid curve is a prediction from theory [8]. It was pointed out by Lindhard [12] that this estimate could be too low by a factor of two as only distant collisions were taken into account. The dashed line is drawn according to this revised prediction. More details of a recent measurement performed with antiprotons at LEAR [11] are presented in a talk given at this school.

At low particle velocities the Bethe-Bloch formula is no longer valid. This has several reasons:

- More and more electrons are too strongly bound to participate in the interaction with the charged particle. Even the valence electrons may no longer be lifted across the band gap.
- The exotic particle no longer moves through the atom on a straight line.
- The Born approximation becomes inadequate.

Fermi and Teller [13] first showed that at low velocities the v^{-2} dependence of the stopping power goes over into a v dependence. They viewed the slowing-down process as the interaction of the charged particle with a degenerate electron gas. The maximum energy transfer ΔE is given by

$$\Delta E = m_e * v_0 * v \tag{3}$$

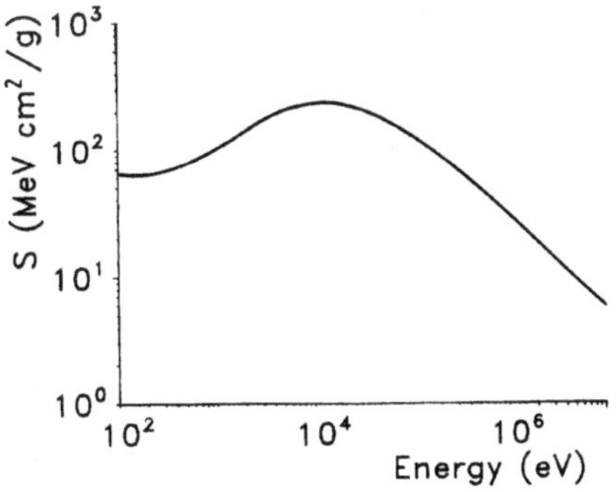

Figure 2. Calculated stopping power for muons in Cu [14].

with m_e, v_0 electron mass and Fermi velocity, respectively. To interact with the slow exotic particle the electrons have to have velocities between $v_0 - v$ and v_0. Their number n is roughly

$$n \approx m_e^3 * v_0^2 * v/\hbar^3. \tag{4}$$

The cross-section σ for interaction is of the order

$$\sigma = [e^2/(m_e * v_0^2)]^2. \tag{5}$$

This finally gives an energy loss per unit time

$$-dE/dt \approx \Delta E * n * \sigma * v_0 \propto v^2. \tag{6}$$

and an energy loss per unit path length

$$-dE/dx \propto v. \tag{7}$$

The charged particle is still assumed to follow a straight line in the medium and a possible acceleration inside the atom is neglected. The latter effect makes the velocity inside the atom - and with it the energy loss per traversed atom - roughly independent from the velocity outside the atom. The stopping power for muons at low energies was calculated by Wilhelm [14]. By following the classical path of a swift muon in the potential of the atom (a Lenz-Jensen potential corrected for solid state effects at the periphery) a periodicity of the stopping power with atomic number was found. The stopping power itself is shown for Cu in Fig. 2.

3 Coulomb capture

The problem of Coulomb capture of exotic particles can be assumed to be solved in principle if one knows the following two quantities:

- the differential cross-section $d\sigma/d\epsilon(E,\epsilon)$ for energy loss ϵ at energy E,

- the capture cross-section σ_{capt}(E,n,ℓ) into a bound level with principal quantum number n and angular momentum quantum number ℓ (these are good quantum numbers only for systems with rotational symmetry).

Competition between slowing-down and capture then determines the fraction of particles reaching energy E. In a complete theory of Coulomb capture this competition has to be taken properly into account.

Unfortunately no way has been found up to now to get direct information about these quantities from experiment. What can we really measure?

- The spectral flux density of particles n(E), i.e. the number of particles with energies between E and E+dE entering per unit time a sphere of radius r, divided by the cross-section of this sphere [15] is accessible to the experiment. Leon [16] has introduced an equivalent quantity, the arrival probability F(E), to characterize the fraction of exotic particles not yet captured during the slowing-down process when reaching energy E.

- The distribution p(ℓ) of exotic particles over the angular momentum levels in an exotic atom can, in principle, be deduced from measurements of the exotic x-ray pattern. Assumptions about the evolution of the atomic cascade have, however, to be made.

- A wealth of data exists on the per-atom capture probability

$$P(Z) = \int n(E) * \sum_{n,\ell} \sigma_{capt}(E, n, \ell) dE. \qquad (8)$$

It will be my task now to describe the theoretical efforts undertaken to get a picture of atomic capture and also to sketch a few of the experiments which measured the quantities relevant to Coulomb capture.

4 Theoretical models on Coulomb capture

4.1 Quantum-mechanical calculations

Attempts to develop theories of Coulomb capture based on a quantum-mechanical treatment of incoming and outgoing particles started already in the fifties [17]. The transition rate for capture by electron ejection is given in first order perturbation theory by

$$w = \frac{2\pi}{\hbar} < f|\frac{e^2}{|\vec{r}_1 - \vec{r}_2|}|i >^2 *\rho(f), \qquad (9)$$

with $\vec{r}_1, \vec{r}_2$ denoting the positions of exotic particle and electron, respectively, and $\rho(f)$ standing for the density of final states f. Plane waves or Coulomb wave functions were taken for the incoming exotic particle and the outgoing electron. Hydrogen-like bound-state wave functions were applied. Later on more elaborate wave functions were used [18]. Table 1 gives an overview over the calculations performed since 1954, most of them for low Z elements.

Table 1. Quantum-mechanical calculations of Coulomb capture. Methods: [a] Born approximation (BA) with plane waves or distorted waves for free particles and hydrogen-like wave functions for bound particles; [b] capture treated as internal conversion; [c] adiabatic approximation; [d] BA, Hartree-Fock functions for bound particles; [e] diabatic treatment; [f] time-dependent Hartee-Fock calculations.

Author	Z	Method	Capture energy	First n	$p(\ell)$
DeBorde, 1954 [17]	any	[a]			
Baker, 1960 [19]	1	[a]	thermal	≈ 15	peaks at $\ell=7$
Mann, Rose, 1961 [20]	6	[a]	8 keV	≈ 7	statistical
Au-Yang,Cohen, 1968 [21]	16 29	[a]	100-500 eV		
Haff,Tombrello, 1974 [22]	1-3	[a]	H: 80 eV He:200 eV Li:240 eV	He: 12 Li: 15 K 15 L	
Korenman, Rogovaya, 1975 [23]	1,2	[a]	H: 50 eV He: 90 eV	≈ 10	statistical
Daniel, 1976 [24]	any	[b]		Zn: 28 Nd: 28	
Baird, 1977 [25]	1	[c]			
Cherepkov, Chernycheva, 1980 [18]	2	[d]	25 eV	peaks at n= 17	statistical
Cohen, Martin,Wadt, 1981 ff. [26]	1	[e]	13 eV	peaks at n=17	statistical, high ℓ cut-off
Garcia, Cohen,Kwong, 1987 [27]	1	[f]		peaks at n=10	

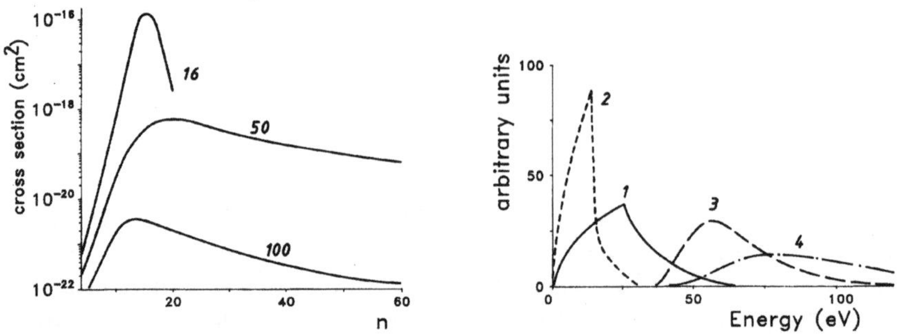

Figure 3. Capture cross-sections and capture energies for μ^- in low-Z elements. Left: dependence of $\sum_\ell \sigma_{capt}(n,\ell,E)$ on the principal quantum number of first bound states for capture from 16, 50 and 100 eV into He [18]. Right: Relative meson-capture probability as a function of energy. Solid line: Results from ref. [18] for He. Short dashed line: Results from ref. [25] for H. Dashed and dot-dashed lines: Results from ref. [23] for H and He, respectively.

What can we learn from these calculations? First of all it seems to be established now that capture takes place at energies roughly corresponding to the first ionization potential of the element, i.e. at energies of some tens of eV. The radius of the first bound orbit of the exotic particle and the radius of the orbit of the electron ejected are roughly equal (the matrix element in eq. (9) may then be expected to reach a maximum). This means e.g. n values around $n = 14$ for capture (only K electron emission) in lowest Z muonic atoms. For higher-Z elements capture by L and later by M electron emission should become predominant; capture to highly excited atomic levels should result. Most authors predict a statistical initial ℓ distribution $p(\ell) \propto 2\ell+1$, which corresponds to equal population of the substates with magnetic quantum number m. Note, however, that also initial distributions with $p(\ell)$ exhibiting a maximum at medium ℓ were derived [19]. Figure 3 shows $\sum_\ell \sigma_{capt}(n,\ell,E)$ as a function of n for different capture energies and the capture probability as a function of energy.

A more detailed description of the most recent quantum-mechanical approaches to Coulomb capture, mostly for lowest Z elements, will be presented by J. Cohen in an invited talk given at this school.

4.2 Semiclassical calculations

There are good reasons to believe that quantum-mechanical calculations are *not* the best choice for Coulomb-capture calculations.

- As the incoming exotic particle is slow, the deBroglie wave-length is small and the exotic-particle wave function has many oscillations within the atom.
- Simple perturbation theory breaks down at low energies as the action integral for the interaction of the exotic particle with the atomic electrons becomes larger than $\hbar$.

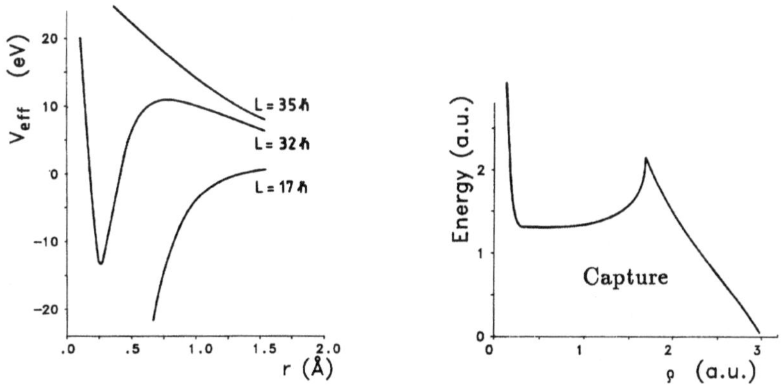

Figure 4. Left: Effective potential for the movement of muons at Z=18 [31]. Right: Capture and escape regions in the (E,ρ) plane for Z= 40 [33](ρ = impact parameter).

After these shortcomings had been detected the semiclassical picture was developed. The following assumptions are made:

- The exotic particle moves on a classical trajectory in a spherically symmetric atomic potential. The movement may be described in one dimension by addition of a centrifugal potential.
- The electrons form a degenerate electron gas in the screened nuclear potential.
- The exotic particle looses energy by lifting electrons to unoccupied states via shielded Coulomb interaction.
- In principle an energy-loss distribution has to be taken into account [28,29]. To make calculations easier most authors used a mean energy loss in their calculations.
- The exotic particle is captured by an atom if its total energy drops below zero during the passage through the electron cloud. Capture is also possible if the particle is trapped behind the centrifugal barrier. This may happen if the atomic potential is flat enough to enable a centrifugal barrier to appear at certain angular momenta (as is shown in Fig. 4a). The resulting regions in the energy-impact parameter plane for capture and escape are shown in Fig. 4b.

The outcome of a typical calculation is sketched in the next Figure. Figure 5a shows the distribution of energies immediately before capture as calculated for muons and Z=40 [28]. Figure 5b shows the angular momentum distribution in μ^-K in the compound KCl after capture and when the muon has reached the electronic K shell [31].

In Table 2 a synopsis of semiclassical calculations performed between 1947 and 1989 is presented.

Table 2. Semiclassical calculations of Coulomb capture: Trajectory:[a] straight line, [b] analytical integration of the equations of motion, [c] numerical integration of the equation of motion. Capture probability: [A] $P(Z) \propto Z^{1/3}\ln(0.57Z)$, [B] $P(Z) \propto Z^{1/3}\ln(0.57Z)/R(Z)$, R(Z) = atomic radius. n(E): [C] depends on the existence of a centrifugal barrier: constant, if no centrifugal barrier exists, $\propto$ E, if a centrifugal barrier is effective.

Author Reference	Exotic particle trajectory	Potential V(r)	Energy loss	Capture probability P(Z)	$p(\ell)$	n(E)
Fermi+Teller [13]	[a]	$1/r^2$	continuous	Z		$\propto$ E
Daniel [30]	[b]	$1/r^2$	continuous	[A]		const.
Vogel et al. [31]	[b]	Lenz-Jensen	continuous	$Z^{7/6}$	$\propto 2\ell+1$	$\propto$ E
Leon+Seki [32,33]	[c]	Thomas-Fermi	continuous		$\propto 2\ell+1$	$\propto$ E
Leon+Miller [28]	[c]	Thomas-Fermi	energy loss distr.		flat or $\propto 2\ell+1$	$\propto$ E
Vogel et al. [29]	[c]	Lenz-Jensen	energy loss distr.	compound dependent		$\propto$ E
Daniel [34]	[b]	$1/r^2$	continuous	[B]		const.
Daniel [35]	[b]	$1/r^2$	continuous			[C]
Daniel [36]	[b]	$1/r^2$	continuous	depends on n(E)	depends on n(E)	
Cohen [37]	Classical trajectory Monte Carlo					
Kwong Garcia Cohen [38]	Classical quantal coupling					

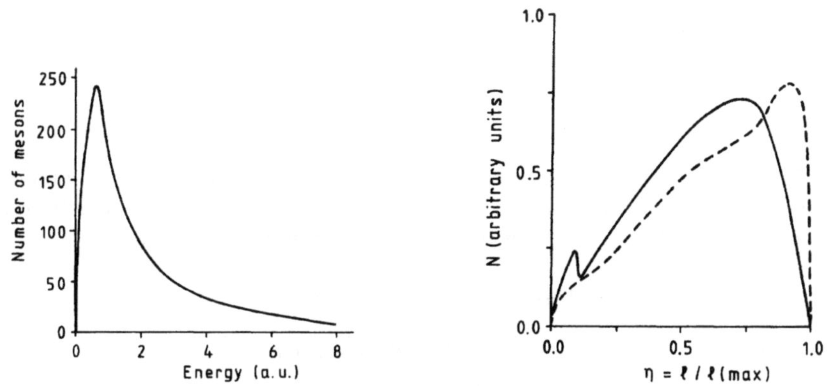

Figure 5. Left: Distribution of energies immediately before capture (μ^- at Z=40). Right: Angular momentum distribution of muons after capture in K^+ in the compound KCl. Solid line: Immediately after capture. Dashed line: At the electronic L shell.

4.3 Semiempirical calculations

In 1978 Schneuwly, Pokrovsky and Ponomarev [39] published calculations of Coulomb capture probabilities based on the model of large mesic molecules developed many years before [40]. This SPP model was very successful in predicting global per-atom capture probabilities but was not intended to give evidence on the details of the Coulomb capture process.

The idea of the SPP model is the following: In a compound an exotic particle can be either captured directly into the constituents or it can be trapped in molecular orbits around the atoms of the compound. As capture takes place by Auger effect the probability of capture is assumed to be proportional to the number of loosely bound electrons. In a binary compound $Z_k Z'_\ell$ there are n core electrons per atom which cause direct capture into element Z, n' core electrons cause direct capture into component Z' and $\alpha_m = (k * \nu + \ell * \nu')$ valence electrons lead to capture into mesomolecular orbits. From these molecular orbits the exotic particle again can be captured into Z and Z' with probabilities ω and ω', respectively. A per-atom capture ratio

$$A(Z/Z') = \frac{\ell}{k} * \frac{k * n + \alpha_m * \omega}{\ell * n' + \alpha_m * \omega'} \tag{10}$$

results. The authors now make the following additional assumptions:

- Only loosely bound core electrons take part in the direct capture process. Which electrons are involved is determined by adjustable parameters.
- Due to the ionicity of the chemical bond a fraction p = (1-σ)/2 of the valence electrons is attributed to element Z, the fraction p' = 1-p is located near Z'. The localization of the exotic particle in the mesomolecular orbits follows this electron distribution.
- The exotic particles are redistributed from their original location with probabilities q and q', which depend symmetrically on Z and Z'.

A semiempirical formula for the per-atom capture probability P(Z) was developed a few years ago [41]). The capture probability may be expressed by

$$P(Z) \propto \sum_i N_i * \sigma_i(E_i, n_i, \ell_i, Z, k_{ex}) \tag{11}$$

where N_i electrons with binding energy E_i are in shell i (characterized by quantum numbers n_i, ℓ_i). σ_i also depends on the momentum k_{ex} of the exotic particle. First order perturbation theory was used to derive an expression for $\sigma(k_e, k_{ex})$

$$\sigma(k_e, k_{ex}) = c_0(k_{ex}) * k_e * \frac{n_i}{Z} \tag{12}$$

with k_e the momentum of the electron ejected from shell i

$$k_e = \sqrt{2(k_{ex}^2/2m_{ex} + E_{ex} - E_e)}. \tag{13}$$

A parametrization of the form

$$\sigma_i = C * \sqrt{1 - E_i/E_0} * Z^a * n_i^b, \quad E_i < E_0 \tag{14}$$

and

$$\sigma_i = 0, \qquad E_i \geq E_0 \tag{15}$$

was suggested. Parameters a, b, C and E_0 are adjusted to yield best agreement with experiments.

5 Experimental tests of the picture of Coulomb capture

5.1 Energy of the exotic particle immediately before capture

A qualitative picture of the exotic-particle energy immediately before capture can be gained from the observation of the angular distribution of exotic-atom x rays. Anisotropy of the emitted radiaton is expected if the exotic particle is captured at such high energies that the correlation of the angular momentum with the original beam direction is not yet destroyed. Lum et al.[42] looked for the anisotropy of pionic x rays from liquid Ar. In a similar experiment Abela et al. [43] investigated the anisotropy of x rays from muonic Se and Sn. Experimental values for the anisotropy coefficient β in the expression

$$W(\Theta) \propto 1 + \frac{\beta}{2} * P_2[cos(\Theta)] \tag{16}$$

(with Θ the angle between beam axis and x-ray direction and P_2 a Legendre polynomial) are given in Table 3.

A small anisotropy is evident indicating a small possibility of capture from higher energies.

More quantitative conclusions can be drawn from experiments on the spectral flux density of muons at very low energies [44]. A beam of low energy muons was directed onto a moderator with the help of a magnet spectrometer and the energy of the

Table 3. Anisotropy of pionic (Ar) and muonic (Se,Sn) x rays

Element	Ar	$Se_{metallic}$	$Se_{amorphous}$	$Sn_{metallic}$
Anisotropy coefficient β %	<5	0.27±0.05	0.26±0.06	<13

muons emerging was measured by a time-of-flight method. The result for the number of detected particles in an energy interval ΔE leaving a thin silver moderator as a function of E is given in Fig. 6.

What can we learn from this result? The spectral distribution of muons follows the relation [45,15]

$$n(E) * S(E) = constant \tag{17}$$

(with S(E) the stopping power), if no particles are lost by atomic capture at higher energies. Any losses at positive energies would result in a deviation from this relation. It can be seen from the Figure that - within the rather large errors - no deviations are visible. Hence we may conclude that atomic capture takes place in Ag only at energies below several tens of eV.

5.2 The dominant capture process

The question of the dominant capture process, radiative or Auger capture, was answered by theory unambiguously: Exotic particles are preferentially Coulomb-captured by electron emission. What can the experiment tell us? Radiative capture would strongly favour dipole transitions to low excited levels in the atom as their rate goes with the third power of the transition energy. Henceforth radiative transitions should be visible in muonic x-ray spectra (where no strong interaction effects obscure the results) in the energy regions above the series limits. Measurements in P and Se [46] and in Mg, Al, Fe, In, and Ho [47] gave the upper limits for radiative Coulomb capture shown in Table 4. From the weakness of these transitions we may conclude that Auger capture is the predominant effect.

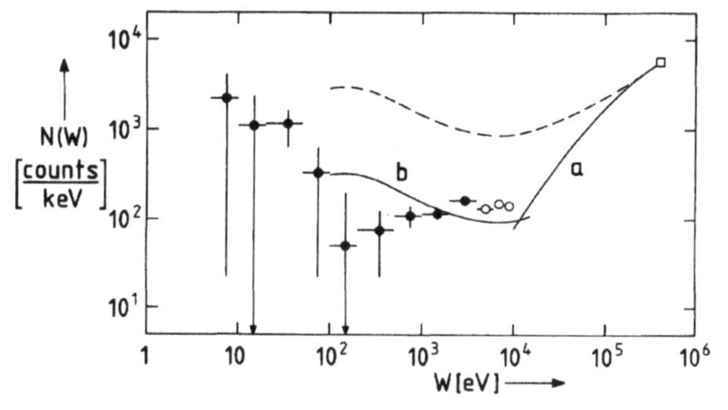

Figure 6. Spectral density N(W) of slow muons after leaving a Ag moderator vs. muon energy W [44]. Open square: Normalization. Dashed line: Expectation without multiple scattering corrections. Solid line: Expectations with multiple scattering corrections - a) in Gaussian approximation, b) relying on Lambert's law.

Table 4. Upper limits for radiative Coulomb capture into different levels of muonic atoms from continuous state energies between zero and one keV (90% confidence).

Element	Mg	Mg	Mg	Al	Al	Al	P	Fe	Se	In	Ho
Lower level	1s	2p	3d	1s	2p	3d	1s	1s	1s	1s	1s
Limit(%)	0.09	0.6	0.12	0.08	0.25	0.12	0.045	0.35	0.35	0.5	0.13

5.3 Principal quantum numbers after capture

The same muonic x-ray spectra used to show the Auger capture to be predominant can tell us something about the first bound state in muonic atoms [47,48]. An intensity of (0.27±0.06) x-ray transitions (20p→1s) per 100 captured muons was seen in μ^-Fe, for instance. If we assume equal population of all ℓ levels (which is corroborated by experiment [48]) and bear in mind that only 6% of the muons leave the 20p level by radiation we find out that all muons captured in iron must pass n=20 and therefore must have been captured into higher levels.

Evidence about the first level after capture was also gained [49] by looking at the electronic x rays emitted during slowing-down, capture and cascade of antiprotons in Kr at low pressure (25 hPa). Electronic K holes are generated by

- ionization of the atoms during slowing-down,
- atomic capture,
- K-Auger effect during the atomic cascade.

Ionization by K-electron emission was not observed because the antiprotons entering the target were too slow. The x rays emitted to fill holes generated during capture are well separated in energy from those emitted after electron depletion by K-Auger effect, because Auger capture takes place to the highest bound levels and the antiprotonic atom is highly ionized in the final stage of the cascade when K-Auger effect becomes possible. Hence the x-ray intensity and the number of K holes can be easily determined. Less than 30% of the antiprotons are captured into Kr by K-electron emission. L-electron emission should lead to capture into orbits around n $\approx$ 85 (cf. section 4.1).

5.4 Angular momentum distribution after capture

The angular momentum distribution of exotic particles after capture cannot be determined directly by experiment. Only the distribution at some intermediate levels may be derived from cascade calculations. Another talk at this school will deal in more detail with the exotic-atom cascade. Let me point out here only that in most cases a statistical initial distribution is a good approximation.

5.5 Ratios of per-atom capture probabilities for μ^-

As mentioned earlier, a huge number of data exist on Coulomb capture ratios for μ^- in all kinds of mixtures, alloys, solutions and compounds. What we can conclude

Table 5. Per-atom capture ratios for μ^- in binary noble gas mixtures.

A(Ar,N)		1.14±0.07	1.53±0.03	1.62±0.04
$k_{Ar}/(k_{Ar}+k_N)$	at%	11.5±0.6	36.6±0.4	65.7±0.3
p	MPa	5.11	5.15	5.11
A(Ar,Ne)		0.91±0.04	1.01±0.03	1.23±0.02
$k_{Ar}/(k_{Ar}+k_{Ne})$	at%	20.2±0.8	50.3±0.6	79.2±0.2
p	MPa	5.11	5.11	5.11
A(Kr,Ar)		1.76±0.03	2.26±0.05	2.67±0.16
$k_{Kr}/(k_{Kr}+k_{Ar})$	at%	21.0±0.2	50.0±0.5	79.5±0.8
p	MPa	5.03	5.12	5.11
A(Xe,Ar)		2.22±0.08	2.33±0.11	2.13±0.22
$k_{Xe}/(k_{Xe}+k_{Ar})$	at%	21.5±0.5	53.7±1.0	84.2±1.3
p	MPa	5.11	5.11	5.19

Table 6. Per-atom capture ratios for μ^- in Nb/V solid solutions as function of the ratio of the stoichiometric ratio Nb/V.

Stoichiometric ratio Nb/V	0.046±0.003	0.182±0.005	0.97±0.02	4.05±0.09	18.5±0.4
A(Z,Z')	1.16±0.09	1.26±0.06	1.17±0.05	1.23±0.06	1.11±0.10

from these experimental data shall be described now.

5.5.1 The question of concentration (in)dependence of the per-atom capture ratio: Noble gases and solid solutions

Slowing-down and atomic capture of exotic particles have to be treated as a whole. P(Z) depends on the spectral density n(E), which may very well change if the composition of the target is changed. Measurements of A(Z,Z') = P(Z)/P(Z') for μ^- in binary noble gas mixtures [50] revealed (cf. Table 5) that for all mixtures with elements lighter than Xe the per-atom capture ratio A(Z,Z') increases with increasing concentration of the heavier element. This was interpreted as an increase of the slope of n(E) with increasing atomic ratio between heavier and lighter element [50]. A concentration dependence of A(Z,Z') was *not* found for solid solutions, however [51]. Table 6 gives details. What may be the solution? Leon [16] pointed out that a concentration dependence of n(E) can be expected (but need not show up) if the exotic particle is trapped behind the centrifugal barrier and atomic capture takes place from positive energies. Such a barrier apparently exists for gases but not for solid solutions with their steep atomic potential in the periphery of the ion cores.

5.5.2 Comparison of experimental results for P(Z) with calculations

By far the most extensive collection of experimental information on Coulomb capture lies in values for capture ratios. Horvath and Entezami [52] collected 321 experimental

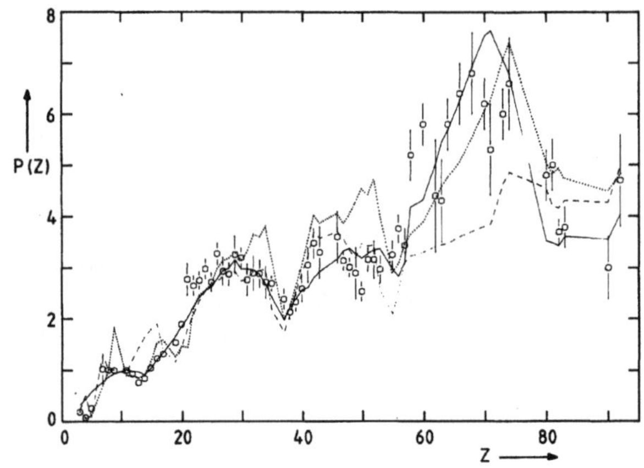

Figure 7. Experimental (open circles [53]) and calculated values for P(Z). Muons in binary compounds. Solid line: Ref. [41]. Dashed line: Ref. [34]. Dotted line: Ref. [39].

ratios and compared them with different calculations. Von Egidy and Hartmann [53] used 146 per-atom capture ratios for μ^- from binary compounds to extract per-atom capture probabilities P(Z) for 64 elements. The possible influence of the chemical structure of the compound on P(Z) was not taken into account as it is in most cases an effect below 10 %. Extracted values [53] for P(Z) are compared in Fig. 7 with the predictions of Schneuwly et al. [39], Daniel [34] and von Egidy et al. [41]. When evaluating the agreement one should keep in mind that in ref.[41] six parameters and in ref.[39] two parameters were adjusted. The pronounced periodic behaviour of P(Z) is reproduced by all models, the agreement with the experiment is reasonable.

5.5.3 Correlation of P(Z) with atomic and solid state properties

Numerous attempts have been made to correlate the per-atom capture ratios with various atomic and solid state properties.

- The strong correlation with the number of core and valence electrons [39,41] and with the atomic radius [34] has already been shown in Fig. 7.
- Stanislaus et al [54] established a correlation between target density and capture ratios for muons in oxides and were even able to deduct a functional dependence of P(Z) from the density.
- Already quite early Kunselman et al. [55] correlated several solid state properties with mesic atom properties. To my opinion the most striking example is a correlation of P(Z) with positron annihilation times in annealed metals. Positron lifetimes depend strongly on the electron density in the periphery of the target atoms. Hence the correlation points into the same direction as the SPP model.
- Capture-ratio results were compared with stopping power data by Hutson et al. [56] and by Evseev at al. [57] and a correlation was found.

- The correlation of A(Z,Z') for muons in compounds of transition elements and of rare-earth elements with the number of possible electronic intra-shell transitions has been pointed out by Naumann et al. [58]. The number of possible transitions peaks for half-filled electron shells; so do the per-atom capture ratios, e.g. in fluorides of rare-earth elements [59].

6 Conclusions

What are the conclusions from this outline of slowing-down and capture of exotic particles for higher Z?

- Slowing-down and atomic capture can be treated similarly for *all* exotic particles.
- Coulomb capture takes place only when the exotic particle has reached low energies (around and below 100 eV)
- The first bound state of the exotic atom is highly excited with principal quantum numbers $n \gg 20$ at medium Z.
- The angular momentum distribution after Coulomb capture is normally statistical ($\propto 2\ell+1$), but flat ℓ distributions are possible.
- The experimental findings for the per-atom capture probability are quite well described by several models. Correlations with atomic and solid state properties have been established.

References

[1] G.P.S. Occhialini, C.F. Powell, Nature 159, 186(1947).

[2] D. Horvath, R.M. Lambrecht, Exotic Atoms - a bibliography 1939 - 1982, Elsevier, 1984.

[3] R. Bacher, P. Blüm, D. Gotta, K. Heitlinger, M. Schneider, J. Missimer, L.M. Simons, Preprint PR-88-14, PSI, Villigen, Switzerland, 1988.

[4] E. Köhler, R. Bergmann, H. Daniel, P. Ehrhart, F.J. Hartmann, Nucl. Instr. Meth. 187, 563 (1981).
H. Daniel, F.J. Hartmann, E. Köhler, U. Beitat, J. Riederer, Archaeometry 29, 1 (1987).

[5] H.A. Bethe, Ann. Phys. 5, 325 (1930). H.A. Bethe, Z. Physik 76, 293 (1932).

[6] F.Bloch, Z. Phys. 81, 363 (1933).

[7] M. Smith, W. Birnbaum, W.H. Barkas, Phys. Rev. 91, 765 (1953).
W.H. Barkas, J.N. Dyer, H.H. Heckman, Phys. Rev. Letters 11, 26 (1963).

[8] J.D. Jackson, R.L. McCarthy, Phys. Rev. B6, 4131 (1972).

[9] H.H. Heckman, P.J. Lindstrom, Phys. Rev. Letters 22, 871 (1969).

[10] W. Wilhelm, H. Daniel, F.J. Hartmann, Phys. Letters 98B, 33 (1981).

[11] L.H. Andersen, P. Hvelplund, H. Knudsen, S.P. Møller, J.O. Pedersen, E. Uggerhøj, K. Elsener, E. Morenzoni, Phys. Rev. Letters 62, 1731 (1989).

[12] J. Lindhard, Nucl. Instr. Methods 132, 1 (1976).

[13] E. Fermi, E. Teller, Phys. Rev. 72, 399 (1947).

[14] W. Wilhelm, Ph. D. thesis, Technische Universität München, 1980.

[15] H. Daniel, Nucl. Instr. Methods 147, 297 (1977).

[16] M. Leon, Phys. Rev. A17, 2112 (1978).

[17] A.H. deBorde, Proc. Phys. Soc. London 67, 57 (1954).

[18] N.A. Cherepkov, L.V. Chernysheva, Yad. Fiz. 32, 709 (1980).

[19] G.A. Baker jr., Phys. Rev. 117, 1130 (1960).

[20] R.A. Mann, M.E. Rose, Phys. Rev. 121, 293 (1961).

[21] M.Y. Au-Yang, M.L. Cohen, Phys. Rev. 174, 468 (1968).

[22] P.K. Haff, T.A. Tombrello, Ann. Physics (New York) 86, 178 (1974).

[23] G.Ya. Korenman, S.I. Rogovaya, Yad. Fiz. 22, 754 (1975).

[24] H. Daniel, Radiat. Eff. 28, 189 (1976).

[25] T.J. Baird, Report LA-6619-T, Los Alamos 1977.

[26] J.S. Cohen, R.L. Martin, W.R. Wadt, Phys. Rev. A24, 33 (1981), Phys. Rev. A27, 1821 (1983).

[27] J.D. Garcia, N.H. Kwong, J.S. Cohen, Phys. Rev A35, 4068 (1987).

[28] M. Leon, J.H. Miller, Nucl. Phys. A 282, 461 (1977).

[29] P. Vogel, A. Winther, V. Akylas, Phys. Lett. 70B, 39 (1977).

[30] H. Daniel, Phys. Rev. Lett. 35, 1649 (1975).

[31] P. Vogel, P.K. Haff, V. Akylas, A. Winther, Nucl. Phys. A 254, 445 (1975).

[32] M. Leon, R. Seki, Phys. Rev. Letters 32, 132 (1974).

[33] M. Leon, R. Seki, Nucl. Phys. A 282, 445 (1977).

[34] H. Daniel, Z. Phys. A291, 29 (1979).

[35] H. Daniel, Ann. Phys. (NY) 129, 303 (1980).

[36] H. Daniel, Z. Physik A302, 195 (1981).

[37] J.S. Cohen, Phys. Rev. A27, 167 (1983).

[38] N.H. Kwong, J.D. Garcia, J.S. Cohen, to be published.

[39] H. Schneuwly, V.I. Pokrovsky, L.I. Ponomarev, Nucl. Phys. A312, 419 (1978).

[40] L.I. Ponomarev, Yad. Fiz. 2, 223 (1965).

[41] T. von Egidy, D.H. Jakubassa-Amundsen, F.J. Hartmann, Phys. Rev. A29, 455 (1984).

[42] G.K. Lum, C.E. Wiegand, G.L. Godfrey, Phys. Letters 65B, 43 (1976).

[43] R. Abela, W.. Kunold, R. Metzner, L.M. Simons, Helv. Phys. Acta 52, 419 (1979).

[44] H. Daniel, G. Fottner, H.Hagn, F.J. Hartmann, P. Stoeckel, W. Wilhelm, Phys. Rev. Lett. 46, 720 (1981).
G. Fottner, H. Daniel, P. Ehrhart, H. Hagn, F.J. Hartmann, E. Köhler, W. Neumann, Z. Phys. A304, 333 (1982).

[45] U. Fano, Ann. Rev. Nucl. Science 13, 1 (1963).

[46] K. Kaeser, T. Dubler, B. Robert-Tissot, L.A. Schaller, L. Schellenberg, H. Schneuwly, Helv. Physica Acta 52, 238 (1979).

[47] F.J.Hartmann, R.Bergmann, H. Daniel, H.-J.Pfeiffer, T.von Egidy, W.Wilhelm, Z. Phys. A305, 189 (1982).

[48] F.J. Hartmann, T. von Egidy, R. Bergmann, M. Kleber, H.-J. Pfeiffer, K. Springer, H. Daniel, Phys. Rev. Lett. 37, 331 (1976).

[49] R. Bacher, Ph.D. thesis, Universität Karlsruhe, 1987.

[50] P. Ehrhart, F.J. Hartmann, E. Köhler, H. Daniel, Phys. Rev. A27, 575 (1983).
P. Ehrhart, F.J. Hartmann, E. Köhler, H. Daniel, Z. Phys. A311, 259 (1983).

[51] R. Bergmann, H. Daniel, T.von Egidy, F.J. Hartmann, J.J. Reidy, W. Wilhelm, Phys. Rev. A20, 633 (1979).

[52] D. Horvath, F. Entezami, Nucl. Phys. A407, 297 (1983).

[53] T. von Egidy, F.J. Hartmann, Phys. Rev. A26, 2355 (1982).

[54] S. Stanislaus, F. Entezami, D.F. Measday, Nucl. Phys. A475, 642 (1987).

[55] R. Kunselman, J. Law, M. Leon, J. Miller, Phys. Rev. Letters 36, 446 (1976).

[56] R.L. Hutson, J.D. Knight, M. Leon, M.E. Schillaci, H.B. Knowles, J.J. Reidy, Phys. Letters 76A, 226 (1980).

[57] V.S. Evseev, T.N. Mamedov, V.S. Roganov, N.I. Kholodov, Yad. Fiz. 35, 513 (1982).

[58] R.A. Naumann, H. Daniel, P. Ehrhart, F.J. Hartmann, T. von Egidy, Phys. Rev. A31, 727 (1985).

[59] F.J. Hartmann, R. Bergmann, H. Daniel, T. von Egidy, G. Fottner, R.A. Naumann, J.J. Reidy, W. Wilhelm, Z. Phys. A308, 103 (1982).

RECENT RESULTS OF EXPERIMENTS WITH SLOW MUONS

W. Schott, H. Daniel, F.J. Hartmann, and W. Neumann[1)]

Physik-Department, Technische Universität München
D-8046 Garching, Germany

ABSTRACT

The ratio n_1/n_2 of measured spectral flux densities and, thus, inverse stopping powers S_2/S_1 of negative muons emerging from a MgF_2 and a Au moderator is larger than $S_2{}^a/S_1{}^a$ for a muon energy T less than 10 keV, with S^a obtained from proton atomic S data which are combined in case of MgF_2. Furthermore, n_1/n_2 is larger than the relation of calculated spectral flux densities using S^a values. The deviation increases with decreasing T. This is caused by the lack of decelerating electrons in the MgF_2 insulator which has an energy gap of 11 eV. Negative muons can be used as a surface probe for elemental analysis. An about monoatomic oxygen layer on a Si surface has been detected by characteristic X rays of μ O atoms which were formed by the capture of μ with $T \approx 20$ eV in O.

1. INTRODUCTION

When negative muons pass through matter, they loose energy mainly by collisions with electrons and are finally captured forming muonic atoms. As the capture takes place at very low energies[1-2)] and the capture process with the succeeding X ray emission gives a signal of the stop, μ^- particles are a useful tool for measuring the energy loss of very slow heavy particles in matter.

a) The spectral flux density n(T), that is the number of particles per unit energy and unit time entering a small sphere divided by the cross section of the sphere at the low μ^- energy T is the appropriate observable which yields the stopping power S(T) using the relation[2)]

$$n(T) \cdot S(T) = \text{const.} \tag{1}$$

The energy loss may be characterized by the collision of "fast" electrons in atomic or crystalline bound states with the slowly moving heavy particle. As only free states can be reached by the colliding electrons the band structure and in particular the gap in a crystal is expected to play a dominant role: a large gap in an insulator is expected to yield a low stopping power while in metals a high stopping power will be found.

[1)]Present address: Eidgenössische Technische Hochschule, Zürich, Switzerland

Electromagnetic Cascade and Chemistry of Exotic Atoms
Edited by L. M. Simons *et al.*, Plenum Press, New York, 1990

An example of an especially small S of MgF_2 is given. Eq.(1) holds, if the number of muons is conserved, the moderator is homogeneous, and multiple scattering and energy straggling are neglected. Proton atomic stopping power data S^a based on experiments and calculations, are available for all elements in a wide energy range[3]. In the energy region where the difference in S of μ^+ and μ^- is neglible they can be used also for muons at a reduced energy in order to match the velocity of the two particles.

b) The capture cross section $\sigma(T)$ ($T \leq 100$ eV) is obtained by n(T) and the differential capture rate

$$dN(T) = n(T) \cdot \sigma(T) \cdot dT.$$

Because of the characteristic X ray emission, the μ^- can be used as a surface probe for elemental analysis. The detection of a thin O larger on a Si surface is discussed.

2. THE STOPPING POWER OF SLOW μ^- IN MgF_2

Averaged stopping power values S^a for compound moderators may be calculated from those for elements properly combined according to the atomic weights. If solid state effects are present which change the electronic structure of the compound, the actual stopping power will be different from the averaged value, especially, in an energy regime where the transitions to single electronic states are important.

The measurement of n(T) for MgF_2 is described which corresponds with eq. (1) for $T \lesssim 10$ keV to an S much smaller than one obtains by the combination method. This is due to the large energy gap[4] (≈ 11 eV) of the MgF_2 insulator.

We have performed a corresponding experiment in 1985 with the highest-intensity beam available at that time, beam μE1 at SIN, now PSI, at Villigen, Switzerland. Similar to a previous experiment[2] muons of an energy of 54 MeV with an energy spread of 12.5 MeV (FWHM) leave the channel, pass through an array of scintillators and a degrader and enter a magnetic spectrometer through a thin window. Particles with the momentum $p \approx 9$ MeV/c and $\Delta p/p = 0.1$ are selected by the spectrometer and transported onto a mylar wedge which makes their energy more homogeneous. Then, they pass the very thin scintillator Sc6 (1.2 mg/cm^2) and leave the spectrometer vacuum chamber through a thin mylar window. On the outside there is attached, either a 0.4 mg/cm^2 thick MgF_2 or a 1 mg/cm^2 Au moderator both yielding the same energy loss. The energy of the muons emerging from the moderator (in the range 250 keV ... 1 keV), is measured by time-of-flight (TOF) with a 5.5 cm long flight path immediately following the moderator. The particles are stopped in a 0.1 mg/cm^2 Pd target on a 0.5 mm thick carbon backing. The resulting Pd 5-4 X rays are measured by a Ge diode. The half maximum acceptance cone angle of the Pd target is $\theta_{max} = 33^o$. The diode pulse is used as the start signal and the delayed Sc6 pulse as the stop signal of the TOF measurement. An event is defined by the coincidence of the scintillator telescope and the Ge detector. At an event rate of about $12 s^{-1}$ and $2 \cdot 10^6$ telescope events for the measurements with the MgF_2 and Au moderators the time for each run was about 45 h.

Fig. 1a shows the TOF spectra for the two moderators. The energy scale is drawn below. At all TOF channels (except the first) there are more counts in the spectrum obtained with the MgF_2 moderator than in the spectrum obtained with the Au moderator. However, only for $T \leq 8$ keV (cf.below) the resulting spectral flux density of MgF_2 ($n_1(T)$) is larger than that of

Au(n_2(T)). This is due to the larger multiple scattering for Au than for MgF_2. The multiple scattering has been calculated by solving a diffusion equation using an ansatz $n \propto \exp(-(\theta/\alpha)^2)$ by means of which the scattering angle α is obtained[5]. The slowing down of the muons in the different materials is taken into account in the diffusion constant which is proportional to $1/(p^2v^2X_o)$, where p and v is the muon momentum and velocity, respectively, and X_o the radiation length of the material. The stopping is calculated using proton S^a values[3] scaled for muon energies (programs SCATT, SCATT2, SCATT3). Using SCATT2, from a given TOF channel or T after the moderator the incident energy into the moderator was calculated and then further in the backward direction the incident energies into the other layers ending finally at the incident energy into the wedge. Knowing all energies α could be determined.

In fig. 1b the multiple scattering correction factors n/I, where I is the measured intensity per unit energy and time, for the MgF_2 and Au moderators are drawn vs T.

Fig. 1c shows the spectral flux densities n for the MgF_2 and Au moderators, as deduced from the TOF spectra (fig. 1a), corrected for multiple scattering, vs T. Both distributions rise towards smaller energies, go through a maximum at about 3 keV for MgF_2 and 5 keV for Au and decrease slowly. For $T \leq 8$ keV $n_1 \gg n_2$ holds.

A computer program (SCATT3) was written by means of which n_1(T) and n_2(T) could be simulated. Starting from a gaussian distribution in muon momentum n^s at the wedge with a center and width which is deduced from the spectrometer current and acceptance, respectively, n^s after the wedge in muon energies is obtained using scaled proton S^a data for calculating the energy loss. The energy straggling is taken into account by the folding of n^s and another gaussian with the variance $\sigma_e = 396.1 \sqrt{Zx/A}$, where x is the thickness in g/cm^2 and Z and A the atomic number and weight of the corresponding target yielding σ_e in keV. σ_e is due to electron collision straggling[6]. The resulting n^s distribution is taken as input for the next material and so on. Fig. 1d shows n^s after the MgF_2 and Au moderators vs T. In the whole considered energy range $n_1^s < n_2^s$ holds which is completely different to n (cf. fig. 1c). The shape of n is roughly reproduced. n^s shows, however, a steeper rise towards smaller T and a maximum at a higher energy of about 10 keV for MgF_2 and 15 keV for Au.

In fig. 1e n_1/n_2, n_1^s/n_2^s and $(S_2^a(1)/S_2^a(0))/(S_1^a(1)/S_1^a(0))$ are drawn vs T, where $S_2^a(1)$ and $S_2^a(0)$ mean the S^a values at the muon energy after and before the Au moderator, etc. The $S^a(1)$ values are divided by the $S^a(0)$ data to eliminate differences in x and A. According to eq. (1) n_1/n_2 is equal to S_2/S_1. It turns out that for $T \leq 10$ keV the measured and corrected ratio n_1/n_2 is different from the relation S_2^a/S_1^a and n_1^s/n_2^s. The deviation increases with decreasing T yielding a factor of about 6 at T = 2 keV. On the other hand, S_2^a/S_1^a and n_1^s/n_2^s agree quite well in the whole energy range.

At low muon energy the measured quotient of the spectral flux densities of MgF_2 and Au n_1/n_2 can neither be described by the inverse relation of the stopping powers S_2^a/S_1^a, nor by a simulation calculation where the stopping powers S^a are used. The reason is that MgF_2 is an insulator with an energy gap of about 11 eV, where an electron must be excited into the conduction band in order to take up momentum for decelerating a muon. Thus, the effective stopping power of MgF_2 is much smaller than that which results from a combination of S^a data for Mg and F giving rise to a much larger n_1 than one would obtain neglecting this solid state effect.

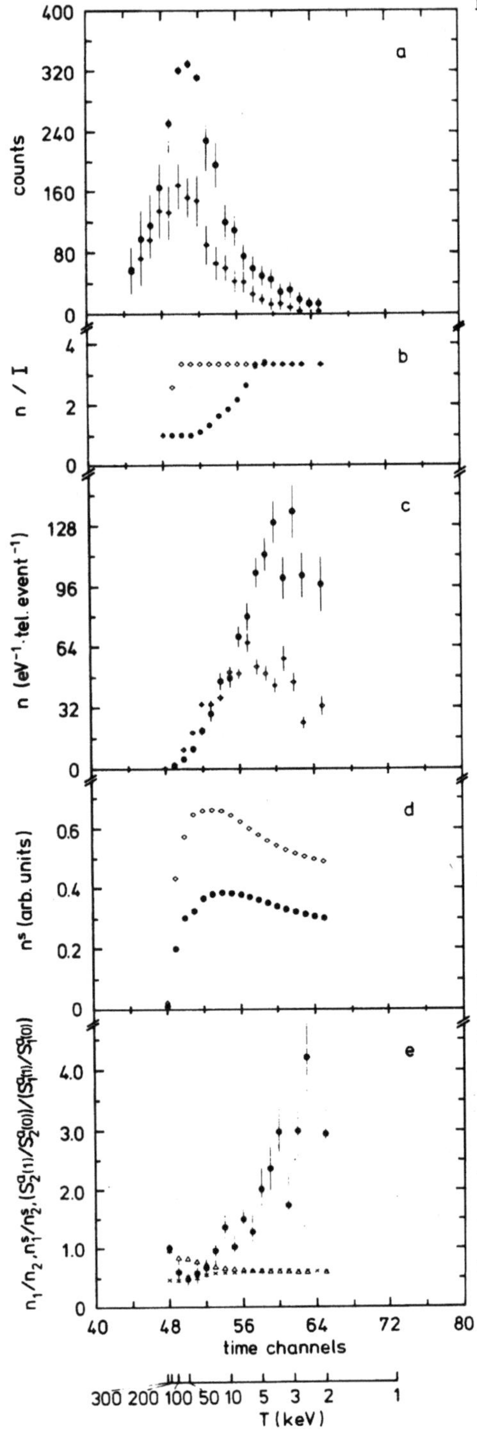

Fig.1 Several spectra and distributions vs TOF channels and muon energy T.
a. TOF spectra for the MgF_2 (full circles) and the Au moderator (diamonds).
b. Multiple scattering correction factor n/I for the MgF_2 (full circles) and the Au moderator (diamonds). n means spectral flux density corrected for multiple scattering. I is the measured intensity per unit energy and time.
c. n for the MgF_2 (full circles) and the Au moderator (diamonds).
d. Computer simulated spectral flux densities n^s for the MgF_2 (full circles) and the Au moderator (diamonds).
e. Ratios of n, n^s and atomic stopping power values S^a. The index 1 means for the MgF_2, the index 2 for the Au moderator. The number 1 and 0 in parenthesis means after and before the moderator, respectively. The full dots denote n_1/n_2, the crosses n_1^s/n_2^s and the triangles $S_2^a(1)/S_2^a(0))/(S_1^a(1)/S_1^a(0))$.

3. μ^- AS A SURFACE PROBE FOR ELEMENTAL ANALYSIS

The application of μ^- for surface analysis is based on the fact that the cross section of atomic capture of very slow μ^- is very large, of the order of the geometrical cross section.

In order to investigate the Coulomb capture a similar setup as described has been used. The experiment was performed at area πE3 of the PSI. As target we used a Si single crystal disk of 7.5 cm diameter with the (1,1,1) surface facing the incoming μ^- beam. The surface was etched with HF acid, heated to 100° C while pumping, received a "shower" of O_2 at 10^{-6} Torr for 20 s and then kept at 10^{-7} Torr for the time of the run (4 d). In this way a thin oxygen layer on Si was produced.

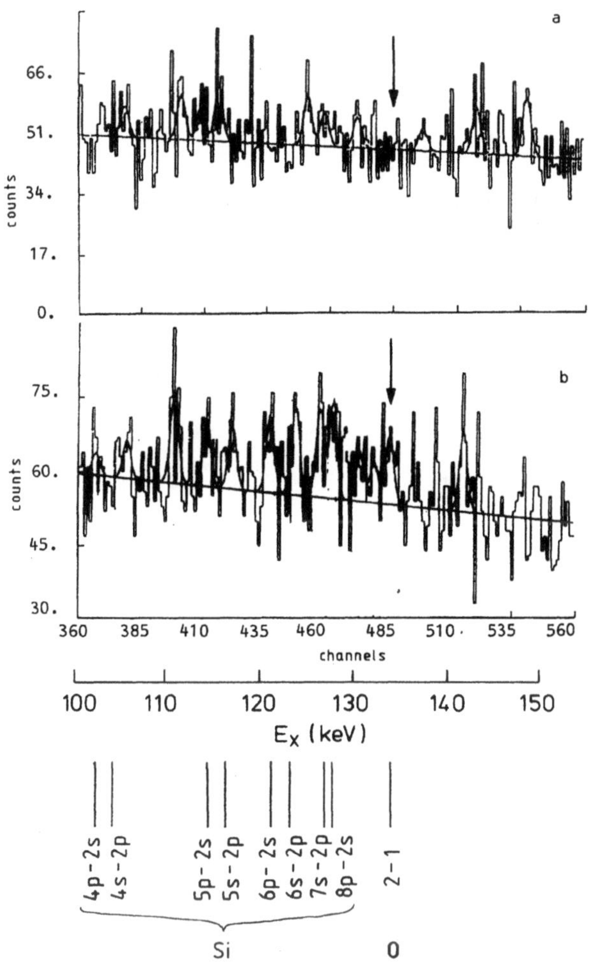

Fig. 2 X ray spectra taken with μ^- TOF corresponding to $0.5 < T \leq 20$ eV. a. "Background", where the spectrometer current I is chosen too large. b. I matched to the T range yielding an indication for the O 2p - 1 s line.

The result of the experiment is shown in fig. 2. The part of the muonic X ray spectrum around the O 2p-1s line is displayed. Fig. 2a shows the spectrum with the μ^- TOF chosen corresponding to $0.5<T\leq20$ eV but with a spectrometer current too large for selecting this μ^- energy range. Thus, fig. 2a shows "background" containing essentially Si lines. No O 2p-1s line is visible. Fig. 2b with $0.5<T\leq20$ eV and the spectrometer current for this range adjusted gives a hint for the O 2p-1s line with an intensity of 58 ± 19 counts. Si lines also appear.

The interpretation of the results is as follows: The very slow μ^- are captured in the first atom they encounter, which is in the majority O. The somewhat faster μ^- pass the O layer and produce Si X rays.

The thickness of the O layer was estimated to be $dx = (-dE)/S = 3.0 \cdot 10^{16}$ atoms/cm^2, where $-dE = 20$ eV is the energy loss at Coulomb capture and S at the corresponding proton energy of 0.18 keV is $6.6 \cdot 10^{-16}$ eV/(atoms/cm^2)[3]. Taking $a_0/2$ for the O radius the thickness of a monoatomic O layer results to be $4.6 \cdot 10^{16}$ atoms/cm^2 which yields dx = 0.7 monoatomic layers.

ACKNOWLEDGEMENTS

We wish to thank the Paul Scherrer Institute for hospitality and support and Prof. Dr. H.S. Plendl, Dr. G. Schmidt, C.A. Schug for discussions.

REFERENCES

1. E. Fermi and E. Teller, Phys. Rev. 72, 399 (1947)
2. H. Daniel et al., Phys. Rev. Lett. 46, 720 (1981)
3. J.F. Ziegler et al., The Stopping and Range of Ions in Solids, Pergamon Press (New York, 1985)
4. H.P.R. Frederikse, in American Institute of Physics Handbook, third edition, edited by D.E. Gray (Mc Graw-Hill Book Company, New York, 1972) p. 9-18
5. F.J.M. Farley et al., Nucl. Instrum. and Methods 152, 353 (1978), W. Wilhelm, Programm SCATT (1980) unpublished, W. Schott, Programs SCATT2, SCATT3 (1988), unpublished
6. N. Bohr, Phil. Mag. 30, 581 (1915)

RECENT PROGRESS IN ANTIPROTON-ATOM COLLISIONS

L.H. ANDERSEN[1], K. ELSENER[2], P. HVELPLUND[1], H. KNUDSEN[1]
S.P. MOLLER[1], E. MORENZONI[3], J.O.P. PEDERSEN[1]
and E. UGGERHOJ[1]

[1] Institute of Physics, University of Aarhus
DK-8000 Aarhus C, Denmark
[2] CERN, CH-1211 Geneve 23, Switzerland
[3] PSI (formerly SIN), CH-5234 Villigen, Switzerland

INTRODUCTION

Deceleration of negatively charged particles in matter plays a decisive role in the field of experimental physics with exotic atoms. Although at velocities much larger than the Bohr velocity, i.e. v>>c/137, it is generally expected that the cross sections for atomic collision processes (ionization, excitation, etc.) are independent of the projectile's charge, care has to be taken at intermediate and low velocities, where the Born approximation is no longer valid.

With the advent of low energy antiprotons at LEAR (CERN), an important step towards the understanding of negative particle interactions with atoms has become possible[1]. The quality of the $\bar{p}$ beam allows to measure energy loss and ionization cross sections to impact energies as low as 50 keV. In this paper, a brief summary of recent results from the LEAR experiment PS194 is given, emphasizing ionization of noble gases and the studies of energy loss in Silicon targets.

IONIZATION

In continuation of earlier experiments[2], the ratio of single to double ionization cross sections for antiproton impact on Helium, Neon and Argon has been measured. Results are now available for the entire range of impact energies from 65 keV to 20 MeV (see Fig. 1). The surprising enhancement for antiproton impact as compared to proton impact on Helium is seen for all the energies below ≈10 MeV.

For energies above 0.5 MeV, this effect was found to result from an enhancement of the double ionization cross section alone. At the lower energies, a reduced single ionization cross section is expected for antiproton impact, yielding again an enhancement in the measured ratio. Similar results as for Helium are obtained for Neon and Argon ionization. Details of the experiment and results are described in [3]).

Electromagnetic Cascade and Chemistry of Exotic Atoms
Edited by L. M. Simons *et al.*, Plenum Press, New York, 1990

ENERGY LOSS AND BARKAS-EFFECT

The theory of energy loss of fast charged particles in matter is based on calculations by Bethe[4], who derived the stopping power in the first Born approximation. Hence the Bethe result is proportional to the projectile charge squared, Z_1^2. It was thus a surprise, when Barkas et al. found that the range of negative pions was longer than that of positive pions of equal momentum, and the effect was due to a difference in the stopping power stemming from the opposite charge of the particles. The reduction in dE/dx of negative particles was later investigated with sigma-hyperons, pions and muons, but these measurements suffer from the poor quality of the particle/antiparticle beams used.

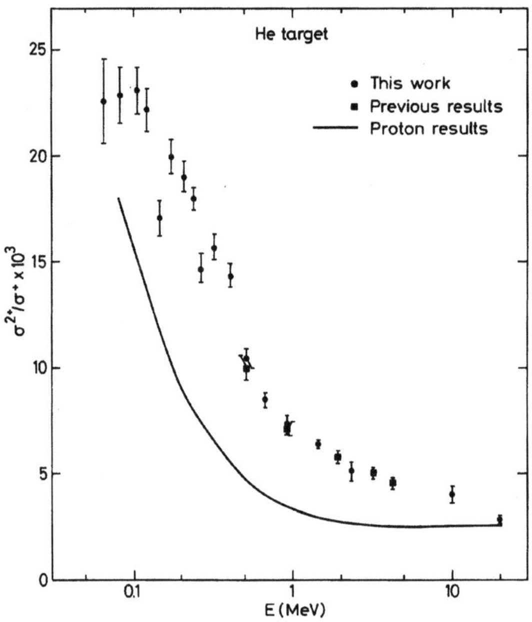

Fig. 1. Measured ratio $R=\sigma^{++}/\sigma^{+}$ for antiproton impact on Helium. Recent data covering the energy range from 65 keV to 20 MeV are compared to a fit through proton data (solid line).

The so-called Barkas effect was interpreted as a polarization effect in the stopping material, appearing as the next term (proportional to Z_1^3) in the implied Born expansion of the energy loss. Similar deviations from a strict Z_1^2 dependence also emerge from an analysis of the stopping power of protons, alpha particles and Li^{3+} projectiles. An overview of the experimental and theoretical status of the Barkas effect has been given by Andersen[5].

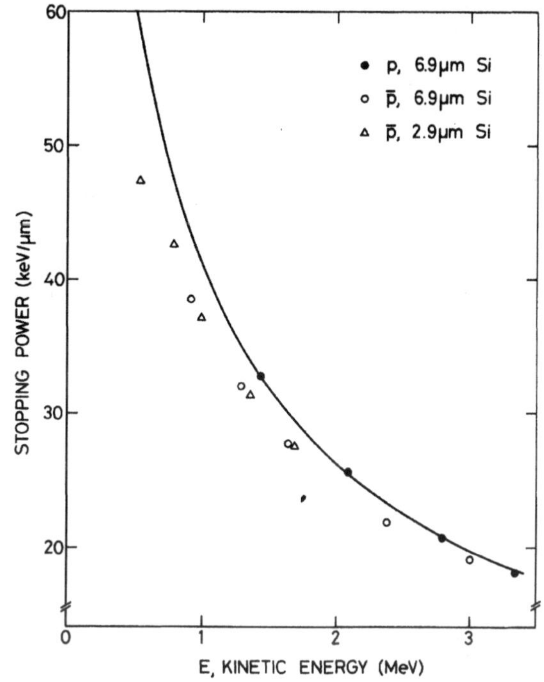

Fig. 2. Stopping power of antiprotons and protons in Silicon. Proton data in a 6.9µm Si surface barrier detector are compared to reference data from Andersen and Ziegler For energies below a few MeV, the lower stopping power for antiprotons in a 6.9 and 2.9µm detector clearly shows the Barkas effect.

We have recently investigated the energy loss of MeV antiprotons and protons at LEAR. A degraded beam was passed through a scintillator telescope to determine the mean beam energy, and the energy loss was determined in a 6.9 μm and a 2.9 μm Si detector, respectively. The experiment is described in detail in [6]). The results for both protons and antiprotons are shown in Fig. 2. The Barkas term L_1, i.e. the strength of the Z_1^3 correction to the energy loss, can be deduced from these data in the energy range of 0.5 MeV to 3 MeV, corresponding to projectile velocities ranging from 5 to 11 times the Bohr velocity. The result is shown in Fig. 3, where it is compared to theoretical estimates by Jackson and McCarthy[7].

This new data on the Barkas effect give accurate information on the difference in dE/dx between positive and negative particles, and provide further indications for possible contributions from close collisions to the L_1 term in the expansion of the stopping power[8].

FUTURE EXPERIMENTS

In a forthcoming run, it is foreseen to study the single ionization of Helium to energies as low as 10 keV. This should provide the first direct measurement of a fundamental cross section for negative particle atom collisions at such low energies. The particular interest in this case is the cross-over of proton- and antiproton induced cross sections at around 35 keV, which has been predicted by several authors[9].

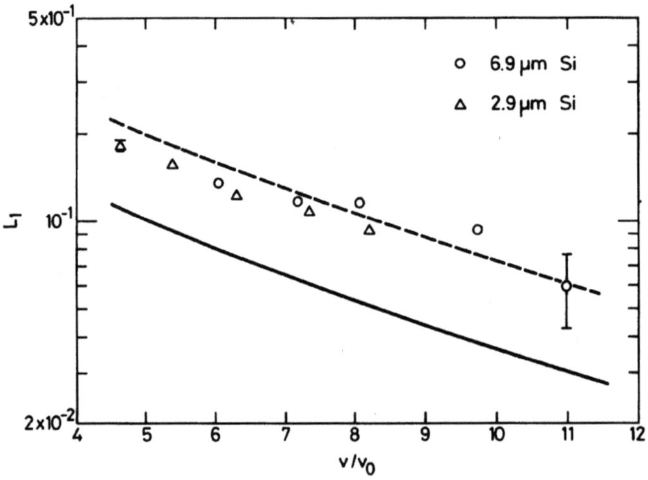

Fig. 3. The Barkas term L_1, deduced from the stopping power data of antiprotons in Silicon (see Fig. 3). The solid line shows the theoretical results of Jackson and McCarthy[7], the dashed line twice their result.

The experiment on the Barkas effect in Silicon will be continued to lower energies. If a good solid state detector of less than 1μm thickness becomes available, such measurements should be feasible down to E≈100 keV, i.e. to the region around the stopping power maximum.

Measurements of inner-shell excitation have started at LEAR, and first results on a 3 MeV $\bar{p}$ beam inducing K-Xrays in Ti, Cu, Se and Nb may soon be available. These measurements would provide a first valuable test of theories developped to describe inner-shell vacancy production at low energy by proton and ion projectiles.

Finally, the first experiment of channeling in single crystals with an MeV antiproton beam is being prepared. This would be the first time, that a slow, heavy, negatively charged particle could be channeled. There is hope that a longstanding obstacle in channeling, stemming from electron/positron channeling experiments, could be removed, and that better insight in the interaction of channeled negative particles with a crystal can be gained.

REFERENCES

1. K. Elsener, Comm. At. Mol. Phys. XXII, 263 (1989).
2. L. H. Andersen, P. Hvelplund, H. Knudsen, S. P. Møller, A. H. Sørensen, K. Elsener, K.G. Rensfelt and E. Uggerhøj, Phys. Rev. A36, 3612 (1987).
3. L. H. Andersen, P Hvelplund, H. Knudsen, S. P. Møller, J. O. P. Pedersen, S. Tang-Petersen, E. Uggerhøj, K. Elsener and E. Morenzoni, Phys. Rev. A, in print.
4. H. A. Bethe, Ann. Phys. (Leipzig) 5, 325 (1930), U. Fano, Ann. Rev. Nucl. Sci. 13, 1 (1963).
5. H. H. Andersen, in "Semiclassical Descriptions of Atomic and Nuclear Collisions", J. Bang et al. (eds.) Elsevier Sci. Publ. (1985), p.409.
6. L. H. Andersen, P. Hvelplund, H. Knudsen, S. P. Møller, J. O. P. Pedersen, E. Uggerhøj, K. Elsener and E. Morenzoni, Phys. Rev. Lett. 62, 1731 (1989).
7. J. D. Jackson and R. L. McCarthy, Phys. Rev. B6, 4131 (1972).
8. J. Lindhard, Nucl. Instr. Meth. 132, 1 (1976). J. C. Ashley, R. H. Ritchie and W. Brandt, Phys. Rev. B5, 2393 (1972).
9. P. D. Fainstein, V. H. Ponce and R. D. Rivarola, Phys. Rev. A36, 3639 (1987).

THE FORMATION AND REACTIONS OF PIONIC HYDROGEN ATOMS

D.F. Measday

Physics Dept.
University of British Columbia
Vancouver, B.C. Canada V6T 2A6

INTRODUCTION

In the last few years a variety of experiments have been undertaken at TRIUMF in order to study the properties of pionic hydrogen atoms. These atoms can be formed in any hydrogenous material and can then undergo reactions in which the pion is transferred to other elements. Such atomic reactions are difficult to study and the measurements do not always have a unique interpretation, but by observing patterns of behaviour, it is possible to build up an understanding of the mechanisms involved.

The original studies were made over many years by Petrukhin and his collaborators at Dubna. An old but still excellent review is that of Ponomarev[1]; more recently Horvath[2] has discussed the chemical effects in complex environments including organic molecules and aqueous solutions. These Russian measurements established many of the fundamental features, but were limited by the techniques then available. At TRIUMF we have been able to take advantage of modern instrumentation to improve on the previous measurements and thereby to quantify the effects more precisely.

The first thing to realize is that pions, muons and other negatively charged particles resemble each other in the manner in which they stop and are captured by atoms in the medium, because the effects are dominated by the Coulomb interaction. However, as they cascade towards the nucleus, the strong interactions come into play and radically affect the final outcome. Thus one can benefit from information gathered with one probe, and apply it to another, so long as care is exercised. Thus the study of pionic atoms can be useful in understanding muonic atoms, and can contribute to our investigations of muon catalyzed fusion.

The dominating characteristic of exotic hydrogen ($\mu p, \pi p, \mathrm{K}p, \overline{p}p,$ etc.) is that the atom is neutral and does not experience the Pauli exclusion principle and so can burrow through neighbouring atoms. For those accustomed to thinking of muonic hydrogen, it is essential to emphasize that the hadronic atoms experience strong absorption in nS states, quite high in the cascade, because of Stark mixing. Thus the overall lifetime is much shorter than the 2μs available to muonic hydrogen atoms. For example, in liquid hydrogen the total cascade time, as given in the review of Burhop[3], is

$$\begin{array}{ll} \pi^- p & \tau = (2.3 \pm 0.6) \times 10^{-12} s \\ K^- p & \tau \sim 4 \times 10^{-12} s \\ \Sigma^- p & \tau \sim 5 \times 10^{-12} s \end{array}$$

(In gases the lifetime is increased significantly.)

Electromagnetic Cascade and Chemistry of Exotic Atoms
Edited by L. M. Simons *et al.*, Plenum Press, New York, 1990

As the cascade times for hadronic atoms are so short, any transfer must occur from excited states and although the rate is higher, the overall effect is a much reduced total transfer. Thus for hadronic atoms, capture on the hydrogen is still observed in most materials (eg. H_2O, C_5H_{12}, LiH etc.), whereas for muons a few percent of another element in hydrogen will result in total transfer.

Ponomarev[1] introduced the useful concept of the large mesic molecule. He suggested that the negatively charged particle undergoes molecular capture first, ejecting loosely bound electrons and passing through a "doorway state" which is a molecular orbital. Then the particle is attracted to one or other of the component atoms. It is not clear whether the captured particle samples the whole molecule but it certainly is aware of several neighbouring atoms, so we shall employ the language of this model as a convenient description. If an exotic hydrogen atom is formed, it is released from the hydrogen site and can travel freely in the local environment. Frequently the negatively charged particle can then be transferred to neighbouring nuclei, whether they be other protons, deuterons, or heavier nuclei. It seems reasonable to suppose that they can be transferred to atoms of the same molecule (internal transfer) as often as to atoms of neighbouring molecules (external transfer), which can be the same type of molecule, or a different type. Our experiments cannot distinguish between internal transfer and prior selection of the heavier atom in the cascade, so there is always a lingering ambiguity. Studies of the mesic X-ray cascade might give some information on this topic however.

EXPERIMENTAL TECHNIQUES

We detect only the final denouement. The measurement consists of determining the relative π^o production when a π^- stops in a variety of substances. In the materials we use, a π^o can come only from pionic absorption on hydrogen, which goes through one of only two reactions, viz.

$$\pi^- + p \rightarrow \gamma + n \quad [B.R. = 39.32\,(13)\%, ref.4]$$
$$\pi^- + p \rightarrow \pi^o + n \quad [B.R. = 60.68(13)\%, ref.4]$$

The only other nuclei which can produce a π^o in any quantity are

$$^3He \quad B.R. \sim 15\%, ref.5$$
$$D \quad B.R. = 1.45\,(19) \times 10^{-4}, ref.6$$

For all other nuclei the reaction at rest is either energetically forbidden (^{12}C,^{16}O, etc.) or strongly suppressed to the level of $\sim 10^{-5}$ or even lower.[7] Charge exchange can also occur in flight but we have deliberately kept our pion beam energy very low (<20MeV), so no in flight reactions have been detected from the target. Thus for all practical purposes, the π^o production comes only from π^- capture on a proton of the target material.

The π^o is detected by its characteristic decay into two gamma rays. The energy of these gamma rays is measured from coincident events in two large NaI crystals in a 180° geometry to maximize the effective solid angle, see Fig.1. To eliminate a background contribution from pions stopping in S_3, this scintillator is made from deuterated plastic. This system is the one that we have used in most of our previous π^o experiments, the only minor modification is that we now define an event by either summing the two gamma-ray energies or multiplying them together. This produces a convenient and precise definition of the events. A typical spectrum taken with a gas target is illustrated in Fig.2.

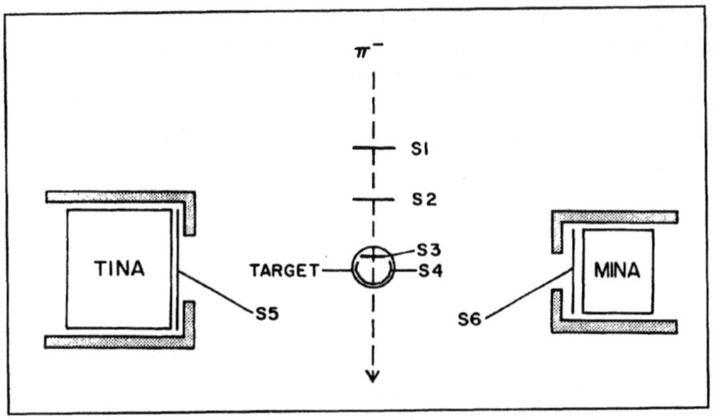

Figure 1 Experimental lay-out of the two large NaI crystals.

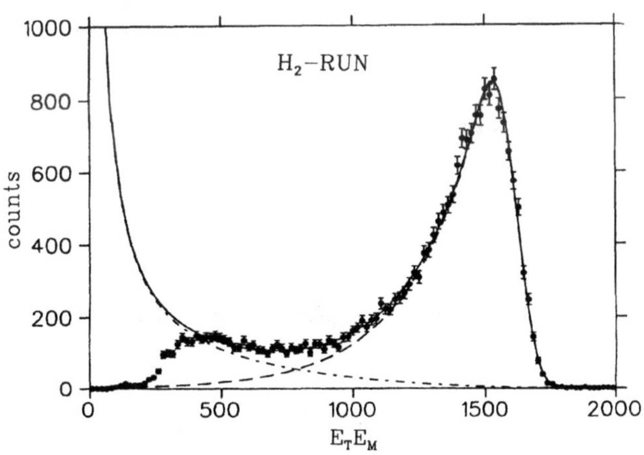

Figure 2 Spectrum of the product of the γ-ray energies for coincident events.

TRANSFER IN GASES

Let us first consider the experiments with gaseous targets. We have studied mixtures of H_2 and D_2 at pressures of 100 atmospheres. This is an extension of our previous work[8] on 50:50 mixtures and on HD. The gases stay as H_2 and D_2 unless subjected to high temperatures or catalysts. We assume that the atomic capture cross-section is the same for H_2 and D_2 molecules, so that on a 50:50 mixture for example, half the pions would be captured by the H_2 molecules. However less than half the pions are finally absorbed by the protons, so transfer must be occurring. In Fig.3 we illustrate the relative π^0 production as the percentage of deuterium is increased. One might anticipate a straight line but the data clearly fall below, indicating preferential transfer of the pions to deuterium. (The two data sets were taken several months apart and analyzed separately to determine our systematic errors.)

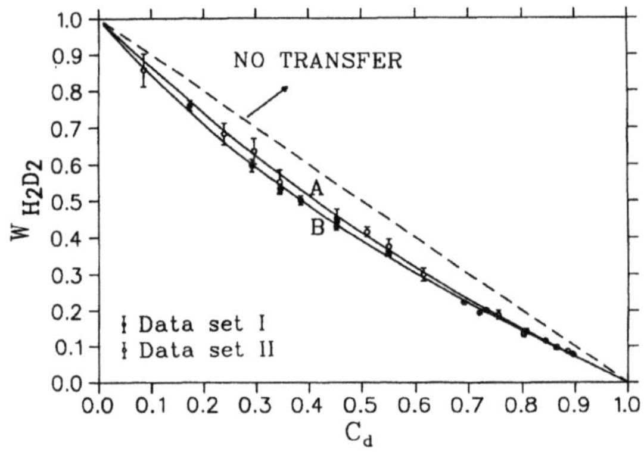

Figure 3 Relative π^0 production in the gas target for $H_2 + D_2$ mixtures versus the atomic percentage of deuterium.

If we define W_H as the probability for capture on the proton and Q as the transfer probability, then

$$W_H = \frac{C_p}{C_p + C_d}(1 - Q(C)) = \frac{1 - Q(C)}{1 + C}$$

where C is the relative concentration of deuterium (i.e. $C = C_d/C_p$). We plot this transfer probability as a function of concentration in Fig.4. Our results are compared to the earlier results of Petrukhin et al.[9] and we see that the agreement is quite satisfactory, however our results are internally much more consistent and cover a wider range of concentration.

We have also included our earlier result[8] as well as the recent results from Gatchina of Kravtsov et al.[10] The curves result from a simple model proposed by Petrukhin and Prokoshkin.[9]

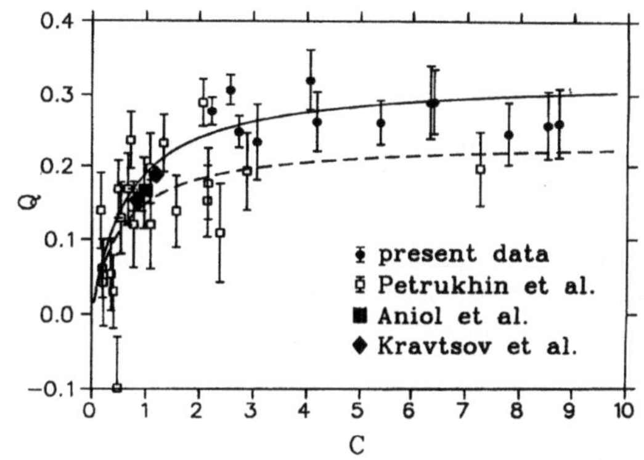

Figure 4 Transfer probably Q as a function of the relative concentration of deuterium and hydrogen. The curves are fits to the model of Petrukhin and Prokoshkin.

The primary idea is that all nuclear captures occur in collisions because of the Stark mixing (only in rarefied gases can the hadron reach the 1s state). Evidence also supports the supposition that transfer occurs one way only, from hydrogen to deuterium. Thus, for absorption to occur on a proton, the hadron must first be captured into a large orbit around the proton and then be absorbed in a collision with another hydrogen, or a deuteron, but not be transferred to the deuteron. Thus defining

1. C_p and C_d as the fractional concentrations of protons and deuterons respectively;
2. $\beta_{pp}C_p$ as the rate for nuclear capture on the proton in πp + H collisions;
3. $\beta_{pd}C_d$ as the rate for nuclear capture on the proton in πp + D collisions;
4. $\lambda_{pd}C_d$ as the transfer rate to the deuteron in πp + D collisions;

then the probability for nuclear capture on the proton is

$$W_H = \frac{C_p}{C_p + C_d} \cdot \frac{\beta_{pp}C_p + \beta_{pd}C_d}{\beta_{pp}C_p + \beta_{pp}C_p + \lambda_{pd}C_d}$$

and the transfer probability is given by:

$$Q = \frac{\lambda_{pd}C_d}{\lambda_{pd}C_d + \beta_{pp}C_p + \beta_{pd}Cd} = \frac{\Lambda \cdot C}{1 + \kappa \cdot C + \Lambda \cdot C}$$

$$\text{where } \kappa = \beta_{pd}/\beta_{pp}$$
$$\Lambda = \lambda_{pd}/\beta_{pp}$$
$$\text{and } \frac{\kappa}{\Lambda} = \frac{\beta_{pd}}{\lambda_{pd}}$$

i.e. κ/Λ is the ratio of nuclear capture probability to transfer probability on (π^-p+D) collisions.

In Fig.4 the full line is our best fit to all data and the dashed line is the fit obtained by Petrukhin and Prokoshkin. The parameters are listed in Table 1. Unfortunately there is some correlation between the parameters κ and Λ, as can be seen by comparing our results to the world fit, so the actual errors are hard to assess.

Table 1. Comparison of the fit parameters: the present measurement is a common fit of both data sets; the world fit comprises our data as well as the data by Aniol et al.[8] and Kravtsov et al.[10]

Parameter	Present Measurement	Petrukhin et al.[9]	World Fit
Λ	0.65±0.07	0.4±0.1	0.45±0.04
κ	1.40±0.22	1.3±0.4	0.93±0.14
κ/Λ	2.2±0.4	3.3±1.3	2.1±0.4
Q(C→ ∞)	(32±3)%	(23±4)%	(33±3)%

In an interesting analysis of these results Kravtsov et al.[10] have shown that most transfers occur from the n=4 and n=5 excited states of the πp system and they also confirmed that backtransfer is negligible. According to their calculations the average kinetic energy of the πp atom during the relevant reactions is about 1eV which is somewhat higher than thought previously, as most de-excitation mechanisms for the πp atom transmit the energy to an electron, so the recoil energy is very small. However more and more evidence is accumulating that the πp system gains a kinetic energy of about 1eV from de-excitation during the cascade to the n = 4 or 5 state from which the transfer and absorption occurs.

We have tested the pressure dependence of the transfer and found no effect in the range 33 to 100 atm. It is difficult to do a much larger range with our equipment. It is interesting to note that an experiment with liquid hydrogen, by Derrick et al., found

Q = (28 ± 13)% at C = 0.84, a result that is compatible with our gas measurements which are at much lower densities.

As a check on backgrounds we filled the target with nitrogen and observed no π^o production (<7 × 10^{-5} at 90% C.L.). The $^{14}N(\pi^-,\pi^o)^{14}C$ reaction at rest is allowed, being exothermic by 3.94 MeV.

TRANSFER IN LIQUIDS

Let us now briefly consider a related system. We investigated mixtures of water (H_2O) and heavy water (D_2O). We observe transfer in a similar manner to that found in gaseous mixtures. To first order the presence of the oxygen is not affecting the reactions between pionic hydrogen and deuterium. Any collision with an oxygen atom is like meeting Pacman, it is all over for the πp (or πd) system, so in pure water πp absorption has a probability of only 0.35%.

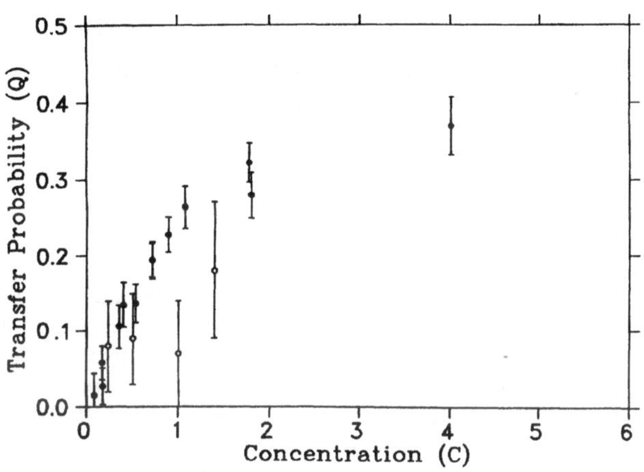

Figure 5 Transfer probability Q for mixtures of H_2O and D_2O.

In Fig.5 we illustrate our results and compare them with the previous data of Petrukhin et al.[12] Our results confirm the effect, however, with greater precision. If we fit these data to the model proposed for gases we obtain similar parameters except that the "transfer" for infinite concentration is higher (54 ±10%) compared to that for gases (33±3%). We suggest that the difference is caused by the fact that in the liquid most hydrogen (especially for higher deuterium concentrations) will be in the form HOD, so that, in the initial capture, some preference could be shown for the deuterium. This confirms our previous comparison of HD and $H_2 + D_2$ in gases.[8]

Table 2

Charge Exchange Probabilities for Alcohols

The data are normalized to a polyethylene value of 12.9×10^{-3}.

Target	Formula	N_{π°	$W_H(\times10^{-3})$
Water	H_2O	3802	3.52±0.07
Methanol	CH_3OH	10438	7.47±0.09
Deuterated methanol	CD_3OH	807	0.845±0.031
Deuterated methanol	CH_3OD	3884	5.55±0.11
Ethanol	CH_3CH_2OH	4305	8.88±0.16
Propan-1-ol	$CH_3(CH_2)_2OH$	5975	9.98±0.17
Butan-1-ol	$CH_3(CH_2)_3OH$	6838	10.70±0.04
Butan-2-ol	$CH_3CH_2CH(OH)CH_3$	7471	10.85±0.17
2-Methyl-propan-2-ol	$(CH_3)_3COH$	6256	10.66±0.18
Pentan-1-ol	$CH_3(CH_2)_4OH$	6569	10.95±0.18
Hexan-1-ol	$CH_3(CH_2)_5OH$	6565	11.03±0.18
Cyclohexanol	$(CH_2)_5CHOH$	3143	8.23±0.18
Heptan-1-ol	$CH_3(CH_2)_6OH$	5796	11.29±0.19
Octan-1-ol	$CH_3(CH_2)_7OH$	7211	11.54±0.17
Decan-1-ol	$CH_3(CH_2)_9OH$	5160	11.96±0.22

In a very illuminating comparison we studied the π° production probability for various alcohols from methanol up to decanol, see Table 2. The π° probability increases monotonically with the size of the molecule. A simple model which interprets this effect quite well is to postulate that the pion capture on the hydrogen depends on the position of the hydrogen in the molecule. This model was suggested by Smith for the anhydrides and a preliminary account of that work was presented in a conference.[13] Thus if we hypothesize that the probability for capture on the hydrogen of the OH group is R_H^O, for a hydrogen atom attached to the neighbouring carbon atom is R_H^α, and for all other hydrogen atoms is R_H^β then, assuming that capture on carbon has a probability R_C and on oxygen R_O, the probability for the final experimental observation of π° is given by

$$W_H = \frac{n_H^\alpha R_H^\alpha + R_H^O + n_H^\beta R_H^\beta}{n_H^\alpha R_H^\alpha + R_H^O + n_H^\beta R_H^\beta + n_C R_C + R_O}$$

where n is the appropriate number of atoms for that particular molecule. It is worth emphasizing that the parameter R encompasses both the initial selection of that atom plus subsequent transfer to and fro. The best fit is given in Table 3 and illustrated in Fig.6, which includes the datum for water.

We see that the model fits very well and seems a reasonable approach. We have confirmed the interpretation with an alternative technique, using deuterated methanols. Thus we measured the hydrogen capture probability for CD_3OH, CH_3OD and CH_3OH. The results are given in Table 4.

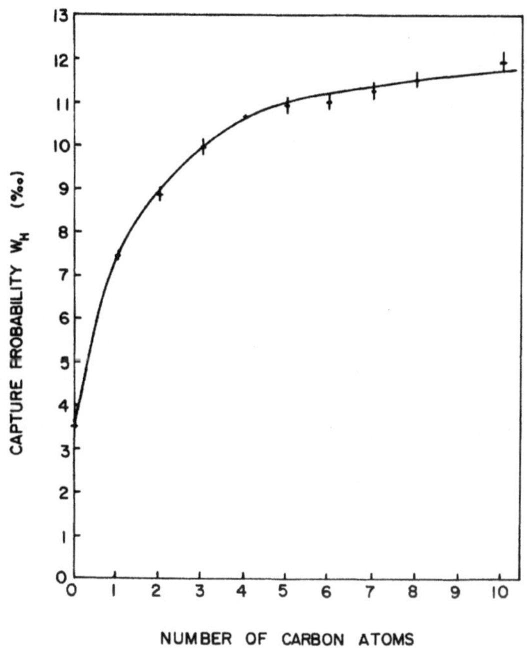

Figure 6 Fit of the Smith model to the capture probability in alcohols (water is also included).

Table 3
Best fit capture ratios for alcohols

Capture Ratio	Best Fit
$R^{\beta}{}_{H}/R_C$	$(6.62\pm0.35)\times10^{-3}$
$R^{\alpha}{}_{H}/R_C$	$(5.73\pm0.59)\times10^{-3}$
$R^{O}{}_{H}/R_C$	$(2.96\pm0.59)\times10^{-3}$
R_H/R_C	1.7 ± 0.2

Table 4

Capture probabilities for deuterated methanols (all in units of 10^{-3})

	W_H(exp)	Corrected (H_2/D_2)	Corrected (H_2O/D_2O)	Alcohol Series
CD_3OH	0.845±0.031	1.1	1.3	1.09±0.15
CH_3OD	5.55±0.11	6.1	6.2	6.32±0.65
CH_3OH	7.47±0.09	(7.2)	(7.5)	7.41

The interesting observation is that the sum of the capture probabilities in CD_3OH and CH_3OD falls far short of the probability for CH_3OH. If we correct for "external transfer", i.e. transfer after the πp atom is formed, then we use the transfer probabilities that we observe for H_2/D_2 gas mixtures and obtain the next column. This is an improvement but not perfect. If we wish to include preference for the deuterium in the initial choice we can correct using the "transfer" probabilities found for H_2O/D_2O. The results are better, although marginally so because of the errors. Now we may compare with the results obtained from the alcohol series and obtain the final column (using the fit from Table 3). The agreement is very satisfactory and supports the whole approach.

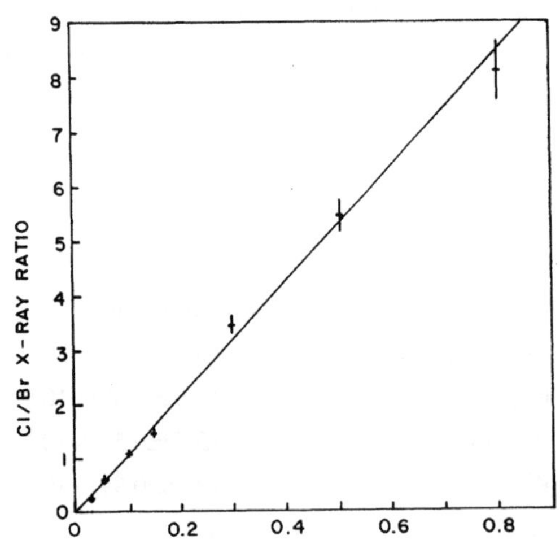

Figure 7 Concentration dependence of the intensity ratio of the Cl(3d→2p) and Br(4f→3d) X-rays.

It is also interesting to discuss our results on external transfer. In this we wish to establish that transfer occurs in liquids and in a manner compatible with observations in solids and gases. Petrukhin et al.[11] attempted a measurement at the same time as they studied H_2O/D_2O mixtures, but they were unable to establish whether it was transfer that was occurring, or whether the data could be interpreted as excessive capture by the added molecule. They chose mixtures of $C_{10}H_{22}$ + CCl_4, as well as mixtures of C_6H_6 + CCl_4. We followed the same philosophy of adding carbon tetrachloride, because it has no hydrogen. For the hydrogenous material, however, we selected bromodecane ($C_{10}H_{21}Br$), so that we could tag the stops on this molecule using mesic X-rays. We used the 4f-3d X-ray in bromine ($E = 223.53 \pm 0.10$ keV) and the 3d-2p X-ray in chlorine ($E = 150.66 \pm 0.10$ keV). The energies for chlorine are consistent with other measurements.[14,15] Apparently there are no previous measurements for bromine.

In Fig.7 we plot the ratio of the X-ray yields against the concentration of the carbon tetrachloride and find a linear relationship. This agrees with previous observations in solids[16,17] for which no effect (<10%) was observed over a concentration range of 400:1, but is different from that observed in gas mixtures for which non-linearities up to about 50% have been observed.[18,19] We can now turn our attention to the π° production as a function of the concentration. We assume that the number of π^- stopping on the bromodecane is proportional to the intensity of the bromine X-ray line, thus the ratio of π° yield is a measure of the transfer of pions from the πp system to the added material, in this case carbon tetrachloride. In Fig.8 we present our data which exhibit a significant drop with carbon tetrachloride concentration, which proves that external transfer is strong and severely depleting the π° production. We have attempted to model this behaviour, but always end up with too many adjustable parameters to make a unique and quantitative interpretation. However it is certainly correct to make the qualitative statement that external transfer to the CCl_4 molecules is very important and must be included in any comprehensive analysis. We have recently confirmed these observations using mixtures of bromodecane with trifluoro-trichloro-ethane (CF_3CCl_3).

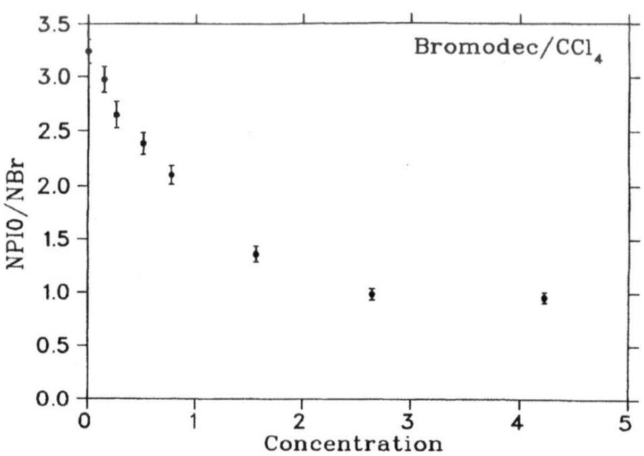

Figure 8 Ratio of π° production to bromine X-ray yield versus carbon tetrachloride concentration in mixtures with bromodecane.

Finally it is worth mentioning that we have confirmed the suppression of π° production when a π^{-} is stopped in benzene or its derivatives. The double bond seems to suppress capture by the hydrogen, even in straight compounds, but in aromatic compounds the effect can be as large as a factor of two. This has practical implications because most scintillators are benzene derivatives, so π° production is quite low for such hydrocarbons. Some typical results are listed in Table 5. It is interesting that the Japanese scintillator SCSN 38 has lower π° production than the North American products. It is probably because SCSN 38 is based on polystyrene, i.e. $(C_6H_5{\cdot}CH{\cdot}CH_2)_n$, whereas Nuclear Enterprises and Bicron use as the base polyvinyltoluene, i.e., $(CH_3{\cdot}C_6H_4{\cdot}CH{\cdot}_2)_n$.

KAONS AND ANTIPROTONS

Another useful comparison is with heavier particles, for example kaons. In the recent experiment a KEK to study hypernuclei,[20] they used a plastic scintillator (SCSN 38) as a live target. They found that the kaons were absorbed on the hydrogen about 1.1% of the time, which is to be compared with 0.36% for pions, so this indicates that kaons do not exhibit such a strong suppression of absorption by the hydrogen.

The only other evidence is for propane because it was used in bubble chambers. Unfortunately there is no direct pion data but if we fit the paraffin series with a model with equal weighting for all hydrogen atoms, we obtain $R_H/R_C = 6.23 \times 10^{-3}$ and then can calculate the capture probability by hydrogen for pions in propane, viz. $W_H = 1.63\%$. (We have renormalized the data in Horvath's review to our normalization of $(CH_2)_n$ as 12.9×10^{-3}, instead of his value of 14.5×10^{-3}). For kaons we have the results of C. van der Velde-Wilcke et al.[21] of (4.0 ± 0.4%) and of Murphy et al.[22] of (3.2 ± 0.4)% which are in reasonable agreement yet significantly higher than for pions.

Table 5

Material	$N_{\pi^{\circ}}$	W_H (units of 10^{-3})
Cyclohexane $[C_6H_{12}]$	1831	12.30±0.35
Benzene $[C_6H_6]$	1417	3.42±-0.10
Toluene $[C_6H_5CH_3]$	1445	4.97±0.15
Ethylbenzene $[C_6H_5C_2H_5)]$	1208	5.86±0.19
p-Xylene $[C_6H_4(CH_3)_2]$	1434	6.30±0.23
Biphenyl $[C_6H_5C_6H_5]$	832	3.19±0.12
SCSN 38	382	3.61±0.20
BC 412	417	5.16±0.27
NE 102	349	4.77±0.27
UVT acrylic	350	4.22±0.30
BBQ	341	3.87±0.22

For antiprotons the value of (11±3)% was found by Pawlewiz et al.[23]. The trend is quite clear and speaks for itself, although further confirmation of these data would be important and is one of the many tasks awaiting a KAON factory. The interpretation given by these authors was that the transfer probability for antiprotons was very small because the cascade time is much quicker. However the evidence that exists, although admittedly of poor quality, indicates that the cascade time for K^- and Σ^- is longer! Because the masses of the KAON and the antiproton are so large, mass effects could easily modify the characteristics of the cascade, so the full answer to the problem may involve both mechanisms.

ACKNOWLEDGEMENTS

The work described here is the result of many years of intense effort by a large number of people. The programme was catalyzed by the sojourn in Vancouver of D. Horváth, who continues to give valuable advice and assistance. The interest and support of Prof. D.F. Jackson has provided valuable womanpower and manpower, first Dr. J.R.H. Smith who suggested the simple molecular model, and more recently Dr. M.R. Harston, who prevented us from poisoning ourselves. The rock around which all this maelstrom swirled was Dr. S. Stanislaus, whose patient dedication has saved us from drowning in our own data. More recently Dr. P. Weber has taken up the torch, whilst Dr. K.A. Aniol continues to fire penetrating salvoes from Los Angeles.

REFERENCES

1. L.I. Ponomarev, Ann. Rev. Nucl. Sci. *23*:395 (1973).
2. D. Horváth, Radiochimica Acta *28*:241 (1981).
3. E.H.S. Burhop in High Energy Physics Vol. III, Ed. E.H.S. Burhop, Academic Press (1969).
4. J. Spuller, D. Berghofer, M.D. Hasinoff, R. MacDonald, D.F. Measday, M. Salomon, T. Suzuki, J-M. Poutissou, and J.K.P. Lee, Phys. Lett. *67B*:479 (1977).
5. G. Backenstoss, M. Izycki, W. Kowald, P. Weber, H.J. Weyer, S. Ljungfelt, U. Mankin, G. Schmidt, and H. Ullrich, Nucl. Phys. *A448*:567 (1986).
6. R. MacDonald, D.S. Beder, D. Berghofer, M.D. Hasinoff, D.F. Measday, M. Salomon, J. Spuller, T. Suzuki, J-M. Poutissou, R. Poutissou, P. Depommier, and J.K.P. Lee, Phys. Rev. Lett. *38*:746 (1977).
7. B. Bassalleck, F. Corriveau, M.D. Hasinoff, T. Marks, D.F. Measday, J-M. Poutissou, and M. Salomon, Nucl. Phys. *A362*:445 (1981).
8. K.A. Aniol, D.F. Measday, M.D. Hasinoff, H.W. Roser, A. Bagheri, F. Entezami, C.J. Virtue, J.M. Stadlbauer, D. Horváth, M. Salomon, and B.C. Robertson, Phys. Rev. *A28*:2684 (1983).
9. V.I. Petrukhin and Yu. D. Prokoshkin, Sov. Phys. JETP *29*:274 (1969).
10. A.V. Kravtsov, A. Yu. Mayorov, A.I. Mikhailov, S. Yu. Ovchinnikov, N.P. Popov, V.M. Suvorov, and A.I. Shchetkovsky, Muon Catalyzed Fusion 2:199 (1988).
11. K. Derrick, M. Derrick, J.C. Fetkovich, T.H. Fields, E.G. Pewitt, and G.B. Yodh, Phys. Rev. *151*:82 (1966).
12. V.I. Petrukhin, V.E. Risin, and V.M. Suvorov, Sov. J. Nucl. Phys. *19*:317 (1974).
13. J.R.H. Smith, A.S. Clough, D.F. Jackson, K.R. Smith, D.F. Measday, F. Entezami, A. Noble, S. Stanislaus, C. Virtue, M. Salomon, and K.A. Aniol, Nucl. Inst. Meth. *A242*:465 (1986).
14. G. Backenstoss, Ann. Rev. Nucl. Sci. *20*:467 (1970).
15. I. Schwanner, G. Backenstoss, W. Kowald, L. Tauscher, H-J. Weyer, D. Gotta, and H. Ullrich, Nucl. Phys. *A412*:253 (1984).

16. R. Bergmann, H. Daniel, T. von Egidy, F.J. Hartmann, J.J. Reidy, and W. Wilhelm, Phys. Rev. *A20*:633 (1979).
17. R.A. Naumann, G. Schmidt, J.D. Knight, L.F. Mausner, C.J. Orth, and M.E. Schillaci, Phys. Rev. *A21*:639 (1980).
18. P. Ehrhart, F.J. Hartmann, E. Köhler, and H. Daniel, Phys. Rev. *A27*:575 (1983).
19. P. Ehrhart, F.J. Hartmann, E. Köhler, and H. Daniel, Z. Phys. *A311*:259 (1983).
20. T. Yamazaki, Nucl. Phys. *A434*:363c (1985).
21. C. van der Velde-Wilquet, G.S. Keyes, J. Sacton, J.H. Wickens, D.H. Davis, and D.N. Tovee, Nuovo Cim. Lett. *5*:1099 (1972).
22. C.T. Murphy, G. Keyes, M. Saha, and M. Tanaka, Phys. Rev. *D9*:1248 (1974).
23. W.T. Pawlewicz, C.T. Murphy, J.G. Fetkovich, T. Dombeck, M. Derrick, and T. Wangle, Phys. Rev. *D2*:2538 (1970).

COLLISION PROCESSES IN THE FORMATION OF LIGHT EXOTIC ATOMS

DEZSŐ HORVÁTH

Central Research Institute for Physics
H-1525 Budapest, P.O. Box 49, Hungary

The formation of exotic hydrogen atoms is discussed on the basis of the available experimental information. Direct evidence shows the significant role of the molecular mechanism and the various collisional processes.

Pionic hydrogen atoms are ideal model systems for the study of exotic atoms due to the unique signal of $p\pi^-$ decay, the charge-exchange reaction following pion absorption in proton

$$p\pi^- \rightarrow n\ \pi^0; \quad \pi^0 \rightarrow \gamma\gamma. \tag{1}$$

Reaction (1) made it possible to study the formation and decay of pionic hydrogen atoms in quite various chemical systems [1-3], gas mixtures, liquids and solid coumpounds. The recent results of pion chemistry have been reviewed by D. F. Measday at this meeting [4]. The revival of interest toward meson transfer is connected to its extreme importance in muon catalyzed fusion [5]. The goal of the present work is to summarize the situation and set the newest models against the experimental results.

The experiments quoted here [6-10] have been performed in Dubna, in Gatchina and in Vancouver with the aim of clarifying the atomic processes involved in pion capture in hydrogen. This information is needed to interpret the pion chemistry experiments. It is well established that the probability with which a stopped pion is absorbed by a proton in a chemical system containing hydrogen is closely related to the electronic environment of the proton in the molecule but we still lack a quantitative description of the elementary processes involved.

The well-known model of *large mesic molecules* (LMM) of Ponomarev [1-3] describes pion capture by proton in three independent, consecutive steps: Coulomb capture of the pion in a mesic molecular orbit via external Auger process by probability P; transition of the pion from a molecular to a $p\pi^-$ atomic state by probability Q; and finally, nuclear capture of π^- by a proton with probability R (Fig. 1). Thus, according to the LMM model the probability that a stopped pion will be absorbed by a proton is

$$W = P\ Q\ R. \tag{2}$$

It is amazing that after more than two decades of intensive experimental study this rather rough model still seems to work quite well. In this time many details of quantities P and R were uncovered by the systematic investigation [6-11] of gas mixtures containing ^{1}H or ^{3}He, the only two isotopes with high enough branching ratios for reaction (1).

Two alternative approaches have been recently suggested to replace the LMM model by H. Daniel *et al.* [12] and D. F. Jackson *et al.* [13-16]. Both have questioned the molecular mechanism of $p\pi^-$ formation and Jackson makes an attempt to describe the de-excitation and decay of the $p\pi^-$ atom via non-collisional processes.

Daniel *et al.* [12] have studied muonic X-ray spectra in aqueous solutions. They claim that (i) there is no evidence for the molecular mechanism of $p\pi^-$ formation and (ii) the muon (and pion) transfer is responsible for all observable effects of meson capture in hydrogen and so the molecular mechanisms are ruled out.

Indeed, although there is a great number of experimental results showing strong molecular effects in pion capture by hydrogen [1-3] many of them can be interpreted in terms of a pure transfer model as assumed in [12]. The exceptions which prove the great importance of the molecular mechanism are effects observed in systems of the same chemical composition but different bond structure. As the pion transfer reaction

$$p\pi^- + Z \longrightarrow p + Z\pi^-, \tag{3}$$

according to the generally accepted picture [1-3], occurs in random collisions of the neutral $p\pi^-$ atom with the nuclei of other atoms of the system it should be independent (apart from some second-order screening effects) of the molecular bonding of the atoms involved. Any influence of the molecular bond shows thereby the role of the molecular mechanism.

The most spectacular effect of this kind is, probably, the many times quoted large difference between the probability of pion capture by proton in hydrazine and in a gaseous mixture of hydrogen and nitrogen [1]:

$$W\left(N_2 + 2H_2\right) / W\left(N_2\,H_4\right) \approx 30. \tag{4}$$

Somewhat less overwhelming but maybe even more convincing is the effect due to a molecular bond between hydrogen isotopes [11] as here even the atomic number is unchanged:

$$W\left(H_2 + D_2\right) / W\left(\;H\,D\;\right) = 1.23 \pm 0.03. \tag{5}$$

For completeness let us mention some pure chemical effects like those due to the disruption of hydrogen bonds in hydrogen-bonded liquids when heated above critical temperature [17, 18]:

$$Q\;(H_2O) = 2.0 \pm 0.1;\quad Q\;(NH_3) = 1.51 \pm 0.12 \tag{6}$$

where

$$Q = W\left(T > T_c\right) / W\left(T < T_m\right)$$

ie. the pion capture probability in hydrogen is much higher in supercritical fluid than in solid phase. Another example is the effect of double bonds in organic compounds of the same chemical composition, e.g., W is higher in polyethylene than in ethylene by 30 % [8].

Apparently, H. Daniel *et al.* were unaware of these results although they have quoted review [1] in their paper [12]. Let us add for the sake of clarity that the molecular mechanism of $p\pi^-$ formation and the significant role of pion transfer (3) do not contradict each other at all as they are consecutive processes in the phenomenological model used for the interpretation of pion capture data [2-4, 6-10, 19].

As opposed to [12], D. F. Jackson [13-16], while accepting an effect of the molecular bond, has tried to discard the collisional processes of deexcitation and transfer. According to her picture (sometimes called the *Surrey model*) the $p\pi^-$ atom is formed via direct atomic capture of the pion on the proton but the freshly formed pionic hydrogen atom may lose its π^- in transfer to the still close-by heavy atom Z via tunnelling along the Z-H bond.

D. F. Jackson's model contradicts the generally accepted picture of mesoatomic processes as the role of radiative meson capture is considered to be negligible on the base of rate considerations as compared to the Auger-capture and at the same time the collisional transfer to be very important, especially for muons. It was also shown in [19] that this model contradicts the evidence available from the pion capture experiments made with gas mixtures and simple compounds. However, as the Surrey group has published papers [15, 16] with the same considerations two years after [19] it is probably worth to briefly return to these problems.

The first crucial statement is that we do see at least one collisional process, namely the transfer reaction (3) from $p\pi^-$ to Z. It is observed in the mixtures of hydrogen with any other gas and also in mixtures C_mH_n + Z with no chemical bond of type H-Z. The second statement is that a possible coexistence of collisional and non-collisional processes should lead to a density dependence of the various measured probabilities - like in the case of muon capture - as the collision rate is proportional to the density and that dependence is cancelled in the pure collisional case. However, all attempts to find any density dependence of $p\pi^-$ capture have been so far unsuccessful. The pion capture probability did not change in ethane [20] or in H_2+D_2 and HD [11] when the gas density was changed by two orders of magnitude. Even condensation or solidification cannot change the measured W probability unless changes in chemical bonding are involved as shown in the case of gaseous and liquid hydrocarbons [20], solid and liquid ammonia [18] and solid and melted aquacomplexes [3]. Thus the role of the process of non-collisional 'inner' transfer can be neglected.

The negligible - if any - role of the assumed 'inner' transfer is further supported by the fact that the concentration dependence of the W pion capture probability is similar in H_2 + D_2 and H_2O + D_2O mixtures [21, 22]. According to the Surrey model the 'inner' transfer should dominate in the water molecule whereas it is naturally impossible in the hydrogen gas mixture.

Another interesting result of this analysis of possible processes is the prevalence of the collisional mechanism in the absorption of the π^- in proton as compared to the radiative process. This was shown by the Dubna group long ago [6, 7] in gas mixture studies.

In conclusion we can state that - in spite of its rather rough assumptions - the model of large mesic molecules of Ponomarev [1, 2] gives still the best description of the processes involved in pion capture in hydrogen bound in molecules. Both new attempts, H. Daniel's *transfer model* and D. F. Jackson's *Surrey model* are inconsistent with the present set of experimental information.

ACKNOWLEDGEMENTS

The author is indebted to D. F. Jackson, D. F. Measday and G. Torelli for illuminating discussions.

REFERENCES

[1] S. S. Gershtein, V. I. Petrukhin, L. I. Ponomarev, Yu. D. Prokoshkin: Usp. Fiz. Nauk **97**, 3 (1969) [Sov. Phys. Usp. **12**, 1 (1970)]
[2] L. I. Ponomarev: Annu. Rev. Nucl. Sci. **23**, 395 (1973)
[3] D. Horváth: Radiochim. Acta **28**, 241 (1981)
[4] D. F. Measday: In the present volume.
[5] See, e.g., the talks of J. S. Cohen, V. E. Markushin, C. Petitjean and L. I. Ponomarev in the present volume.
[6] V. I. Petrukhin, Yu. D. Prokoshkin, V. M. Suvorov: Zh. Eksp. Teor. Fiz. **55**, 2173 (1968) [Sov. Phys. JETP **28**, 1155 (1969)]
[7] V. I. Petrukhin, V. M. Suvorov: Zh. Eksp. Teor. Fiz. **70**, 1145 (1976) [Sov. Phys. JETP **43**, 595 (1976)]
[8] V. I. Petrukhin, V. E. Risin, I. F. Samenkova, V. M. Suvorov: Zh. Eksp. Teor. Fiz. **69**, 1883 (1975) [Sov. Phys. JETP **42**, 955 (1976)]
[9] V. M. Bystritskii, V. A. Vasilyev, A. V. Zhelamkov, V. I. Petrukhin, V. E. Risin, V. M. Suvorov, B. A. Khomenko, D. Horváth: In *Mesons in Matter*, Proc. Intern. Symp. on Meson Chemistry and Mesomolecular Processes in Matter, edited by V. N. Pokrovskii, (Dubna, USSR, 1977), p. 223.
[10] A. V. Bannikov, B. Lévay, V. I. Petrukhin, V. A. Vasilyev, L. M. Kochenda, A. A. Markov, V. I. Medvedev, G. L. Sokolov, I. I. Strakovsky, D. Horváth: Nucl. Phys. **A403**, 515 (1983)
[11] K. A. Aniol, D. F. Measday, M. D. Hasinoff, H. W. Roser, A. Bagheri, F. Entezami, C. J. Virtue, J. M. Stadlbauer, D. Horváth, M. Salomon, B. C. Robertson: Phys. Rev. **A28**, 2684 (1983)
[12] H. Daniel, F. J. Hartmann, R. A. Naumann, J. J. Reidy: Phys. Rev. Lett. **56**, 448 (1986)
[13] D. F. Jackson, C. A. Lewis, K. O'Leary: Phys. Rev. A **25**, 3262 (1982)
[14] D. F. Jackson: Phys. Lett. **95A**, 487 (1983)
[15] C. Tranquille, D. F. Jackson: Phys. Rev. **A34**, 742 (1986)
[16] D. F. Jackson, J. R. H. Smith: Phys. Rev. **A34**, 763 (1986)
[17] A. K. Kachalkin, Z. V. Krumshtein, A. P. Minkova, V. I. Petrukhin, V. M. Suvorov, D. Horváth, I. A. Yutlandov: Zh. Eksp. Teor. Fiz. **77**, 26 (1979) [Sov. Phys. JETP **50**, 12 (1979)]
[18] D. Horváth, A. V. Bannikov, A. K. Kachalkin, B. Lévay, V. I. Petrukhin, V. A. Vasilyev, I. A. Yutlandov, I. I. Strakovsky: Chem. Phys. Lett. **87**, 504 (1982)
[19] D. Horváth: Phys. Rev. **A30**, 2123 (1984)
[20] V. I. Petrukhin, Yu. D. Prokoshkin: Dokl. Akad. Nauk SSSR **160**, 71 (1965) [Sov. Phys. Doklady **10**, 33 (1965)]
[21] P. Weber, D. S. Armstrong, D. F. Measday, A. J. Noble, S. Stanislaus, M. R. Harston, K. A. Aniol, D. Horváth: Submitted to Phys. Rev. A.
[22] S. Stanislaus, D. F. Measday, D. Vetterli, P. Weber, K. A. Aniol, M. Harston, D. S. Armstrong: Submitted to Phys. Lett. A.

II

CASCADE STUDIES IN EXOTIC HYDROGEN ATOMS

Alice was puzzled. "In *our* country," she remarked, "there is only one day at a time." The Red Queen said "That's a poor thin way of doing things. Now *here*, we mostly have days and nights two or three at a time and sometimes in the winter we take as many as five nights together - for warmth, you know."

Lewis Carroll: Through the Looking-Glass

CASCADE PROCESSES IN EXOTIC ATOMS WITH Z=1

Valery MARKUSHIN

Kurchatov Atomic Energy Institute
Pl.Kurchatova, Moscow 123182, USSR

ABSTRACT

Basis cascade processes in lightest exotic atoms are considered. Recent experiments and their implications for the theory are discussed.

INTRODUCTION

When a negative particle stops in hydrogen, an exotic hydrogen-like atom is formed in high excited state[1]. Thus, studying the interaction of stopped $\mu^-, \pi^-, \bar{p}$, etc. with hydrogen and deuterium we deal with the atomic cascade. Some attractive features immediately come out, such as to make use of possibility to select the initial states from which the nuclear reaction in hadronic atoms takes place. For some while the theory of the atomic cascade was mainly considered as a tool for getting the information about the low energy interaction and static particle parameters. The progress in experimental technique which makes it possible to use the good quality beams of slow muons and antiprotons (such as at PSI(SIN) and LEAR/CERN) and to study the cascade processes in a wide range of external conditions ("muon bottle", "cyclotron trap") and with high energy resolution (crystal spectrometer), as well as the fast developments in new fields, e.g., muon catalyzed fusion, have brought about the necessity of more detailed investigations of the atomic cascade. The physics of atomic cascade proves to be very rich in different

Electromagnetic Cascade and Chemistry of Exotic Atoms
Edited by L. M. Simons *et al.*, Plenum Press, New York, 1990

phenomena having the significance of their own.

The exotic atoms with Z=1 have an important peculiarity that distinguishes them from all the others. They have no own electrons and being electroneutral can easily penetrate close to the nuclei of surrounding atoms where the strong electric field essentially affects the atomic cascade[2,3] which becomes rather sensitive to the external conditions.

In the first part of this lecture we will consider the basis cascade processes, their relative importance under different conditions, and some theoretical and experimental results in the X-ray yields from muonic and antiprotonic hydrogen. The second part will be devoted to the problem of acceleration and deceleration of the atoms during the cascade. Some problems to be solved will be outlined.

STANDARD CASCADE MODEL

The first detailed calculations of the cascade in exotic hydrogen were performed by Leon and Bethe[4] for π^-p and K^-p atoms. The further theoretical studies of the lightest exotic atoms[5-7] have resulted in the present-day version of the cascade model which we shall refer to as the standard cascade model (SCM). The basic cascade processes taken into consideration in the SCM are listed in Table 1.

Table 1. The processes considered in the standard cascade model (π^-p atom is taken as an example).

Process	The dependence on the collision energy	Ref.
Radiative deexcitation		
$(\pi^-p)_i \longrightarrow (\pi^-p)_f+\gamma$	—	see 4
External Auger effect	no (BA)	4
$(\pi^-p)_i+H_2 \longrightarrow (\pi^-p)_f+e^-+H_2^+$	weak (EA)	8,9
Chemical deexcitation		
$(\pi^-p)_i+H_2 \longrightarrow (\pi^-p)_f+H+H$	?	4
Stark mixing		
$(\pi^-p)_{nl}+H_2 \longrightarrow (\pi^-p)_{nl'}+H_2$	moderate	4,10
Coulomb collision		
$(\pi^-p)_i+p \longrightarrow (\pi^-p)_f+p$	strong	5,9
Nuclear absorption		
$(\pi^-p)_{nS} \longrightarrow \pi^0+n,\ \gamma+n$	—	

The initial population of the atomic states is determined by the competition between the slowing down and the atomic capture, this problem is discussed in the lecture by Cohen[11] (see also Refs. 12–14). In the first approximation one can consider the initial orbit of the captured particle to be close to that of the displaced electron, i.e. the exotic atom to be formed in the state with principal quantum number $n \sim \sqrt{M/m_e}$ where M is the reduced mass of the particles forming the atom, m_e is the electron mass.

Radiation vs. Collisional Deexcitation

All the deexcitation processes but the radiation are the collisional ones, i.e., their rates are proportional to the density N. The radiative deexcitation rates are determined by the well-known formula for E1 transitions

$$\Gamma^{\gamma}_{if} = \frac{4}{3} \alpha R^2_{if} \omega^3 \tag{1}$$

where R_{if} is the dipole matrix element and $\omega_{if}=E_i-E_f$ is the transition energy. It follows from (1) that for given initial state i the transitions with the maximum change in principal quantum number $\Delta n=n_i-n_f$ are favorable. With n_i increasing the radiative deexcitation rate drops very fast, the following estimation for the total rate averaged over the sublevels with different orbital angular momentum l being valid

$$\Gamma^{\gamma}_{n} = \frac{1}{n^2} \sum_{n_f, l_f, l} (2l+1)\, \Gamma^{\gamma}_{if} \simeq \frac{4}{3} M \alpha^5 \frac{\ln n}{n^5} \tag{2}$$

To get the estimation of the relative importance of the collisional deexcitation due to the external Auger effect and the radiative deexcitation one can use the well-known formula for the probability of the electron conversion in E1 transition, $\beta(E1)$, that gives the ratio of corresponding rates as following

$$\frac{\Gamma^{e}_{if}}{\Gamma^{\gamma}_{if}} = \beta(E1)\, N_e \pi a_o^3 \tag{3}$$

where

$$\beta(E1) = \frac{\alpha}{2} \left(\frac{2m_e}{\omega_{if}}\right)^{7/2} \tag{4}$$

Here factor $N_e \pi a_o^3$ is the electron density $N_e=N$ normalized to that seen by the nucleus in hydrogen, a_o being the electron Bohr radius ($N_e \pi a_o^3 \sim 10^{-2}$ at liquid hydrogen density N_o). At $\omega_{if} \simeq 10$ keV $\beta(E1) \simeq 1$, and the smaller is the transition energy ω_{if}, the higher

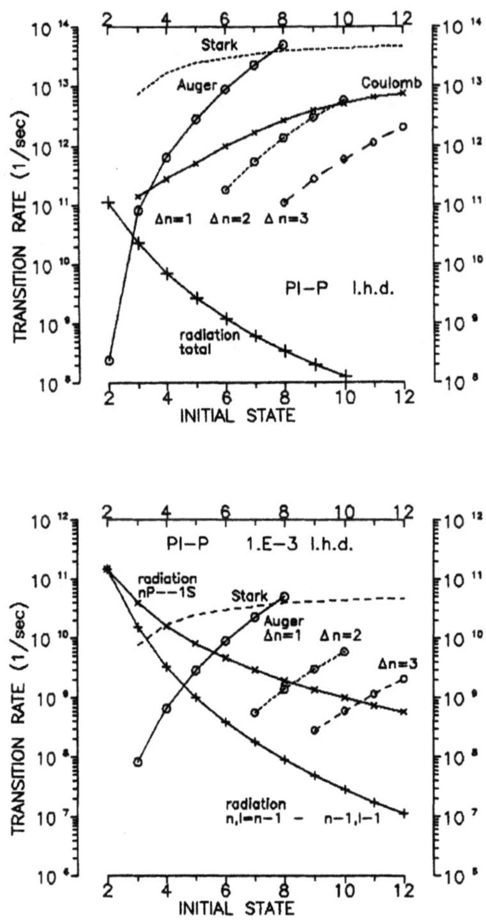

Fig.1. The rates of the cascade processes for π^-p atom in (a) liquid and (b) gaseous ($N=10^{-3}N_o$) hydrogen. For the radiative deexcitation shown are the total average rates (a) and the rates of nP→1S and circular transitions (b). The Auger deexcitation rates are calculated in the Born approximation[4]. The Stark mixing rates are calculated in the Leon-Bethe model[4] at T=1eV.

the conversion coefficient $\beta(E1)$ becomes. It means that the radiation can dominate the deexcitation only for the low-excited states or/and at low density.

The Auger deexcitation rates were calculated in the Born approximation[4] (BA) and in the eikonal approximation[8]. In contrast to the energy independent BA result the latter gives a weak energy dependence of the rates. At atomic kinetic energy $T \sim 1 eV$ and $n \approx 5 \div 7$ the rates calculated with both methods differ by a factor of about two. The relationship between the radiative and Auger deexcitation at different densities is demonstrated in Fig.1. The most important are the Auger transitions with minimum change of the principal quantum number Δn which is sufficient for the electron ionization.

The rates of Coulomb deexcitation calculated by Menshikov[9] at $n>7$ are by an order of magnitude smaller than those obtained by Bracci and Fiorentini[5], the difference appears to be due to using the dipole approximation in Ref.5. Formally extrapolating formulas from Ref.9 in the region $n=3,4$ one gets the results of the same order of magnitude as the earlier calculations.

Stark mixing

The Stark mixing plays an important role when the small and electrically neutral exotic hydrogen atom moves through the electric field inside the hydrogen-atom electron cloud[2-4], that can be easily seen from the following estimation . The frequency ω_{St} of Stark transitions in a homogeneous electric field $E=\alpha^{1/2}a_o^{-2}$ is given by the formula

$$\omega_{St} = \langle nl|\alpha^{1/2}\vec{r}E|n,l-1\rangle = \alpha a_o^{-2}\langle nl|r|n,l-1\rangle = n(n^2-l^2)^{1/2}\cdot 2\cdot 10^{14}s^{-1}$$

($\mu^- p$ is taken as example) that should be compared with the characteristic time of the collision $\tau_{St} \sim a_o/v \sim 0.5\cdot 10^{-14}s$ (kinetic energy $T \approx 1eV$). Since $\omega_{St}\tau_{St} > 1$, there is enough time for the atom to have many transitions between different states nl with given n. The cross section of the Stark mixing must be in order of magnitude comparable with the geometrical size of the hydrogen atom, i.e. the rate of the Stark mixing has the scale $Nv\pi a_o^2 \sim 5\cdot 10^{12}(N/N_o)\ s^{-1}$. The Stark mixing rates were calculated with the impact-parameter method in Refs.4,10, the atom being assumed to move along a straight-line trajectory. The main results are as following.

1. There is a critical value ρ_n of the impact parameter such that for $\rho>\rho_n$ the probability of the Stark transition is negligibly small and for $\rho<\rho_n$ the mixing is complete so that the mixing rate can be defined as following

$$\Gamma_n^{St} = Nv\pi\rho_n^2 \qquad (6)$$

Among all the collisional processes the Stark mixing is the most rapid one (see Fig.1).

2. The Stark mixing provides the statistical population of the sublevels nl for given n, if the levels can be considered as degenerated in the Stark splitting scale. This property is realized either in the scheme of the complete mixing[7], when it is assumed that after the Stark collision the atom "forgets" its initial state, or in the shuffling model[6] where the dipole-like mixing is assumed.

3. In hadronic atoms the strong interaction shifts of the nS states lift the degeneracy and, thus, play an important role resulting in some reduction of the rate of the absorption via the Stark mixing[4,6].

4. The Stark mixing rate depends on the kinetic energy T. For the sake of simplicity used are the effective rates calculated for some value of T.

Cascade in muonic hydrogen

In the absence of the absorption, that is the case for the muonic atoms of hydrogen isotopes, the final stage of the cascade is determined mainly by the radiative and Auger de-excitation. At high density (l.h.d.) the Stark mixing maintains the statistical population of the nl-sublevels with the same n for all the states. With density decreasing the radiative deexcitation gains more importance and starting from some state n_c dominates in the cascade, the Stark mixing being effective only for $n>n_c$. The role of the radiative de-excitation in liquid and gaseous (STP) hydrogen is demonstrated in Table 2.

The X-ray yields for muonic hydrogen have been measured in a wide density range[15-20] and are in a good agreement with the results of the calculations within the SCM[6,7,20,21], as it is shown in Fig.2. Thus, we can conclude that the standard cascade model provides a rather good description of the relative importance of the radiative and collisional de-excitation at the final stage of the cascade.

Table 2. The X-ray yields Y_{if} and populations p_f of the states n_f of $\mu^- p$ atom in liquid and gaseous ($N=10^{-3}N_o$) hydrogen (Ref.7).

liq.	final state		
	n_f=1	2	3
n_i=4	0.002	0.001	0.001
3	0.06	0.047	
2	0.89		
$\sum_i Y_{if}$	0.953	0.049	0.001
p_f	1.	0.937	0.988

gas	final state		
	n_f=1	2	3
n_i=4	0.21	0.17	0.20
3	0.17	0.26	
2	0.5		
$\sum_i Y_{if}$	1	0.50	0.26
p_f	1.	0.50	0.43

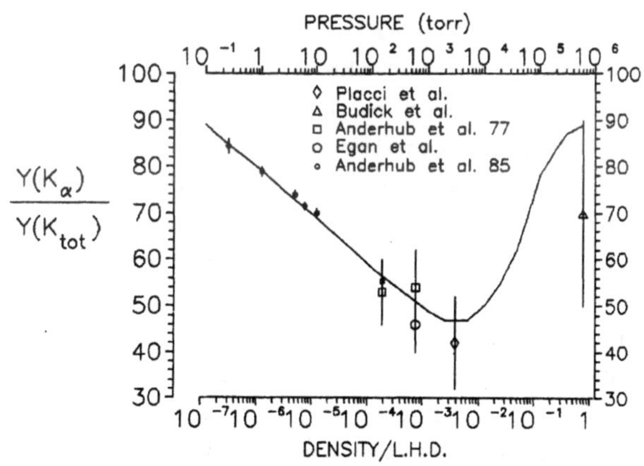

FIG.2. The density dependence of the K_α-line relative yield for $\mu^- p$ atom. The experimental data are from Refs.15-18,20, the curve is the SCM result[6,7,20]

Nuclear Absorption in Hadronic Atoms

With the absorption involved the cascade becomes more complicated. The rates of absorption from different atomic states nl are proportional to the overlap of the atomic wave function with the nuclear interaction region, the most important the nuclear interaction being in the nS-states and, in some cases such as $\bar{p}p$ and $\bar{p}d$, the nP-states. There are two mechanisms of the absorption, one corresponds to the direct absorption following the deexcitation to the nS (nP) state, the other is due to the Stark mixing[2-4] which feeds in the nS (nP) states where the nuclear interaction takes place (Day-Snow-Sucher mechanism). Being a peculiar feature of the exotic atoms with Z=1 the DSS mechanism results in the intensive absorption from the excited states that makes the observation of the radiative transitions between the low-lying levels an extremely difficult experimental problem.

The rates of the absorption via the Stark mixing were calculated by Leon and Bethe[4]. In view of simplifying assumptions used in this calculation (the straight line trajectory approximation, the fixed direction of the electric field) some adjustment of the Stark mixing rate is required . The tuning is performed with the help of the so called Stark factor k_{STK} which multiplies all the Stark mixing rates

$$\Gamma_n^{St} = k_{STK} N \, v \, \pi \rho_n^2 \tag{7}$$

Here $\rho_n = \rho_n(T)$ is the radius of the Stark collisions for the states with principal quantum number n. The kinetic energy T is also considered as a fitting parameter that provides the necessary flexibility of the cascade model. Some examples of the sensitivity of the results of the cascade calculations to the Stark parameter k_{STK} can be found in Ref.6. The more refined calculations of the Stark mixing rates for pionic hydrogen[10], where the effects of the electric field rotation have been taken into account, give parameter k_{STK} depending on n and varying between 1.5 and 4.

Among the exotic atoms with Z=1 the best studied are the antiprotonic hydrogen and deuterium, the X-ray yields have been measured in a wide density range in the experiments[22-26] at LEAR. The experimental data have been analyzed within the SCM in Refs.23,26 and generally, reasonable fits to the yield values were obtained for $\bar{p}p$ atom. The experimentally measured shift and width of the 1S state were used in these calculations. In Ref.23 the data were fitted by adjusting the Stark mixing parameter, the kinetic energy being varied from the thermal one (0.026 eV) to 2 eV. Over this wide range the value of χ^2 remained constant and the parameter k_{STK} was found to be strongly correlated with the value of T (in particular, k_{STK}=1.55±0.09 at T=1eV). In the analysis performed in Ref.26 the best fit to similar sample of data were obtained at T= =$1.0^{+0.3}_{-0.2}$ eV and k_{STK}=2.2±0.2. Figure 3 shows the experimental data on the L-line yields in comparison with the SCM calculations.

For the $\bar{p}d$ results, there are some yields the fit to which is unsatisfactory (see for discussion Ref.23). One of the problems in this case is the lack of the strong interaction shift and width measurement so that it is necessary to use theoretical predictions. More experimental data are required for further theoretical analysis.

Approach to the cascade calculations[27], which is somewhat different from the SCM, is based on the simulation of the collisions between $p\bar{p}$ and H_2 with the classical trajectory Monte Carlo method as it was suggested earlier by Landua and Klempt for exotic helium[28] (see also Ref.29). It allows one to include,

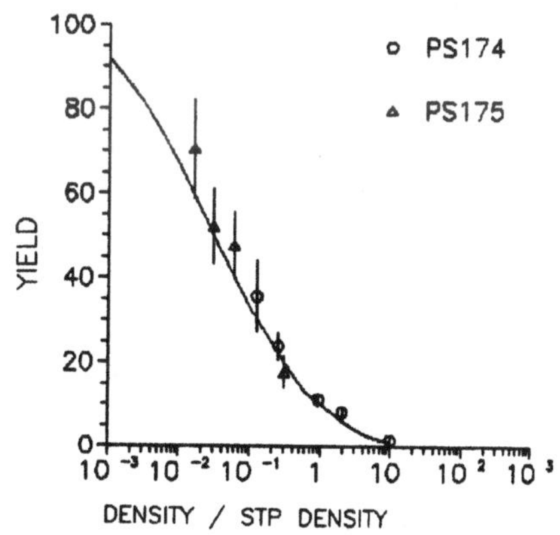

Fig.3. Absolute yield of the L_α-line for antiprotonic hydrogen vs. density (STP conditions correspond to $N=1.3\cdot10^{-3}N_o$). Experimental data are from Ref.23 (o) and Ref.26 (Δ). The curve shows the best fit[26] obtained with the Borie-Leon version of the SCM.

in principle, the acceleration and deceleration processes and thus to get rid of any free cascade parameters. With the present-day version of the model[27] the calculations were done for kinetic energy 1 eV and 0.026 eV and in both cases the results obtained are in reasonable agreement with the experimental data on the X-ray yields from antiprotonic hydrogen. The cascade time of antiprotons in gaseous hydrogen at 1 atm measured in the experiment[30] $\tau^{exp} = (5.1 \pm 0.7)$ ns is also in a fair agreement with the calculation[27] $\tau^{th} = 3$ ns.

The experimental data on $\pi^- p$, $K^- p$, and $\Sigma^- p$ atoms are rather restricted, measured are the capture and cascade time in liquid hydrogen[31-36] and the X-ray yields for $\pi^- p$ in gaseous hydrogen[37-39].

KINETIC ENERGY DISTRIBUTION PROBLEM

Until now we have treated the atomic cascade as if it developed without any interplay between the internal and external degrees of freedom of the exotic atom, the Coulomb nl states and c.m.s. motion, respectively, the kinetic energy T being considered as the cascade parameter. This is probably a suitable approximation for some stage of the cascade but definitely not true for the whole life history of the atom. In this section we concern the problem of the evolution of the kinetic energy distribution during the cascade. After presentation of the relevant experimental data we discuss some possible mechanisms of the acceleration in the atomic transitions.

Muonic hydrogen/deuterium diffusion

The experiment on the muonic hydrogen diffusion[40,41] (see references therein for the earlier works) provides the direct information on the kinetic energy distribution of the atoms ($\mu^- p$ or $\mu^- d$) reaching the ground state after the atom formation when muons are stopped in H_2 or D_2 gas filling the interstices of an array of planar equidistantly spaced foils.

In the recent experiment performed by MISSISSIPI-MUNICH-PSI(SIN)-VIENNA-WILLIAM&MARY collaboration at PSI(SIN)[40,41] the diffusion of muonic atoms was studied in wide density range (pressure 0.094 - 8 bar for D_2 and 0.047 - 8 bar for H_2), so that both the kinetic energy distribution and the rescattering effects can be investigated by measuring the time elapsing between the atom formation coincident with the muon incoming and its arrival at the foil. The example of the experimental time distribution is shown in Fig.4 . Striking is the near coincidence of the hydrogen data with that taken for deuterium with the doubled gap. The data analysis leads to the following conclusions.

1. The velocity distribution of μd atoms in the ground state has a mean kinetic energy T=1.8±0.1 eV.

2. The mean kinetic energy for μp is about twice as much as for μd.

3. The spectra of different shapes are consistent with the observed distribution, for example, Maxwellian and rectangular. More data are needed for the spectrum shape discrimination.

To derive the interpretation of these results we use the language of the SCM. The schematics of the cascade at $N = 10^{-4} N_0$ (0.08 atm) is shown in Fig.5. The cascade is ended up with the fast radiative deexitation, so that at $n \leq 5$ neither acceleration nor deceleration take place. The preceding stage of the cascade ($n>5$) is determined by the collisional processes where the transitions with minimum change in principal quantum number ($\Delta n=1$) are dominant. From this we conclude that the observed kinetic energy distribution with the mean value T=1.8 eV for μd and

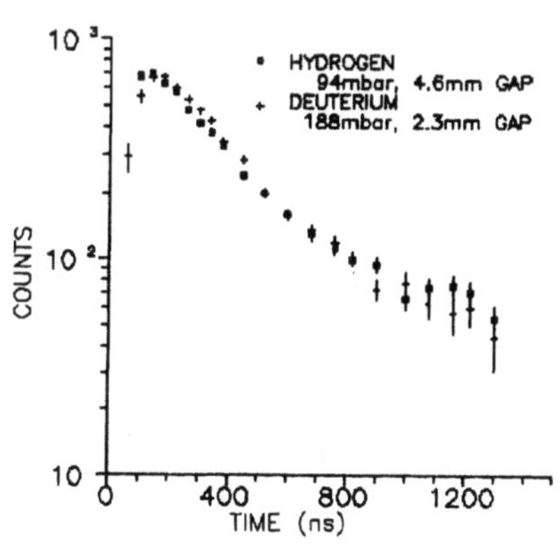

Fig.4. The time distribution for $\mu^- p$ and $\mu^- p$ atoms arrival at the foil after diffusion in gas at low pressure.

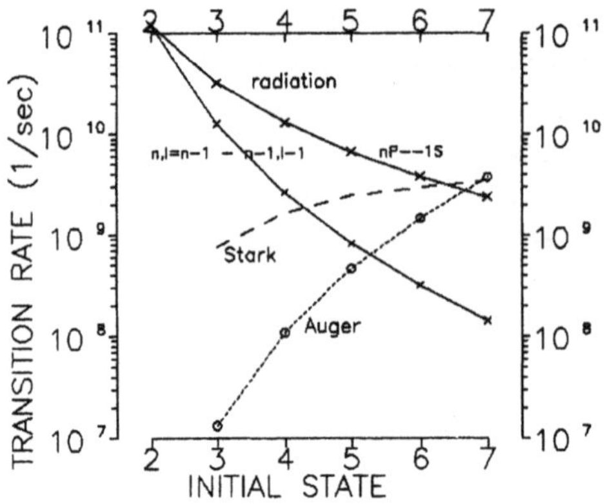

Fig.5. The rates of cascade processes in $\mu^- d$ atom at $N=10^{-4}N_o$.

$T \approx 4$ eV for μp does correspond to the one in the instance of reaching the edge of the collisional deexcitation stage, i.e., the state with n=5.

The search for metastable 2s state of muonic hydrogen

One can get a piece of information about the kinetic energy of $\mu^- p$ atom by measuring the lifetime of the 2s state. If the kinetic energy is greater than the 0.3 eV threshold for the Stark collision $(\mu^- p)_{2s} + H_2 \longrightarrow (\mu^- p)_{2p} + H_2$, then the $(\mu^- p)_{2s}$ is rapidly quenched, otherwise it appears to be the long-lived (metastable) state which could be detected by observing delayed $2p \longrightarrow 1s$ photons from slow collisional quenching[42]. The results of the experiments[17,18] to search for the metastable 2s state in gaseous target are consistent with the assumption that $(\mu^- p)_{2s}$ atoms are formed dominantly with kinetic energy greater than 0.3 eV, the following upper limits for the long-lived fraction f_{2s} with the lifetime τ_{2s} having been obtained

$f_{2s} < 2\%$ at 0.8 atm ($100\text{ns} < \tau_{2s} < 1000\text{ns}$) Ref.18

$f_{2s} < 4\%$ at 0.2 atm (assuming $\tau_{2s} = 2\mu s$) Ref.17

These results are in an agreement with the kinetic energy distribution observed in the $\mu^- p$ diffusion experiment[40].

Doppler broadening of the reaction product spectrum

An elegant way to measure the kinetic energy distribution was used in the experiment on $\pi^- - \pi^o$ mass difference[43,44] performed by PSI(SIN)-VIRGINIA collaboration where the reaction

$$(\pi^- p)_{at} \longrightarrow \pi^o + n$$

was studied in liquid hydrogen. The kinetic motion of the pionic atoms results in the Doppler broadening of the neutron lines in the time-of-flight spectra measured at different distances from the target. The experimental data evidence that, besides the expected broadening with about 1 eV characteristic energy, there is the high energy tail in the kinetic energy distribution. The data are consistent with birectangular distribution

$$f(E) = \begin{cases} C_1, & \text{if } 0<E<T_1, \\ C_2, & \text{if} T_1<E<T_2, \\ 0, & \text{if } \quad E>T_2 \end{cases} \qquad (8)$$

with 56% contribution from the low energy component (T_1=0.94±0.13 eV) and 44% from the high energy one (T_2=71.5±6.1 eV), the mean energy being $\bar{T}$=16.2±1.3 eV.

The life-history of π^-p atom in liquid hydrogen[4] is sketched in Fig.6. The bulk of absorption takes place at the states n=4 and n=3 which are mainly populated by the Δn=1 deexitation. Thus, we conclude that the energy distribution (8) is produced before or during the transitions 5⟶4 and 4⟶3.

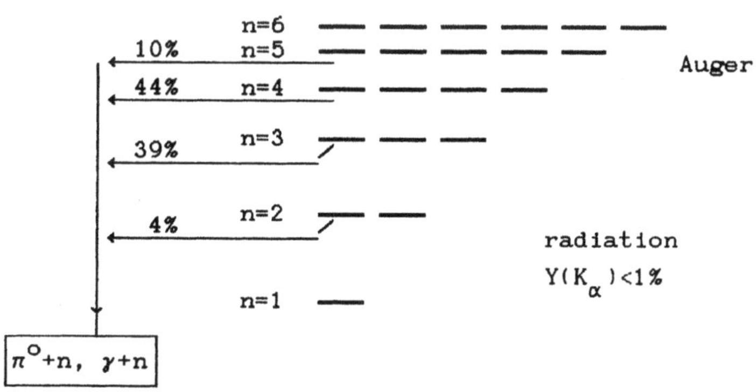

Fig.6. Capture schedule for π^-p atom in liquid hydrogen.

Pion transfer in H-D mixture

Cascade in the mixture of hydrogen isotopes (Fig.7) provides an additional possibility to get some information on the kinetic energy distribution[45,46]. The charge exchange reaction $\pi^- p \rightarrow \pi^0 + n$ can be used as the very clean signature of pion capture on proton since in deuterium the branching ratio for π^0 production from stopped π^- is negligibly small. Disregarding the molecular structure and isotope effects the relative probabilities of the initial muon capture by H and D correspond to the isotope concentration ratio $C_p:C_d$. Due to the isotope mass difference the irreversible pion transfer from hydrogen to deuterium is possible during the deexcitation and, according to calculations[45,46], at n<5 it can significantly influence the cascade.

The transfer rates being strongly energy dependent (see Fig.8), the fraction of pions, W, undergoing the reaction $\pi^- p \rightarrow \pi^0 + n$ after being captured by protons is the function of C_d and the kinetic energy distribution. The theoretical analysis[45] of the experimental data on the pion transfer in H-D mixture, obtained at TRIUMF[47] and LINP[46], suggested that the calculations can be put in agreement with the experiment at the mean energy T≈1eV. The assumption that the $\pi^- p$ atoms are thermalized gives too low value of fraction $W_{\gamma\gamma}$ and, thus, should be rejected. According to the SCM this conclusion concerns the energy distribution at $n=3\div5$.

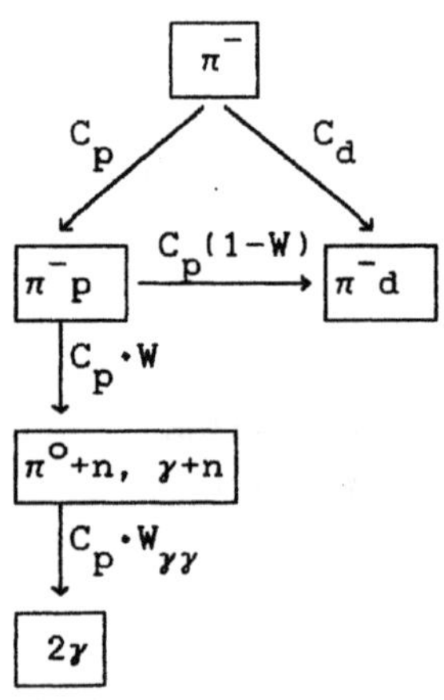

Fig.7. The schematics of the pion capture, transfer, and absorption in H-D mixture.

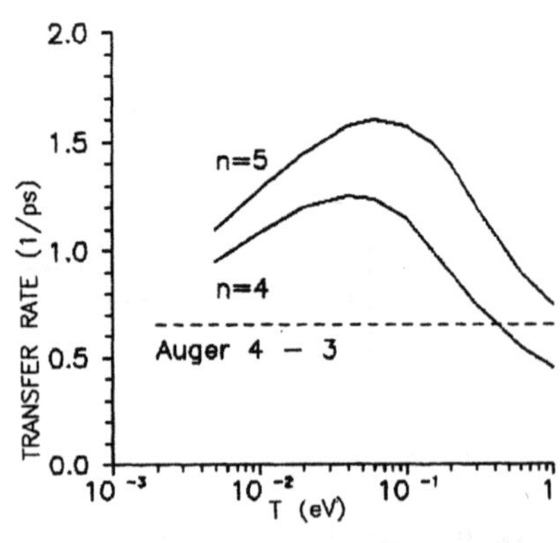

Fig.8. The energy dependence of the normalized rates of the pion transfer (Ref.46) $(\pi^- p)_n + d \longrightarrow (\pi^- d)_n + p$.

Muon transfer in D-T mixture

The schematics of the muon cascade in deuterium-tritium mixture is shown in Fig.9. The probability, q_{1S}, for the muon captured by deuteron to reach the ground state, being the function of density and tritium concentration C_t, is one of the important characteristics of the muon catalysis in D-T mixture (see Ref.48 and references therein). The muon transfer rates were calculated in Refs.45,49-51 and some of the results of the cascade calculations are shown in Fig.10. The data available on the muon catalyzed fusion[53-56] are not compatible with the assumption that the muonic deuterium atoms are thermalized during the cascade. For example, the LINP experiment[56] gives $q_{1S} \geq 0.97$ at $N=0.1 \cdot N_o$ and $C_t=1.24\%$ that requires $\bar{T} \geq 1$ eV. The discussion of the muon transfer at high density can be found in Refs.48,51,53-55.

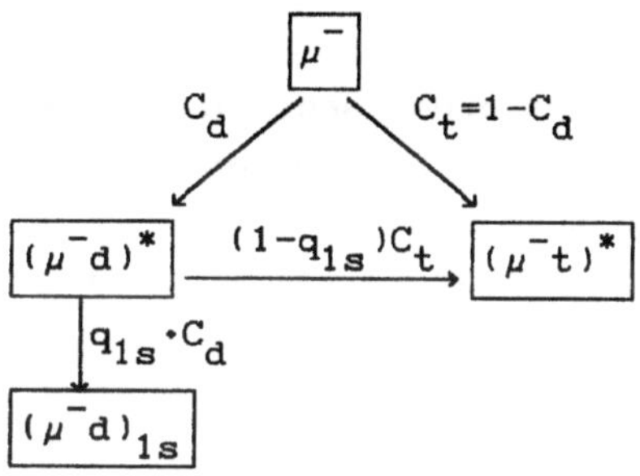

Fig.9. The muon capture and transfer during the cascade in D-T mixture.

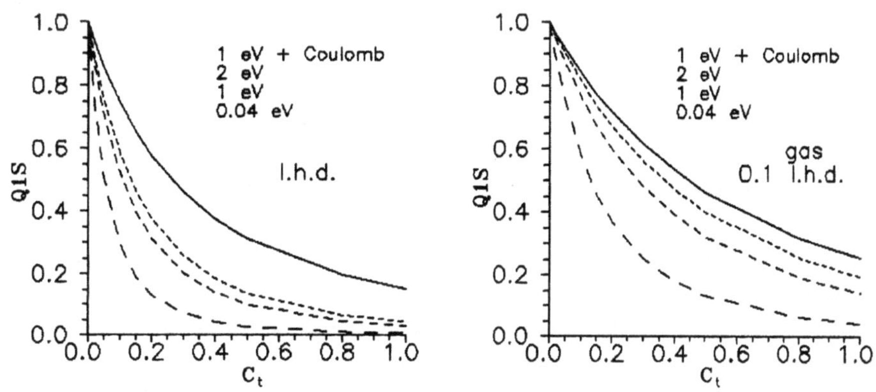

Fig.10. Muon transfer parameter q_{1s} vs. tritium concentration C_t calculated for different values of kinetic energy T (Ref.51) and with the Coulomb acceleration taken into account[52].

The deceleration rates

As we have demonstrated, the energy distribution of the exotic atoms with Z=1 during the cascade is not the thermal one. This is a rather remarkable fact because the atom undergoes a lot of elastic collisions during the cascade. Taking into account that the relative energy loss in the elastic scattering of the atom with mass m_a and kinetic energy T_i from at-rest target with mass m_t is determined by formula

$$\frac{T_i - T_f}{T_i} = \frac{2m_a m_t}{(m_a + m_t)^2}(1-\cos\vartheta) \qquad (9)$$

where ϑ is the c.m.s. scattering angle, one can estimate the rate of the deceleration with formula

$$\lambda_n^{dec} = N\cdot v\cdot \frac{2m_a m_t}{(m_a + m_t)^2}\sigma_n^t \qquad (10)$$

Here $\sigma_n^t = \int(1-\cos\vartheta)d\sigma_n(\vartheta)$ is the transport cross section, $d\sigma_n(\vartheta)$ being the differential elastic cross section for the atom in the state n. The transport cross sections were calculated with WKB method in Ref.50, for kinetic energy $T \geq 1eV$ they can be estimated with formula

$$\sigma_n^t \sim \frac{\pi(n^2-1)}{MT} \qquad (11)$$

The deceleration rates for $\mu^- d$ atoms, calculated according to Ref.50, are shown in Fig.11 in comparison with the collisional deexcitation rates. The deceleration rates are essentially energy dependent, at T>1eV they behave like $T^{-1/2}$.

Atomic cascade with deceleration

In considering the kinetic energy distribution problem it is useful to distinguish following stages of the cascade:

1) The initial stage where the collisional transitions with $\Delta n=1$ are forbidden by the energy conservation (n>7 for $\mu^- d$);

2) The intermediate stage where the rapid collisional transitions $n \longrightarrow (n-1)$ become possible;

3) The final stage of the radiative deexcitation (for $\mu^- d$ $n \leq 2$ in liquid D_2 and $n \leq 4$ in gas at STP).

The initial stage is badly explored. In the SCM it is assumed that the chemical deexcitation plays an important role at this stage and as a result the atom gets kinetic energy about 1eV. The importance of the chemical deexcitation was doubted in Ref.9 where alternative mechanism, the two-step Auger transitions with the excitation of the hydrogen molecule taken into account, was suggested. In order to get a correct cascade time one has to suppose[7] that in liquid hydrogen the atom reaches the next stage in less than 0.4 ps.

The second stage commences with the rapid $\Delta n=1$ Auger transitions (for μ^-d it corresponds to $n\leq7$). With n decreasing the other collisional processes, besides the Stark mixing and Auger transitions, become essential, first of all, deceleration, and, probably, Coulomb deexcitation (see Fig.11). The relative importance of the deceleration is strongly energy dependent, at $T\geq2$eV its rate becomes comparable with that of deexcitation at $n\sim4$, whereas at $T\leq0.5$eV one has to take it into account from the very beginning.

The stage of the radiative deexcitation starts at some n_c depending on density. In liquid hydrogen it involves only the $2\rightarrow1$ transition, but becomes rather lengthy in gas at low pressure. At $\rho<0.1\rho_{STP}$ ($n_c=5$) the deceleration rates are much smaller than radiative ones, so that the cascade at the final stage $n<n_c$ almost preserves the kinetic energy distribution.

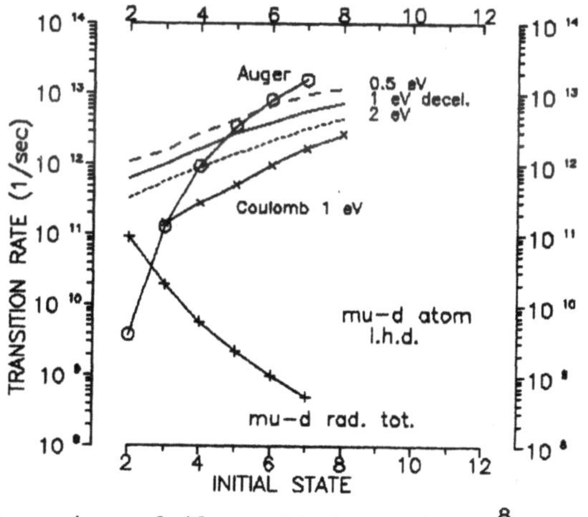

Fig.11. The rates of the radiative, Auger[8], and Coulomb[5] deexcitation and deceleration[50] for μ^-d atom in liquid deuterium.

Coming back to the analysis of the experimental information on mean kinetic energy from the experiments on the muonic atoms diffusion and the search for the metastable 2s state, which were performed at $\rho<\rho_{STP}$, we can put the question what the stage is where the kinetic energy distribution is basically formed. In two extreme cases the cascade could be developed as following.

1) If the atom reaches the intermediate stage with $\bar{T}\geq 2eV$, then at $\rho<\rho_{STP}$ the atom can retain the most of its kinetic energy till the end of the cascade.

2) If the atom enters the intermediate stage with $\bar{T}\leq 0.5eV$, then, without any acceleration accounted, the noticeable energy loss would take place before reaching the final stage. Thus, some acceleration mechanism at the intermediate stage has to form the observed kinetic energy distribution.

Kinetic energy distribution for π^-p in liquid hydrogen

The atomic cascade in π^-p in liquid hydrogen at n>5 is similar to that in μ^-d in gas at STP (with the appropriate time rescaling due to the change in density) because in both cases neither the absorption nor the radiative transitions are yet significant at this stage. Thus, taking into account the information from the μ^-d diffusion experiment[40,41] one can expect that the π^-p has mean kinetic energy about 2eV at n=5. Then follows rather fast transition $5\longrightarrow 4$, which is mainly due to the Auger deexcitation, but some contribution comes from the

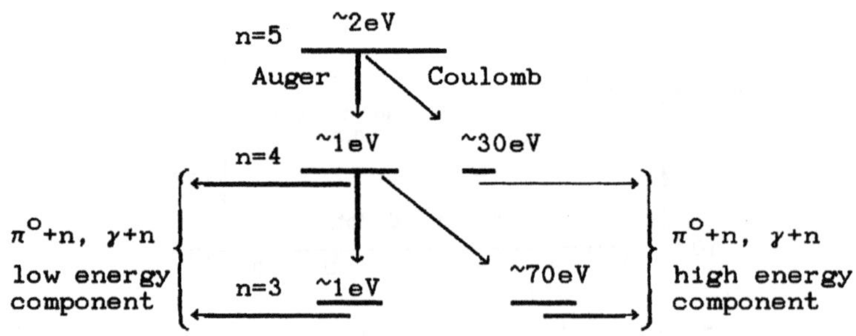

Fig.12. The final stage of the cascade for π^-p atom in liquid hydrogen.

Coulomb collision with the energy release Q=74 eV, so that some fraction of atoms at n=4 gets kinetic energy about 30 eV (see Fig.12). The low energy fraction looses some energy at n=4 before the nuclear absorption or the transition 4⟶3 take place. The Coulomb transition 4⟶3 (Q=157eV) produces the high energy fraction with T≈ 70 eV. The cascade is terminated by the strong absorption at n=3. The high energy tail of the kinetic energy distribution observed in experiment[40,41] on studying the reaction $\pi^- p \to \pi^0 + n$ in liquid hydrogen can be considered as an evidence in favour of importance of the Coulomb deexcitation at n≤5.

Acceleration mechanisms

The acceleration due to recoil is essential only if there are no light particles (e^- or γ) in the final state, as it is the case for Coulomb or chemical deexcitation. The Coulomb transitions may be important even if they do not dominate the deexcitation since the effect of the epithermal component production can be accumulated. In each transition of the chain n⟶(n-1)⟶(n-2)⟶... some fraction of atoms is accelerated in the Coulomb collisions. Being small for a single transition n⟶(n-1) (see Figs.1,11) the fraction of accelerated atoms is growing up while the atom cascades down. The epithermal component can survive till the end of the cascade because the deceleration rates drop with the increase of the kinetic energy.

In case of the Auger transitions the atom can get only small fraction of the energy release Q because of the recoil ($T \sim Qm_e/m_a$ ≪1eV). However there is another acceleration mechanism[9] which is due to the interaction of the atomic dipole moment with the electric field of the systems H_2, H_2^+, and p+p participating consequently in the two-step Auger transitions. This mechanism can produce the epithermal component with characteristic energy T ≈ 0.2÷0.6 eV.

The molecular effects in muonic hydrogen cascade were suggested in Ref.57 to be important at high densities and n≤4 as they result in the atom binding to a hydrogen molecule or molecular ion and influence the kinetic energy distribution.

Taking into account the existence of the quasibound mesic molecule states localized below thresholds $(\mu^- d)_n + d$ one may suggest the following mechanism of the acceleration ($n \geq 3$)

$$(\mu^- d)_n + D_2 \longrightarrow \left[(d\mu d)_{nJv} d\, e^-\right]_f + e^-$$
$$\hookrightarrow (\mu^- d)_{n-1} + d$$

which goes via the resonance formation in Auger process followed by the decay of Coulomb type.

CONCLUSION

The cascade processes in exotic atoms with Z=1 turn out to involve a wide range of phenomena which have to be considered thoroughly in order to get the firm theoretical background for the light exotic atoms studies. The present day cascade models provide a fair description of the populations of the low-lying states and the X-ray yields in a wide density range for $\mu^- p$ and $p\bar{p}$ atoms. More experimental data are required to verify these models at different intensities of nuclear absorption, including $\bar{p}d$ and $\pi^- p$ atoms. As for the problem of the kinetic energy distribution for the atoms during the cascade, it is still to be solved. This can be done by developing the cascade model which incorporates simultaneously the description of the evolution of both internal and external degrees of freedom.

The more advanced theoretical studies of the following problems will be of great importance for this purpose.

1. The elastic scattering of the atoms in excited states.
2. The Coulomb deexcitation and its relative importance among the other collisional processes.
3. The acceleration due to the atomic dipole moment interaction with the electric field of the molecules and ions.
4. New acceleration mechanisms.
5. The cascade features due to the epithermal and high energy components in the kinetic energy distribution.

The experimental possibilities to get new information on the cascade processes are rather promising, the following is to be pointed out.

1. The measurements of the energies and yields of X-ray transitions in pionic and antiprotonic atoms of hydrogen isotopes with crystal spectrometer[58] in wide density range.
2. The search for the density dependence of the kinetic energy distribution in the experiment on muonic hydrogen atoms diffusion.
3. The μ^-, π^-, $\bar{p}$ transfer in the mixture of hydrogen isotopes.

4. The study of the reaction $(\pi^- p) \rightarrow \pi^0 + n$ in gas (the measurements of the Doppler line broadening).

5. The muon transfer to Z>1 elements.

The further progress in the field considered will provide the necessary background for related studies (strong interaction effects, muon catalyzed fusion, etc.).

I would like to express my gratitude to Dr.L.Simons, Dr.R.Frosch, Dr.F.Kottmann, Dr.P.Kammel and K.Heitlinger for fruitful discussions and to thank all my hosts and, especially, Dr.C.Petitjean for the warm hospitality at Paul Scherrer Institute during my visits in 1988-89. Support by PSI is gratefully acknowledged.

REFERENCES

1. E.Fermi and E.Teller, The capture of negative mesotrons in matter, Phys.Rev. 72 : 399 (1947).
2. T.B.Day, G.A.Snow, and J.Sucher, Suppression of P-state capture in (K^-,p) atoms, Phys.Rev.Lett. 3 : 61 (1960).
3. T.B.Day, G.A.Snow, and J.Sucher, High-orbital S-state capture of π^- mesons by protons, Phys.Rev. 118 : 864 (1960).
4. M.Leon and H.A.Bethe, Negative meson absorption in liquid hydrogen, Phys.Rev. 127 : 636 (1962).
5. L.Bracci and G.Fiorentini, Coulomb de-excitation of mesic hydrogen, Nuovo Cimento 43A : 9 (1978).
6. E.Borie and M.Leon, X-ray yields in protonium and mesic atoms, Phys.Rev. A21 : 1460 (1980).
7. V.E.Markushin, Light μ^- atoms in liquid and gaseous hydrogen and deuterium, Sov.Phys.JETP 53 : 16 (1981).
8. A.P.Bukhvostov and N.P.Popov, Depolarisation of negative muons in hydrogen, Sov.Phys.JETP 55 : 12 (1982).
9. L.I.Menshikov, Mesic atoms acceleration mechanisms in cascade transitions, Muon Catalyzed Fusion, 2 : 173 (1988).
10. J.-L.Vermeulen, A solution of the Leon-Bethe model for the Day-Snow-Sucher effect by dynamical group method, Nucl.Phys. B12 : 506 (1969).
11. J.S.Cohen, Formation of low-Z exotic atoms, this volume.
12. J.S.Cohen, R.L.Martin, and W.R.Wadt, Diabatic-state treatment of negative-meson moderation and capture, Phys.Rev. A24 : 33 (1981).
13. J.S.Cohen, Slowing down and capture of negative muons by hydrogen: Classical trajectory Monte Carlo calculation, Phys.Rev. A27 : 167 (1983).
14. G.Ya.Korenman and V.P.Popov, Slowing-down and Coulomb capture of negative muons in molecular hydrogen, Muon Catalyzed Fusion (Sanibel Island, FL 1988), AIP Conference Proceedings 181, pp.145-160, Eds. S.E.Jones, J.Rafelski, and H.J.Monkhorst, AIP, NY, (1989).
15. A.Placci, E.Polacco, E.Zavattini, K.Ziok, G,Carboni, U.Gastaldi, G.Gorini, and G.Torelli, Observation of muonic X-rays in hydrogen gas, Phys.Lett. 32B : 413 (1970).
16. B.Budick, J.R.Toraskar, and I.Yaghoobia, Muonic X-rays from liquid hydrogen, Phys.Lett. 34B : 539 (1971).

17. H.Anderhub, H.Hofer, F.Kottmann, P.LeCoultre, D.Makowiecki, O.Pitzurra, B.Sapp, P.G.Seiler, M.Walchli, D.Taqqu, P.Truttmann, A.Zehnder, and Ch.Tschalar, Search for the metastable 2s state in muonic hydrogen, Phys.Lett. 71B : 443 (1977).
18. P.O.Egan, S.Dhawan, V.W.Hughes, D.C.Lu, F.G.Mariam, P.A.Souder,J.Vetter, G.zu Putlitz, P.A.Thompson, and A.B.Denison, Search for long-lived 2s muonic hydrogen in H gas, Phys.Rev. A23 : 1152 (1981).
19. H.Anderhub, J.Bocklin, M.Devereux, F.Dittus, R.Ferreira Marques, H.Hofer, H.K.Hofer, F.Kottmann, O.Pitzurra, P.G.Seiler, D.Taqqu, J.Unternahrer, M.Walchli, and Ch.Tschalar, Slowing-down of negative muons and formation of muonic hydrogen in gas below 1 torr, Phys.Lett. 101B : 151 (1981).
20. H.Anderhub, H.P.von Arb, J.Bocklin, F.Dittus, R.Ferreira Marques, H.Hofer, F.Kottmann, D.Taqqu, and J.Unternahrer, Measurement of the K-line intensity ratios in muonic hydrogen between 0.25 and 150 Torr gas pressures, Phys.Lett. 143B : 65 (1985).
21. M.Leon, X-ray yields in muonic hydrogen, Phys.Lett. 35B : 413 (1971).
22. M.Ziegler,S.Ahmad, C.Amsler, R.Armenteros, E.G.Auld, D.A.Axen, D.Bailey, S.Barlag, G.A.Beer, M.Bizot, M.Botlo, M.Comyn, W.Dahme, B.Delcourt, M.Doser, K.D.Duch, K.L.Erdman, F.Feld-Dahme, U.Gastaldi, M.Heel, B,Howard, R.Howard, J.Jeanjean, H.Kalinowsky, F.Kayser, E.Klempt, C.Laa, R.Landua, G.M.Marshall, H.Nguyen, N.Prevot, J.Reidlberger, R.Rieger, L.P.Robertson, C.Sabev, U.Schaefer, O.Schreiber, U.Straumann, P.Truoel, H.Vonach, P.Weidenauer, B.L.White, and W.R.Wodrich, Measurement of the strong interaction shift and broadening of the ground state of the $p\bar{p}$ atom, Phys.Lett. B206 : 151 (1988).
23. C.A.Baker, C.J.Batty, S.A.Clark, J.Moir, S.Sakamoto, J.D.Davies, J.Lowe, J.M.Nelson, G.J.Pyle, A.Selvarajah, G.T.A.Squier, R.E.Welsh, R.G.Winter, E.W.A.Lingeman, G.W.E.van Eijk, R.W.Hollander, D.Langerveld, W.J.C.Okh, and A.Zoutendijk, Measurement of X-rays from antiprotonic hydrogen and deuterium, Nucl.Phys. A483 : 631 (1988).
24. L.M.Simons, Results from antiprotonic atoms at LEAR, in Physics at LEAR with low energy antiprotons (the 4-th LEAR Workshop, Villars-sur-Ollon, Switzerland, 1987), pp.703-716 Eds. C.Amsler, G.Backenstoss, R.Klapish, et al., AP, Harwood (1988).
25. G.W.E.van Eijk, R.W.Hollander, D.Langerveld, W.J.C.Okh, A.Zoutendijk, R.Ferreira-Marques, C.A.Baker, C.J.Batty, S.A.Clark, J.Moir, S.Sakamoto, J.D.Davies, J.Lowe, J.M.Nelson, G.J.Pyle, G.T.A.Squier, R.E.Welsh, and R.G.Winter, Nucl.Phys. A486 : 604 (1988).
26. R.Bacher, P.Blum, D.Gotta, K.Heitlinger, D.Rohmann, M.Schneider, J.Egger, L.M.Simons, K.Elsener, Measurement of Balmer and Lyman X-rays in antiprotonic hydrogen isotopes at pressure below 300 hPa, Preprint PR-89-01, Paul Scherrer Institute, April 1989.
27. G.Reifenrother and E.Klempt, The cascade of antiprotonic hydrogen, in Contributed Papers from Few Body XII (12th International Conference on Few Body Problems in Physics), Ed. B.K.Jennings, TRI-89-2, p.I12, Vancouver, Canada (1989).
28. R.Landua and E.Klempt, Atomic cascade of muonic and pionic helium atoms, Phys.Rev.Lett. 48 : 1722 (1982).
29. G.Reifenrother, E.Klempt, and R.Landua, Cascade of muonic helium atoms, Phys.Lett. B191 : 15 (1987).
30. G.Reifenrother, S.Ahmad, C.Amsler, R.Armenteros, E.G.Auld, D.A.Axen, D.Bailey, S.Barlag, G.A.Beer, M.Bizot, M.Botlo,

M.Comyn, W.Dahme, B.Delcourt, M.Doser, K.D.Duch, K.L.Erdman, F.Feld-Dahme, U.Gastaldi, M.Heel, B,Howard, R.Howard, J.Jeanjean, H.Kalinowsky, F.Kayser, E.Klempt, C.Laa, R.Landua, G.M.Marshall, H.Nguyen, N.Prevot, J.Reidlberger, L.P.Robertson, C.Sabev, U.Schaefer, O.Schreiber, U.Straumann, P.Truoel, H.Vonach, P.Weidenauer, B.L.White, W.R.Wodrich, and M.Ziegler, Cascade time of antiprotons in gaseous hydrogen, Phys.Lett. 214B : 325 (1988).

31. T.H.Fields, G.B.Yodh, M.Derrick, and J.G.Fetkovich, Cascade time of π^- in liquid hydrogen, Phys.Rev.Lett. 5 : 69 (1960).
32. E.Bierman, S.Talor, E.L.Koller, P.Stamer, and T.Huetter, Absorption time of negative pions in liquid hydrogen, Phys.Lett. 4 : 351 (1963).
33. J.H.Doede, R.H.Hildebrand, M.H.Israel, and M.P.Pyka, Moderation and absorption times of negative pions in liquid hydrogen , Phys.Rev. 129 : 2808(1963).
34. R.Knop, R.A.Burnstein, and G.A.Snow, Absorption time of negative K^- mesons in liquid hydrogen, Phys.Rev.Lett. 14 : 767 (1965).
35. M.Cresti, L.Limentani, A.Loria, L.Peruzzo, and R.Santangelo, Lower limits on the absorption rate of K^-'s at rest in liquid hydrogen, Phys.Rev.Lett. 14 : 847 (1965).
36. R.A.Burnstein, G.A.Snow, and H.Whiteside, Absorption time of negative Σ hyperons in liquid hydrogen, Phys.Rev.Lett. 15 : 639 (1965).
37. J.Bailey, A.Placci, E.Polacco, E.Zavatttini, K.Ziok, G.Carboni, U.Gastaldi, G.Gorini, G.Neri, and G.Torelli, Measurements of pionic X-rays from hydrogen, Phys.Lett. 33B : 369 (1970).
38. E.Bovet, L.Antonuk, J.-P.Egger, G.Fiorucci, K.Gabathuler, and J.Gimlett, A new determination of the π^-H and π^-D 2P-1S strong interaction shifts using crystal spectrometer, Phys.Lett. 153B : 231 (1985).
39. A.Forster, E.Bovet, J.Gimlett, H.E.Henrikson, D.Murray, R.J.Powers, P.Vogel, F.Boehm, R.Kunselman, and P.L.Lee, Determination of the 2p-1s transition energy and strong interaction shift in pionic hydrogen using crystal diffraction, Phys.Rev. C28 :2374 (1983).
40. W.Breunlich, M.Cargnelli, G.Chen, P.P.Guss, F.Hartmann, P.Kammel, J.B.Kraiman, J.Marton, C.Petitjean, J.Reidy, R.Siegel, W.Vulcan, R.Welsh, H.Woolverton, A.Zehnder, and J.Zmeskal, Diffusion of muonic hydrogen atoms, SIN Newsletter 20 : 58 (1988); Contributed Papers from Few Body XII, TRI-89-2, p.A31, Vancouver, Canada (1989); PSI Newsletter (1989).
41. J.B.Kraiman, G.Chen, P.P.Guss, R.Siegel, W.Vulcan, R.Welsh, W.Breunlich, M.Cargnelli, P.Kammel, J.Marton, J.Zmeskal, F.Hartmann, C.Petitjean, A.Zehnder, J.Reidy, and H.Woolverton, Diffusion of muonic deuterium in D_2 gas, submitted to Phys.Rev.Lett.
42. R.O.Muller, V.W.Hughes, H.Rosenthal, C.S.Wu, Collisional quenching of the metastable 2S state of muonic hydrogen and muonic helium ion, Phys.Rev. A11 : 1175 (1975).
43. J.E.Crawford, M.Daum, R.Frosch, B.Jost, P.-R.Kettle, R.M.Marshall, B.K.Wright, and K.O.H.Ziock, Precision measuement of the mass difference $m_{\pi^-} - m_{\pi^o}$, Phys.Lett. 213B : 391 (1988).

44. J.E.Crawford, M.Daum, R.Frosch, B.Jost, P.-R.Kettle, R.M.Marshall, B.K.Wright, and K.O.H.Ziock, Measurement of the kinetic energy distribution of pionic hydrogen atoms in liquid hydrogen, this volume.
45. L.I.Menshikov and L.I.Ponomarev, The influence of process $d\mu(n)+t \rightarrow t\mu(n)+d$ on the kinetics of muon catalysis in D-T mixture, JETP Lett. 39 : 663 (1984).
46. A.V.Kravtsov, A.Yu.Mayorov, A.T.Mikhailov, S.Yu.Ovchinnikov, N.P.Popov, V.M.Suvorov, and A.I.Shchetkovsky, Quasiresonance pion transfer in the protium-deuterium mixture, Muon Catalyzed Fusion 2 : 199 (1988).
47. K.A.Aniol, D.F.Measday, M.D.Hasinoff, W.H.Roser, A.Bagheri, F.Entezami, C.Virtue, J.M.Stadlbauer, D.Horvath, M.Salomon, and B.C.Robertson, Phys.Rev. A28 : 2684 (1983).
48. L.I.Ponomarev, Summary talk, Muon Catalyzed Fusion 3 : 629 (1988).
49. L.I.Menshikov and L.I.Ponomarev, Charge exchange of hydrogen isotopes mesic atoms in excited states in triple collisions with molecules, JETP Lett. 42 : 13 (1985).
50. L.I.Menshikov and L.I.Ponomarev, Quasiresonance charge exchange of hydrogen isotopes mesic atoms in excited states, Z.Phys.D2 : 1 (1986).
51. A.V.Kravtsov, A.I.Mikhailov, S.Yu.Ovchinnikov, and N.P.Popov, Quasiresonance charge exchange of excited mesic hydrogen in the mixture of hydrogen isotopes, Muon Catalyzed Fusion 2 : 183 (1988); Preprint LINP-1 (1988).
52. V.Markushin, Recent progress in the theory of the cascade processes in light exotic atoms, talk given at PSI Workshop, June 1988.
53. S.E.Jones, A.N.Anderson, A.J.Caffrey, C.DeW.Van Siclen, K.D.Watts, J.N.Bradbury, J.S.Cohen, P.A.M.Gram, M.Leon, H.R.Maltrud, and M.A.Paciotti, Observation of unexpected density effects in muon catalyzed d-t fusion, Phys.Rev.Lett. 51 : 588 (1986).
54. W.H.Breunlich, M.Cargnelli, P.Kammel, J.Marton, N.Naegele, P.Pavlek, A.Scrinzi, J.Werner, J.Zmeskal, J.Bistirlich, K.M.Crowe, M.Justice, J.Kurck, C.Petitjean, R.H.Sherman, H.Bossy, H.Daniel, F.J.Hartmann, W.Neumann, and G.Schmidt,Phys.Rev.Lett. 58 : 329 (1987).
55. P.Kammel, P.Ackerbauer,W.H.Breunlich, M.Cargnelli, M.Jeitler, J.Marton, N.Naegele, A.Scrinzi, J.Werner, J.Zmeskal, C.Petitjean, J.Bistirlich, K.Crowe, M.Justice, R.H.Sherman, H.Bossy, H.Daniel, F.J.Hartmann, H.Plendl, and W.Schott, μCF at low tritium concentration, Muon Catalyzed Fusion 3 : 483 (1988).
56. D.V.Balin, A.I.Ilyin, V.P.Maleev, E.M.Maev, G.E.Petrov, L.B.Petrov, G.A.Ryabov, G.G.Semenchuk, Yu.V.Smirenin, V.A.Volchenkov, A.A.Vorobyov, and An.A.Vorobyov, Analysis of dt μCF data, Muon Catalyzed Fusion 2 : 163 (1988).
57. D.Taqqu, Molecular effects in muonic hydrogen cascade, Muon Catalyzed Fusion (Sanibel Island, FL 1988), AIP Conference Proceedings 181, , pp.217-222 Eds. S.E.Jones, J.Rafelski, and H.J.Monkhorst, AIP, NY, (1989).
58. G.L.Borchert, D.Gotta, O.W.B.Schult, L.M.Simons, K.Elsener, K.Rashid, J.J.Reidy, High resolution spectroscopy of X-rays from antiprotonic hydrogen and helium isotopes using a crystal spectrometer, to be published.

PROTONIUM: The Mainz Cascade Model

G. Reifenröther and E. Klempt
Institut für Physik der Universität Mainz
D-6500 Mainz, Germany

1. INTRODUCTION

Recent experiments at LEAR have studied extensively the properties of antiprotonic hydrogen, often also called protonium.

The main motivation of the experiments was to determine the strong interaction shift and width of the $1S$ ground state and the hadronic width of the $2P$ state. In addition, the intensities of the Balmer and Lyman lines have been measured. The line intensities decrease significantly with increasing target pressure. This indicates important atomic effects in the time between capture and annihilation. Also, the question of the fraction of S and P wave annihilation depends strongly on details of the atomic cascade. This ratio has played an important role in the discussion of $\bar{p}p$ annihilation at rest. The experimental information on protonium, the strong interaction effects and the measured cascade parameters are reviewed in reference 1.

In vacuum the cascade of protonium atoms would be determined only by radiative transitions and nuclear capture, and we would get a "circular" cascade. The $\bar{p}p$ atom would deexcite along the circular l states and annihilation would occur mostly from P wave because the hadronic width of the $2P$ state is much larger than the radiative width of the $2P$ state. But the experiments work at target densities corresponding to some 10^{-2} atm and to that of liquid hydrogen. Therefore the $\bar{p}p$ undergoes many collisions during its life.

In a collision the $\bar{p}p$ experiencies high electric fields which lead to the Stark effect which reshuffles the level population away from the circular l states and makes nuclear capture possible from high nS and nP levels. In liquid H_2, the collisional frequency is much higher than the P wave annihilation rate, and annihilation from S states is dominant. In hydrogen gas both, S and P wave capture are important. The $\bar{p}p$ atom experiences in addition high electron densities leading to Auger ejection of hydrogen electrons and to deexcitation of the $\bar{p}p$. This effect is called the external Auger effect. Because the hydrogen target consist of two-atomic molecules also chemical effects like dissociation and vibrational or rotational excitation have to be considered.

In this paper we present results of a model describing these processes in a microscopic approach. Details of the work and further results are presented in a forthcoming publication[2].

Electromagnetic Cascade and Chemistry of Exotic Atoms
Edited by L. M. Simons *et al.*, Plenum Press, New York, 1990

2. THE CASCADE

2.1. Collision Processes

A $\bar{p}$ with kinetic energy of some MeV moving in a hydrogen target will loose its energy by ionisation of hydrogen atoms. If it is decelerated to some eV it can be captured via Auger effect. We have simulated this process by following the classical trajectory of the three particles. The $\bar{p}$ approaches the hydrogen atom with a kinetic energy of a few eV, $5\,eV$ were chosen for the collision shown in Fig.1.

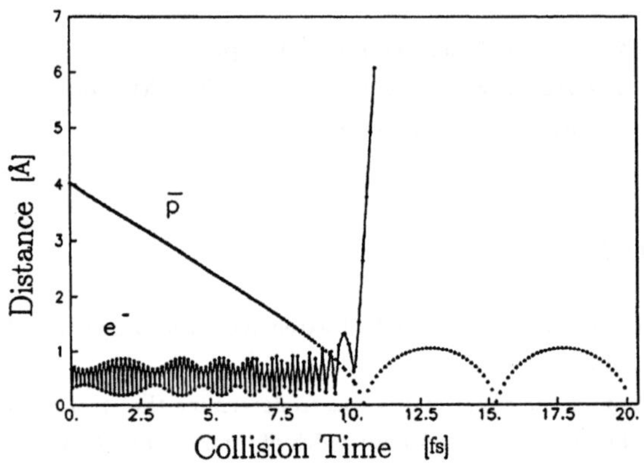

Fig.1. Simulation of $\bar{p}$ capture by a proton via Auger effect.

Plotted are the distances between electron and proton and between antiproton and proton in dependence of the collision time. Obviously the electron is ejected with high velocity. The antiproton is captured by the proton into a highly excited orbit and a neutral $\bar{p}p$ atom is formed. This atom collides with the molecules of the H_2 target.

The collision processes between protonium and hydrogen atoms were simulated by the Classical-Trajectory-Monte-Carlo method[3] (Fig.2). In this simulation the hydrogen atom was treated as a particle with an electric field given by

$$E = \frac{1}{r^2}(1 + 2r + 2r^2)e^{-2r}$$

Thus the problem is reduced from a four-body-problem to a three-body-problem which is solved numerically.

The initial state of the $\bar{p}p$ atom is defined by the choice of five random variables, its path is determined as solution of the three-body problem by the Classical-Trajectory-Monte-Carlo method. The results of these simulations are:

1. The cross sections for five different typical collisions defined by the maximal electric field seen in the collision.

2. The time dependence of the electric field and its directional change $\dot{\theta}$. These functions were used to calculate the Stark mixing.

3. The electron density seen by $\bar{p}p$ atoms in a collision. These electron densities are necessary to calculate the external Auger effect.

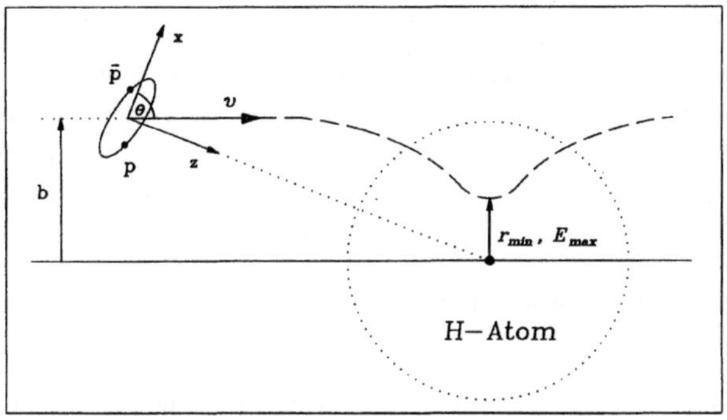

Fig.2. Initial state of a collision process simulation by the Classical-Trajectory-Monte-Carlo method.

2.2 Stark Mixing

The $\bar{p}p$ feels intense electric fields in a collsion with a hydrogen atom. These external fields disturb the spherical symmetry of the Coulomb potential and lead therefore to transitions between states of different angular momenta. The transition probability can be calculated by solving the time dependent Schrödinger equation

$$i\hbar \cdot \frac{\partial \Psi(t)}{\partial t} = H(t)\,\Psi(t) \qquad\qquad H(t) = H_0 + H_I(t)$$

$H_I(t)$ describes the interaction betweeen the induced electric dipole moment $\vec{\mu}$ and the electric field $\vec{E}$. We choose the z-axis in direction of the electric field, the x-axis in the collision plane and θ as the angle between the x-axis and the direction of motion. We expand the wavefunction into hydrogen eigenfunctions Ψ_j and exploit the selection rules:

Electric field induced transitions $\quad \Delta l = \pm 1; \quad \Delta m = 0$

$\dot{\theta}$ induced transitions $\quad \Delta l = 0; \quad \Delta m = \pm 1$

Then a system of differential equations is arrived at which describe the time dependence of the level population[4].

$$\begin{aligned} i\hbar \cdot \dot{b}_{lm}(t) &= \frac{3}{2} \cdot E \cdot n \cdot \{A(n,l,m) \cdot b_{l+1,m} + B(n,l,m) \cdot b_{l-1,m}\} \\ &\quad - i \cdot \hbar \cdot \dot{\theta} \cdot \{C(l,m) \cdot b_{l,m-1} - D(l,m) \cdot b_{l,m+1}\} \end{aligned}$$

The solution yields transition probabilities from every angular momentum l to every angular momentum l' of the same principal quantum number.

2.3. External Auger Effect

In a collision between a protonium and a hydrogen atom the atomic electron can be ejected leading to deexcitation of the antiprotonic hydrogen atom

$$(\bar{p}p)_{nl} \quad + \quad H \quad \longrightarrow \quad (\bar{p}p)_{n'l'} \quad + \quad p \quad + \quad e^-$$

In this model the external Auger transition probabilities were determined in three steps:

- First the electron density integrated over a collision was determined by the collision process simulations. The Auger transition rates are known for electron densities corresponding to those of normal hydrogen atoms[5]. Since the Auger process leads to an exponential decay of the population of an excited level only the integrated electron densities are needed to calculate the Auger transition rates.
- In the second step the transition probabilities of the internal Auger effect for the $1S$ level were calculated.
- Finally the internal Auger transition probabilities were scaled by the integrated electron density seen by the $\bar{p}p$ in a collision.

2.4. Chemical Effects

In very high Rydberg states the $\bar{p}p$ can also deexcite into lower n levels by exciting vibrational or rotational states or by dissociation of hydrogen molecules.

We have not simulated this process quantitatively, instead we assume that the cross section for these molecular effects is given by the size of the $\bar{p}p$ atom

$$\sigma_{chem} = \pi a_{n_i}^2$$

Changes of this cross section by a factor of 1/2 or 2 change the results only by 1 or 2%.

2.5. Radiative Transitions and Nuclear Capture

For radiative transitions we have used the known equations as can be found in the famous book by Bethe and Salpeter[6].

The strong interaction shifts the energy levels of the antiprotonic hydrogen and leads to an additional width. Energy shift and width of the $1S$ ground state and the hadronic width of the $2P$ state have been measured[1] at LEAR. The strong interaction in D states can be neglected. For these calculations we have used the shift and width of

$$\Gamma_{1S} = 1.2\, keV \qquad \Delta E_{1S} = -0.7\, keV \qquad \Gamma_{2P} = 35\, meV$$

The strong interaction effects in higher S and P states can be calculated by use of these scaling laws:

$$\Gamma_{nS} = \frac{1}{n^3}\Gamma_{1S} \qquad \Delta E_{nS} = \frac{1}{n^3}\Delta E_{1S} \qquad \Gamma_{nP} = \frac{32}{3}\left(\frac{1}{n^3} - \frac{1}{n^5}\right)\Gamma_{2P}$$

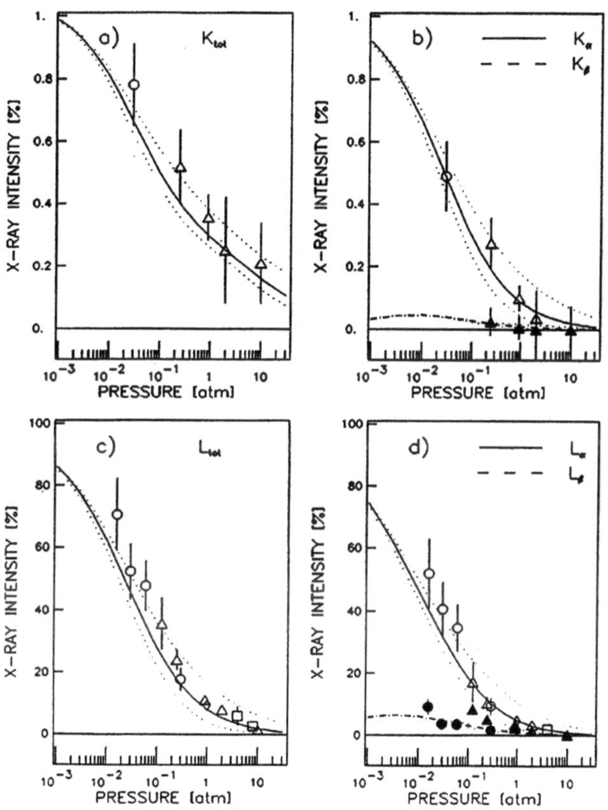

Fig.3. Xray intensities of K and L lines in dependence of target pressure.

2.6. Cascade Calculations

The cascade calculation starts by assuming an initial population of the protonium states. We assume that the $\bar{p}p$ atom is formed in states with $n = 30$ since for this quantum number the Bohr radius of the antiprotonic hydrogen atom and a ground state hydrogen atom have the same size. This guarantees a high capture probability. The population of the angular momentum states is chosen to be a statistical one. Depopulation occurs continuously by radiative decays and by annihilation. Collision take place with collision frequencies determined by:

$$\Gamma_i = N \cdot v \cdot \sigma_i \qquad\qquad i = 0, \ldots, 5$$

Here is σ_0 the chemical cross section and $\sigma_1, \ldots, \sigma_5$ are cross sections leading to five different electric field strengths and electron densities. Note that the size of the cross sections depend on the initial state which is treated. Hence a collision reduces the population of every level. Reshuffling of the population of different angular momentum states is due to Stark mixing. Changes of the principal quantum number are caused by Auger and chemical transitions. All radiative decays during the cascade are registered. Annihilations are monitored for S and P states separately. The collision sequence is stopped when the residual population of all levels with the given principal quantum number is below 10^{-3}.

The calculations were carried out for target densities corresponding to pressures from $10^{-3}\,atm$ up to liquid hydrogen.

3. RESULTS

Fig.3 shows the intensities of the Balmer and Lyman lines as function of target pressure. The experimental data cover a target pressure from 30 $mbar$ to 10 atm. The data are from $PS171$, $PS174$ and $PS175$ at LEAR. The calculated results are represented by the solid lines. We also show (dotted lines) the results which were obtained when the Auger rates were multiplied by an arbitrary factor of 0.5 or 2 (upper and lower curve, resp.). Changes of the chemical cross section by a factor of 0.5 or 2 or of the kinetic energy of the $\bar{p}p$ atom from 1 eV to thermal energies led to even smaller changes of the final results. The density dependence and also the absolute intensities are reproduced very well. This shows that all cascade processes are described with the required accuracy.

4. CONCLUSION

We have developed a model of the cascade of antiprotonic hydrogen. This model allows to calculate the cascade of $\bar{p}p$ atoms over the whole pressure range from $10^{-3}\,atm$ gas up to liquid hydrogen. The calcualtions were made without use of any free parameter. The results agree fully with the data of present experiments.

REFERENCES

1. Ch. Batty, "Antiprotonic Hydrogen", to be published in Rep. on Progr. in Phys. (1989)
2. G. Reifenröther and E. Klempt, "Antiprotonic Hydrogen: From Capture to Annihilation" Nucl. Phys. B (accepted for publication)
3. R.E. Olson and A. Salop, Phys. Rev. A 16 (1977) 531
   J.S. Cohen, Phys. Rev. A 26 (1982) 3008

4. M. Leon and H.A. Bethe, Phys. Rev. 127 (1962) 636
5. G.R. Burbidge and A.H. de Borde, Phys. Rev. 89 (1953) 189
6. H.A. Bethe and E.E. Salpeter, "Quantum Mechanics of One- and Two-Electron-Systems"

DETERMINATION OF K X-RAY YIELDS IN PIONIC HYDROGEN AS A FUNCTION OF GAS PRESSURE

A.J. Rusi EL Hassani, W. Beer, M. Bogdan,
J.F. Gilot, P.A.F. Goudsmit, H.J. Leisi, St. Thomann, W. Volken

Institute for Intermediate Energy Physics (IMP)
ETH Zürich, CH-5232 Villigen PSI, Switzerland

D. Bovet, E. Bovet, J.P. Egger, G. Fiorucci, J.L. Vuilleumier

Institut de Physique de l'Université de Neuchâtel
CH-2000 Neuchâtel, Switzerland

K. Gabathuler, L. Simons
Paul Scherrer Institute, CH-5232 Villigen PSI, Switzerland

INTRODUCTION

The S-wave scattering lengths a_1 and a_3 are among the most interesting quantities in pion-nucleon physics. An experiment is planned[1] at PSI to determine these scattering lengths from a determination of the strong shift (ε_o) and width (Γ) of the 1s level in pionic hydrogen with a crystal spectrometer. As a first step we have measured in the present work the absolute X-ray yields in pionic hydrogen as a function of gas pressure with the aim of obtaining an improved understanding of the atomic cascade processes (Stark mixing, Auger effect, Coulomb deexcitation etc...).

EXPERIMENTAL SET-UP

A major difficulty of measuring X-rays in pionic hydrogen is caused by the Stark mixing making it necessary to work with gazeous hydrogen. We have to stop a large amount of pions in a target with very small density. In our experiment we therefore used the so called cyclotron trap[2]. A schematic layout of the experiment carried out at the $\pi E3$-beam at PSI is shown in Fig. 1.

After entering the trap, pions pass through a thin plastic scintillator (S1) which is used to define the incoming particles. Approximately 10% of these incoming pions are stopped in the target cell located at the center of the trap after passing through a series of degraders.

Electromagnetic Cascade and Chemistry of Exotic Atoms
Edited by L. M. Simons *et al.*, Plenum Press, New York, 1990

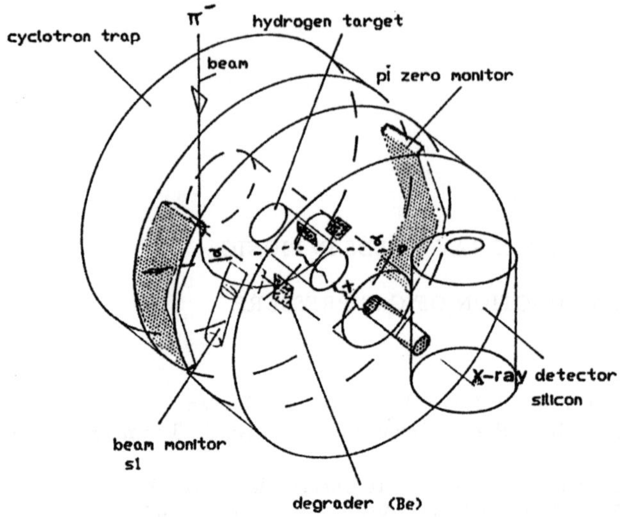

Fig. 1. Schematic layout of the experiment.

The target cell consists of a mylar cylinder of 60 mm diameter and a length of 120 mm, equipped with a Kapton exit window. The hydrogen gas can be cooled down to liquid hydrogen temperatures. For the present experiment the cell was used with the equivalent (room temperature) pressures of 3, 15 and 40 bar.

The formation of a π^- hydrogen atom is identified by the decay of a π^o produced in the charge exchange reaction:

$$\pi^- p \longrightarrow \pi^o n \quad \pi^o \rightarrow \gamma\gamma$$

In this desintegration two gamma rays are emitted back-to-back and are detected by the newly built π^o detector.

This detector[3] consists of a 10 layer thick sandwich of 1 mm thick tungsten plates and 1 mm diameter scintillation fibres. The detector incorporates four modules.

The X-rays were detected with a Si(Li) detector. The relative efficiency of this detector had been determined previously in an experiment[4] on X-rays in antiprotonic hydrogen at LEAR. The absolute intensity calibration in our set-up was obtained from a measurement of the 14.4 keV γ-ray from an absolutely calibrated ^{57}Co source.

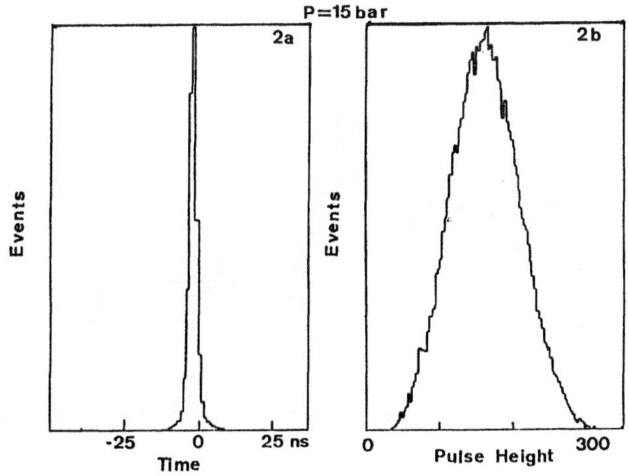

Fig. 2. Spectra from the π^o detector (see text).

The number of X-rays registred in this detector equals

$$N_{k_x} = N_{\pi^- stop} \cdot Y_{k_x} \cdot \varepsilon \cdot \Omega \cdot T$$

where:

$N_{\pi^- stop}$ is the number of π^- stopped in the target cell
Y_{k_x} is the absolute yield
$\varepsilon\Omega$ is efficiency times solid angle of the Si(Li) detector
T is the transmission through the Kapton window

Since the number of stopped pions ($N_{\pi^- stop}$) is not known with sufficiently high accuracy, we determined the yield with a coincidence method. The X-rays in the Si(Li) detector were measured in coincidence with the signal from the π^o detector. The number of such coincident events is

$$N(\gamma\gamma X) = N_{\pi^- stop} \cdot \frac{P}{1+P} \cdot \varepsilon_{\gamma\gamma} \cdot Y_{k_x} \cdot \varepsilon\Omega \cdot T$$

where:

P is the Pano fsky ratio
$\varepsilon_{\gamma\gamma}$ is the efficiency of the π^o detector.

We now define the ratio:

$$R = \frac{N(\gamma\gamma X)}{N(\gamma\gamma)} = Y_{k_x} \cdot \varepsilon\Omega \cdot T$$

where $N(\gamma\gamma)$ is the rate of the π^o detector.

In order to study the background effects, we have performed separate measurements with deuterium instead of hydrogen (essentially no π^o formed) and with an empty target cell.

We now define the appropriate ratio:

$$R' = \frac{N(\gamma\gamma X)_H - N(\gamma\gamma X)_B}{N(\gamma\gamma)_H - N(\gamma\gamma)_B} = Y_{k_x} \cdot \varepsilon\Omega \cdot T$$

where the index B represents the background measurement (deuterium or empty target).

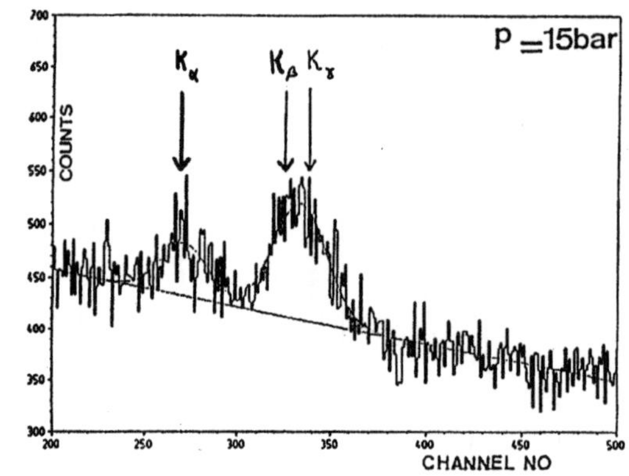

Fig. 3. Spectrum of X-rays triggered with the signal for incoming pions.

The desired yield is then immediately obtained as:

$$Y_{k_x} = R'/\varepsilon\Omega \cdot T \quad .$$

Results:
In Fig. 2 we show some data from the π^o detector. Figure 2a shows the time difference between the two γ-rays falling onto the upper and lower parts of the π^o detector. In Fig. 2b the total pulse hight from both detector halves is shown for the coincident $\gamma\gamma$ events.

The spectrum of X-rays measured in the Si(Li) detector in coincidence with a signal from the incoming pions (S1) is shown in Fig. 3. The background due to the contribution of electrons in the incoming beam was reduced by making use of the difference in the time-of-flight between pions and electrons. The K_α, $K_{\beta,\gamma,...}$ peaks are clearly visible in a non negligi ble background.

The corresponding spectrum of X-rays coincident with the π^o signal is shown in Fig. 4. Evidently, the peak-to-background ratio is improved very much.

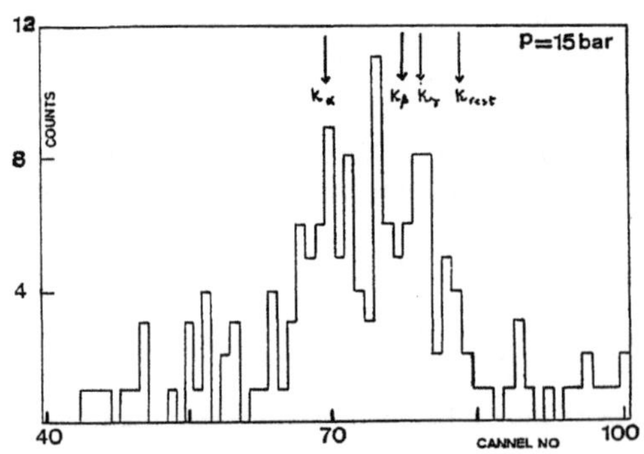

Fig. 4. Spectrum of X-rays coincident with the π^o signal.

Table 1. Total K X-ray yields, preliminary results.

H_2 Pressure [bar]	R'x10^{+5}	$Y_{k_{tot}}$
3	4.41 ± .44	0.21 ± 0.02
15	2.07 ± .22	0.097 ± 0.011
40	1.32 ± .29	0.062 ± 0.014

The resulting ratios R' and the yields obtained from the three H_2-pressures are listed in Table 1. Please note that these data should be considered as preliminary.

Finally, in Fig. 5 we have compared the total yields with a calculation of Borie and Leon[5] and the previously measured value at 4 bar by Bailey[6].

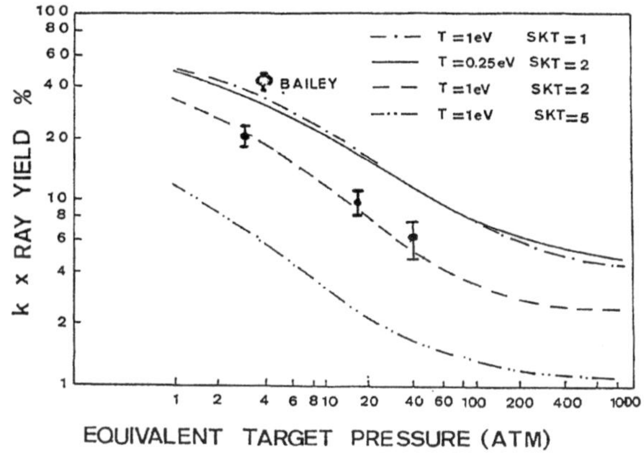

Fig. 5. Total K X-ray yields as function of equivalent target pressure.

ACKNOWLEDGMENTS

We would like to thank L. Knecht and B. Leoni for solving many technical problems and K. Heitlinger, J. Missimer and R. Backer for their support in operating the trap.

Further, we would like to thank the operators of the PSI synchrocyclotron for providing us excellent beam conditions.

REFERENCES

1. D. Bovet et al., Crystal diffraction of pionic hydrogen and deuterium, Proposal SIN R-86-05.1;
D. Bovet et al., Addendum to SIN experiment R-86-05.1, 20.11.1987.
2. L. Simons, Phys. Scripta 22:90-95, 1988.
3. A.J. Rusi El Hassani et al., 'A scintillating fiber detector for π° identification', to be published.
4. L. Simons et al., to be published.
5. E. Borie and M. Leon, Phys. Rev. A21:1460 (1980).
6. J. Bailey et al., Phys. Lett. 33B:369 (1970).

MEASUREMENT OF THE KINETIC ENERGY DISTRIBUTION OF PIONIC HYDROGEN ATOMS IN LIQUID HYDROGEN

J.F. Crawford, M. Daum, R. Frosch, B. Jost, and P.-R. Kettle

Paul Scherrer Institute
CH-5232 Villigen-PSI, Switzerland

R.M. Marshall, B.K. Wright, and K.O.H. Ziock

Physics Department, University of Virginia
Charlottesville, VA 22901, USA

INTRODUCTION

In this experiment[1,2] time-of-flight distributions of neutrons produced by the charge exchange reaction

$$\pi^- \ p \to \pi^0 \ n \tag{1}$$

of $\pi^- p$ atoms in liquid hydrogen were measured. The data were used to determine the pion mass difference $m_{\pi^-} - m_{\pi^0}$ and, as a side result, the kinetic energy distribution of the pionic hydrogen atoms just before the charge exchange reaction.

The resulting neutron velocity for $\pi^- p$ atoms at rest is $v_n = 0.894266 \pm 0.000063 \ cm/ns$; this corresponds to the pion mass difference[2]

$$m_{\pi^-} - m_{\pi^0} = 4.59366 \pm 0.00048 \ MeV/c^2 \ . \tag{2}$$

TIME-OF-FLIGHT DISTRIBUTIONS OF NEUTRONS PRODUCED BY CHARGE EXCHANGE

The measured time-of-flight flight distributions of the neutrons produced by the charge exchange reaction (1), after background subtraction,[2] are shown by histograms in Fig. 1. These three time-of-flight spectra were recorded at neutron flight distances of 3.2 m, 7.9 m, and 18.1 m. The tails to the right of the peaks in Fig. 1 are due not only to the finite kinetic energies of the $\pi^- p$ atoms, but also to neutrons which have reached the neutron detector after scattering in the materials in and around the flight

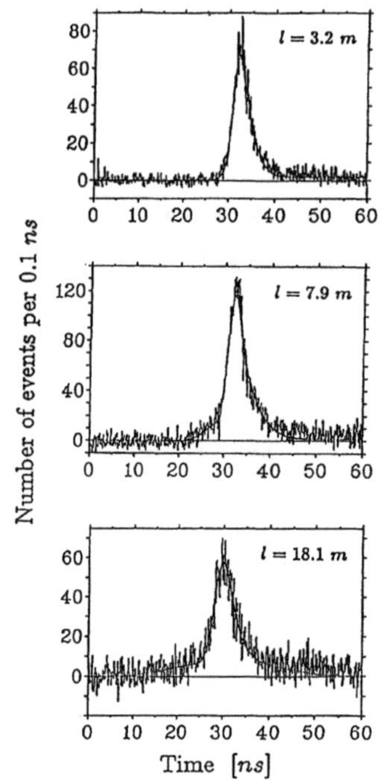

Fig. 1. **Experimental time-of-flight spectra of neutrons from the charge exchange reaction $\pi^- \ p \rightarrow \pi^0 \ n$, after background subtraction, for flight paths of 3.2 m, 7.9 m, and 18.1 m. Time, in ns, is from an accidental photon peak[1] about 30 ns before the neutron peak.**

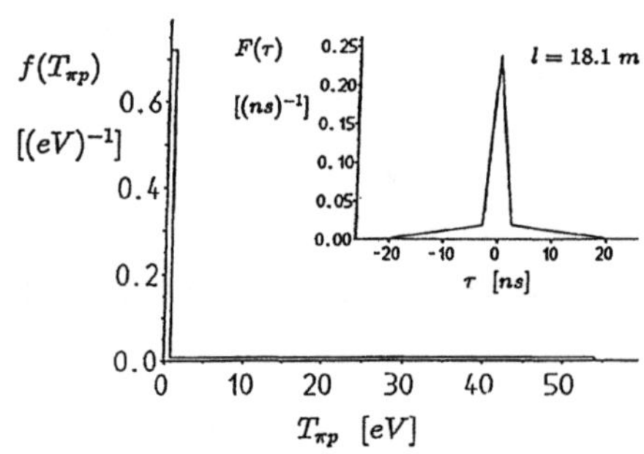

Fig. 2. **Distribution function $f(T_{\pi p})$ found to agree with the neutron time-of-flight spectra of Fig. 1; $T_{\pi p}$ is the kinetic energy of the $\pi^- p$ atoms just before reaction (1); $F(\tau)$ is the corresponding neutron time-of-flight distribution for a flight path of 18.1 m.**

channel.[1] In contrast, the tail to the left is not contaminated by neutron scattering; this tail, not visible at 3.2 m, extends to about 10 ns (20 ns) before the peak at 7.9 m (18.1 m).

The smooth curves in Fig. 1 were obtained by fitting nine free parameters[2] simultaneously to all three experimental spectra. Among these nine parameters were the three free parameters of the distribution function $f(T_{\pi p})$, where $T_{\pi p}$ is the kinetic energy of the pionic hydrogen atom just before the charge exchange reaction (1).

A kinetic-energy distribution function $f(T_{\pi p})$ found to agree with the data is shown in Fig. 2, together with the corresponding neutron time-of-flight distribution $F(\tau)$ calculated[3] for a fixed flight distance of 18.1 m. The model used for the function $f(T_{\pi p})$ was:

$$\begin{aligned} f(T_{\pi p}) &= \eta/T_1\,, \text{ if } 0 < T_{\pi p} < T_1\,; \\ f(T_{\pi p}) &= (1-\eta)/(T_2 - T_1)\,, \text{ if } T_1 < T_{\pi p} < T_2\,; \\ f(T_{\pi p}) &= 0\,, \text{ if } T_2 < T_{\pi p}\,. \end{aligned} \tag{3}$$

The three free parameters of this model are η, T_1 and T_2. The integral of the function defined by Eq. (3) is $\int_0^\infty f(T_{\pi p})\, dT_{\pi p} = 1$.

The analysis[2] of the experimental time-of-flight spectra shown in Fig. 1 yielded the following parameters of the function $f(T_{\pi p})$:

$$\begin{aligned} \eta &= 0.559 \pm 0.013\,; \\ T_1 &= 0.94 \pm 0.13 \; eV; \\ T_2 &= 71.5 \pm 6.1 \; eV. \end{aligned} \tag{4}$$

The corresponding mean kinetic energy of the $\pi^- p$ atoms is

$$\overline{T_{\pi p}} = T_1/2 + (1-\eta)\cdot T_2/2 = 16.2 \pm 1.3 \; eV. \tag{5}$$

The distribution function $f(T_{\pi p})$ presented in Fig. 2 and in Eqs. (3) and (4) consists of two components, the first of which reaches from zero to $T_1 = 0.94 \pm 0.13 \; eV$ and contains $\eta = 55.9 \pm 1.3$ percent of the $\pi^- p$ atoms, whereas the second component, containing the remainder of the $\pi^- p$ atoms, reaches from T_1 to $T_2 = 71.5 \pm 6.1 \; eV$. In this model both of the two components of the kinetic energy distribution are assumed to be uniform. Since the experimental time spectra are of limited statistical accuracy, they do not exclude $T_{\pi p}$ distributions which differ somewhat from that defined by Eqs. (3) and (4); however, the extension of the time-of-flight tails in Fig. 1 implies model-independently that the $T_{\pi p}$ distribution extends up to $\sim 70 \; eV$.

The high-energy component of the distribution $f(T_{\pi p})$ may be due to Coulomb de-excitation[4,5] of the $\pi^- p$ atom near one of the protons of the surrounding liquid hydrogen. In this process the de-excitation energy is partly transformed into kinetic energy of the $\pi^- p$ atom; the energy $T_{\pi p}$ of $\pi^- p$ atoms which have just undergone the $5 \rightarrow 4$ ($4 \rightarrow 3$) Coulomb de-excitation is around 30 eV (70 eV).

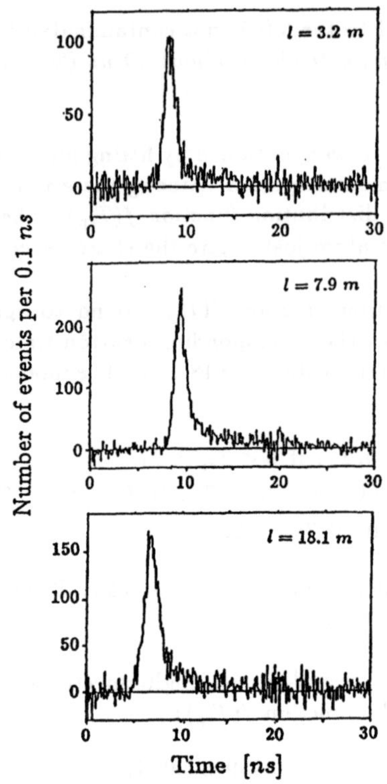

Fig. 3. Experimental time-of-flight spectra of neutrons from the radiative capture reaction $\pi^- \, p \rightarrow \gamma \, n$.

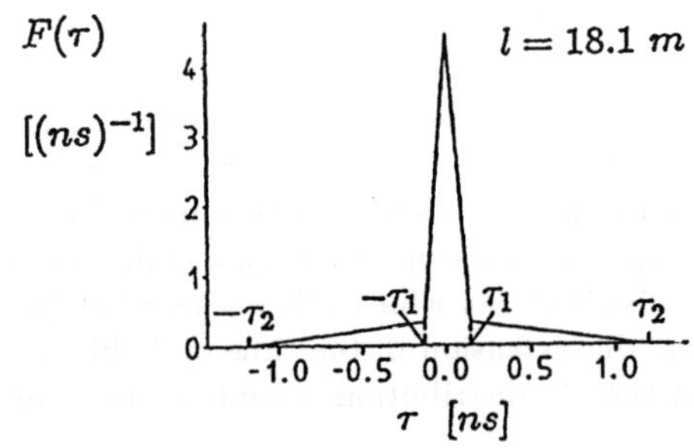

Fig. 4. Time-of-flight distribution of neutrons from radiative capture, for a flight path of 18.1 m. This function $F(\tau)$ was calculated from the function $f(T_{\pi p})$ shown in Fig. 2.

TIME-OF-FLIGHT DISTRIBUTIONS OF NEUTRONS PRODUCED BY RADIATIVE CAPTURE

As a test of the experimental method, time-of-flight distributions of the neutrons from the radiative capture reaction

$$\pi^- \; p \to \gamma \; n \tag{6}$$

were recorded simultaneously with the spectra of the neutrons from charge exchange. The measured distributions, after background subtraction,[6] are shown in Fig. 3.

The resulting neutron velocity[2,6] for $\pi^- p$ atoms at rest, $v_{n'} = 4.09090 \pm 0.00075 \; cm/ns$, leads to a π^- mass value of $139.587 \pm 0.027 \; MeV/c^2$, which is consistent with the world average,[7] $139.56755 \pm 0.00033 \; MeV/c^2$.

The kinetic energy distribution of the $\pi^- p$ atoms defined by Eqs. (3) and (4) corresponds, for neutrons from radiative capture, to the time-of-flight distribution function $F(\tau)$ shown in Fig. 4. This function was calculated[3] from Eqs. (3) and (4) for a neutron flight distance of 18.1 m. At 7.9 m, the times τ_1 and τ_2 indicated in Fig. 4 are equal to 0.06 ns and 0.52 ns, respectively. These time distribution functions $F(\tau)$ are narrower than those for the neutrons from charge exchange by a factor of $(v_n/v_{n'})^2 = 0.048$. The time distributions obtained[6] by folding the functions $F(\tau)$ with the time distributions due to the sizes of the liquid hydrogen target and the neutron detector, to neutron scattering and to the time jitters of the electronics are consistent with the experimental spectra shown in Fig. 3.

References

1. J. F. Crawford *et al.*, Precision Measurement of the Mass Difference $m_{\pi^-} - m_{\pi^0}$, Phys. Rev. Lett. 56:1043 (1986).
2. J. F. Crawford *et al.*, Precision Measurement of the Mass Difference $m_{\pi^-} - m_{\pi^0}$, Phys. Lett. B 213:391 (1988).
3. R. Frosch, Kinetic Energy of $\pi^- p$ Atoms in Liquid Hydrogen, internal report TM-37-21 (1985), Paul Scherrer Institute (unpublished).
4. L. Bracci and G. Fiorentini, Coulomb De-excitation of Mesic Hydrogen, Nuovo Cim. 43A:9 (1978).
5. L. I. Men'shikov, Mesic Atom Acceleration Mechanisms in Cascade Transitions, Muon Catalyzed Fusion 2:173 (1988).
6. J. F. Crawford *et al.*, to be published.
7. Particle Data Group, Review of Particle Properties, Phys. Lett. B 204 (1988).

RESULTS OF X-RAY MEASUREMENTS IN ANTIPROTONIC HYDROGEN (PS 175)

R. Bacher[4], R.Badertscher[4], P. Blüm[3], J. Eades[1], J. Egger[4], K. Elsener[1], D. Gotta[2], K. Heitlinger[3], L.M. Simons[4]

[1] CERN, CH-1211 Geneva 23, Switzerland
[2] KfA, D-5170 Jülich, Germany
[3] KfK/Universität Karlsruhe, D-7500 Karlsruhe 1, Germany
[4] PSI, CH-5232 Villigen, Switzerland

Abstract

At the LEAR storage ring at CERN, Balmer and Lyman X-rays in antiprotonic hydrogen were measured. The measurements have been performed using the cyclotron trap together with Si(Li)-semiconductor detectors at a beam injection momentum of 105 MeV/c. At 30 hPa target pressure, a stop efficiency of 85 % could be obtained. Strong interaction parameters (2p-width, 1s-shift and width) could be determined with good precision.

1 Introduction

Some basic properties of antiprotonic hydrogen ($\bar{p}p$) can be shown in the framework of the Bohr model. Due to the large mass of the antiproton the binding energy in $\bar{p}p$

$$E_n \approx -\frac{\mu c^2}{2}(\frac{\alpha}{n})^2 \approx -12.49keV/n^2 \quad (1)$$

is increased by a factor of about 1000 compared to the e^-p-system. The transition energies between the low lying levels are in the keV region and therefore measurable with Si(Li)-semiconductor detectors. The Bohr radii given by

$$r_n \approx \frac{\hbar}{\mu c}\frac{n^2}{\alpha} \approx 57.64fm \cdot n^2 \quad (2)$$

are reduced by a factor of about 1000 resulting in a drastically increased overlap of the wave functions of the nucleus and the low lying atomic states of the $\bar{p}p$-sytem. The consequence is that the atomic levels are shifted in energy and broadened. The ground state is expected to have a shift of about -500 eV and a strong width of about 1000 eV while the 2p-state should be broadened by an amount of about 40 meV. Due to hyperfine interaction the ground state is composed of two states, the

Electromagnetic Cascade and Chemistry of Exotic Atoms
Edited by L. M. Simons *et al.*, Plenum Press, New York, 1990

so-called triplet- and singlet-state; the relative energy shift between these two states is theoretically expected to be smaller than 300 eV [1].

The observables of the measurement are therefore a shift and width of the ground state (spin averaged or for the two components separately), which can be obtained by a direct measurement of the Lyman-transitions (np $\rightarrow$ 1s). The strong width of the 2p-state is too small to be resolved with our detection method and must therefore be inferred from an intensity balance. This includes the measurement of the population of the 2p-level (Balmer transitions).

The value, 40 meV, for the strong width of the 2p-state is about 100 times larger than its radiative decay width. This reduces drastically the intensity of the following Lyman α-transition to about 1 % of the population of the 2p level. Due to pressure-dependent deexcitation processes in the upper region of the level scheme (described in [2] and [3]), the population of the 2p level is pressure-dependent. In the pressure region between 10 hPa and 1000 hPa the 2p population (and also the following Lyman α-transition) is expected to be inversely proportional to pressure.

Therefore, as a first step it is necessary to measure the population of the 2p-level (i.e absolute yields of the Balmer transitions) as a function of pressure. The comparison of the measured absolute Balmer X-ray yields with a cascade code of Borie and Leon [2] should lead to a better understanding of the cascade processes in the upper region of the level scheme. As a second step, the measurement of the Lyman transitions can be performed under optimal conditions.

2 The Experimental Set-Up

The experiment was performed with the cyclotron trap, a superconducting magnetic system and Si(Li)-semiconductor detectors. The principle of the cyclotron trap was described in former Erice proceedings [4] and [5]. The characteristics of the Si(Li)-detectors used for this experiment are listed in Tab.1. The determination of the response function of the detectors was described in [6].

Table 1. Characteristics of the Si(Li)-semiconductor detectors used for this experiment. Si(Li) I was build in a guardring configuration, that means, a 30 mm^2 detector is surrounded by a Si(Li)-ring detector of 200 mm^2. Such a configuration can be used for background reduction.

	active area /mm^2	in-beam resolution at 6.4 keV /eV	Be-window thickness /μm	crystal thickness /mm
Si(Li) I	30/200	280	35	6.0
Si(Li) II	300	590	40	5.0

3 Results

3.1 Balmer X-Rays and Cascade Calculations

Balmer X-rays have been measured at 16 hPa, 30 hPa, 60 hPa, 120 hPa and 300 hPa, both in 1986 and 1988. Fig.1 shows as an example measured spectra of $\bar{p}p$ at 16 hPa

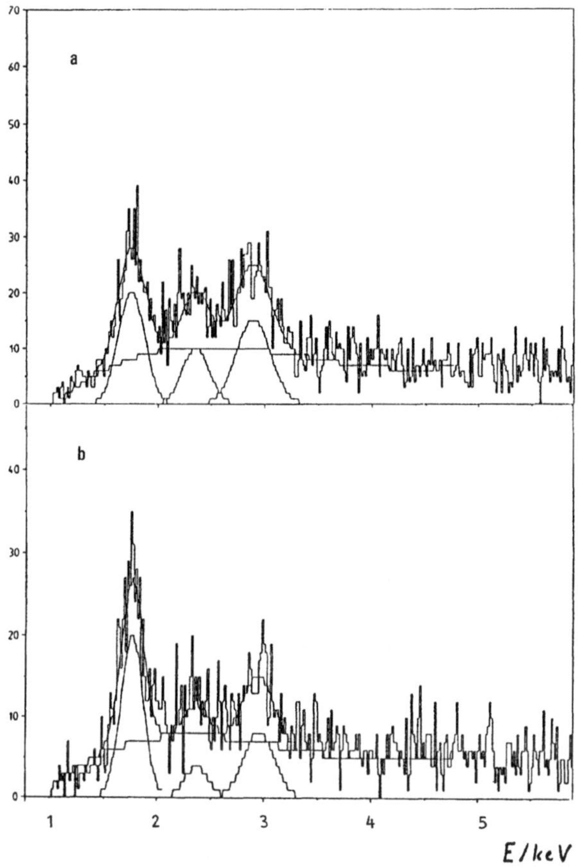

Figure 1: Balmer X-ray spectra of antiprotonic hydrogen measured with Si(Li) I at 120 hPa (1.a.) and 16 hPa (1.b.) together with a fit to the Balmer transition lines: Balmer α at 1.74 keV, Balmer β at 2.34 keV and the series transitions (fitted with one gaussian line).

and 120 hPa (1988) together with a fit to the Balmer transitions. From the measured intensities and precise knowledge of the response function of the detectors, absolute yields can be calculated. They can be compared with the results of a cacade code of Borie and Leon [2]. A fit of the calculations to the measured yields is shown in Fig.2. A successful adaption can be gained by varying only two parameters of the cascade code : the kinetic energy T_{kin} of the $\bar{p}p$-system and the so-called Stark mixing parameter k_{Stk}. T_{kin} is kept constant for all steps of the cascade calculations, which means that acceleration or deceleration processes during the cascade are not taken into account. The Stark mixing parameter is an arbitrary number multiplying all Stark mixing rates globally. This factor is necessary to compensate some approximations in the cascade model of Leon and Bethe [7]. Fig.2 shows that with these two parameters the pressure dependence of the Balmer X-ray yields is in principle understood. The obtained values for the cascade parameters are $T_{kin} = (1.0^{+0.3}_{-0.2})$ eV and $k_{Stk} = (2.16^{+0.23}_{-0.20})$.

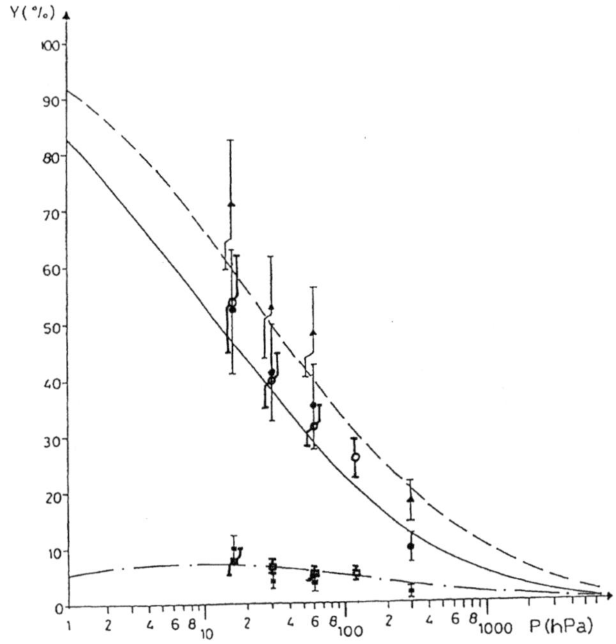

Figure 2. Absolute yields in percent versus pressure for antiprotonic hydrogen. The curves are the best fit to the measured yields as obtained from the cascade code of Borie and Leon. The curves denote: — — — Balmer *tot*, ——— Balmer α and —·— Balmer β.

Balmer *tot* denotes the sum of all Balmer transitions. The notation of the measured yields is : ▲ : Balmer *tot* (1986), ● : Balmer α (1986), ○ : Balmer α (1988), □ : Balmer β (1986), □ : Balmer β (1988).

4 Lyman X-Rays

Lyman X-rays were measured with two different detectors at a pressure of 30 hPa hydrogen. For background studies, antiprotonic deuterium spectra were also recorded. Fig.3 shows two spectra recorded with Si(Li) I : antiprotonic hydrogen (Fig.3a) in the energy range from about 5 keV to 18 keV and antiprotonic deuterium (Fig.3b) in the same energy range. Except for antirotonic oxygen, contamination lines due to a water contamination in the target chamber, the background behaviour in the $\bar{p}$d-spectrum is flat and could be fitted with a polynomial of 3rd oder. The $\bar{p}$p-spectrum shows in addition a broad structure between 8 keV and 10 keV. This structure could be fitted with a lorentzian line profile and identified as Lyman α-transition. A similar structure with the same line profile around about 12 keV can be associated with the Lyman series transitions.

Table 2. Strong interaction parameters for antiprotonic hydrogen obtained from the fits of the different X-ray spectra.

	$-\epsilon_{1s}$ /eV	Γ_{1s} /eV	Γ_{2p} /meV	Detector
spin	690 ± 35	1060 ± 140	36 ± 4	Si(Li) I
averaged	770 ± 30	1290 ± 120	–	Si(Li) II
triplet state	850 ± 40	770 ± 150	–	Si(Li) I

Fig.4 shows the antiprotonic hydrogen spectrum recorded with Si(Li) II. Again the contamination lines stem from antiprotonic oxygen and, in addition, a small contribution of pionic aluminium could be found. The lorentzian line profiles of the Lyman α and the higher Lyman transitions can be clearly identified in this spectrum.

The results of the evaluation of both spectra are given in Tab.2. The comparison of the population of the 2p level and the intensity of the Lyman α transition yields the strong width of the 2p state. As already mentioned in the beginning, the 1s ground state consists of two hyperfine states. Taking the antiprotonic hydrogen spectrum recorded with Si(Li) I and fitting the broad structure between 8 keV and 10 keV with the lorentzian line profiles of two lines, the result shown in Fig.5 is obtained. A comparison with a corresponding Monte Carlo simulated spectrum allows to extract the position and strong width of the larger of the two components (triplet-state) from the fit. For the singlet state, only upper limits can be given.

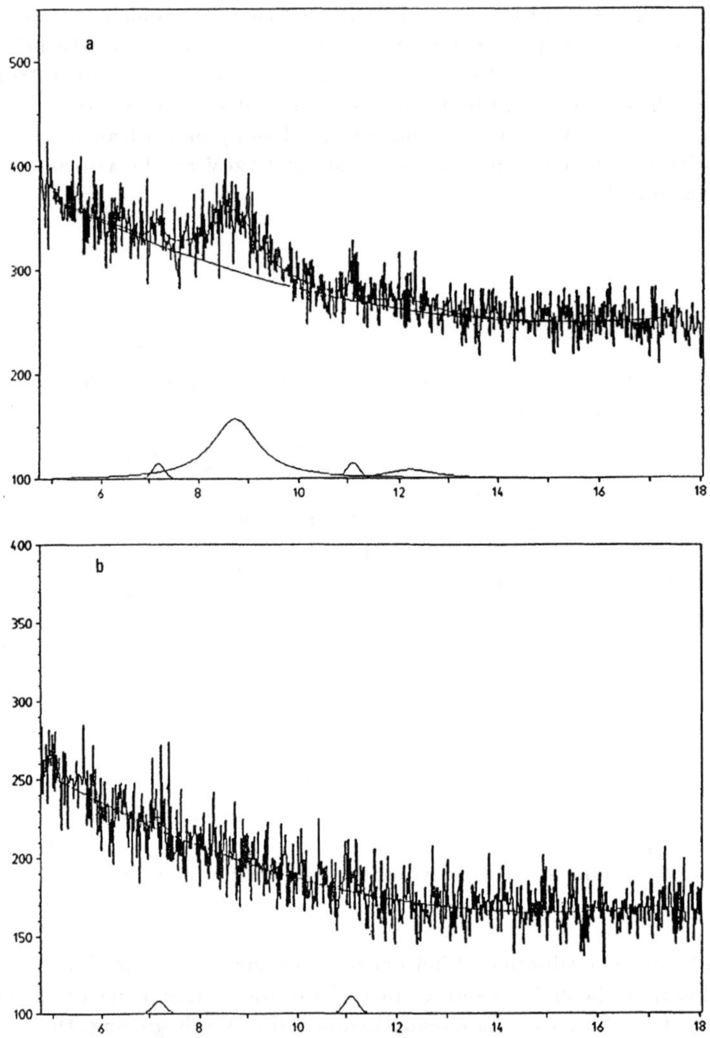

Figure 3. X-ray spectra measured with Si(Li) I :
a) antiprotonic hydrogen measured at 30 hPa
b) antiprotonic deuterium measured at 30 hPa
The binning for both spectra is about 16 eV/channel.

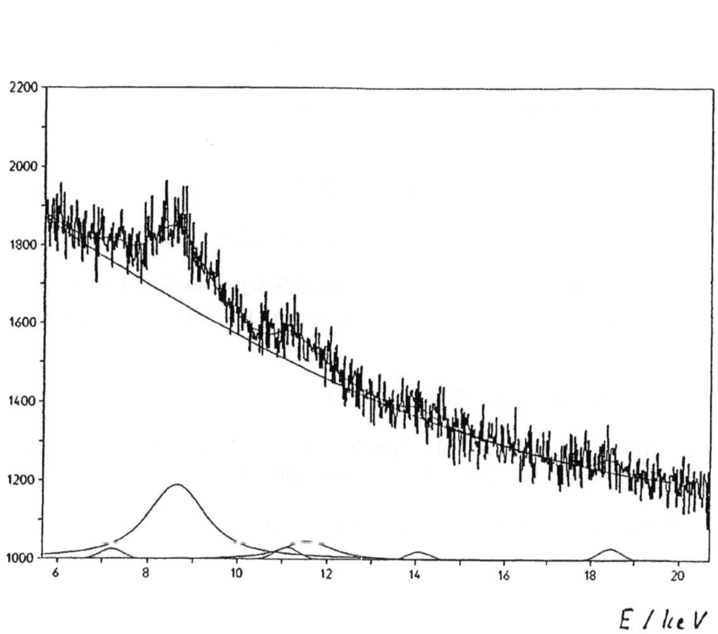

Figure 4. X-ray spectrum of antiprotonic hydrogen measured with Si(Li) II at 30 hPa.

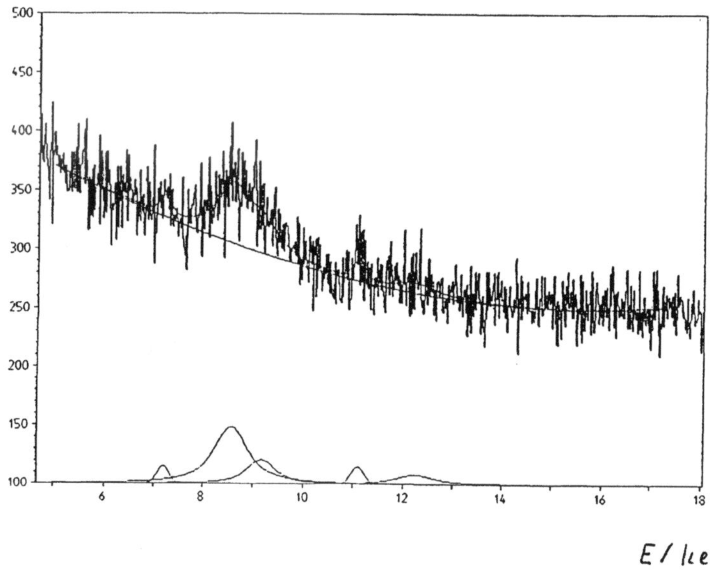

Figure 5. A fit to the hyperfine states in spectrum 3.a. T denotes the triplet state, S the singlet state.

References

[1] Batty, C.J.: 'Antiprotonic-Hydrogen Atoms', accepted for publication in "Reports on Progress in Physics"

[2] Borie, E., Leon, M.: Phys. Rev. **A21**,1460(1980)

[3] Bracci, L., Fiorentini, G.: Il Nuovo Cim. **43A**,9(1978)

[4] Simons, L.M.: Ettore Majorana International Science series, Fundamental Symmetries, **31**,89(1987), eds. Bloch, P., Pavlopoulos, P., Klapisch, R., Plenum Press, New York

[5] Simons, L.M.: Phys. Scripta, **T22**,90(1988)

[6] Bacher, R., Blüm, P., Gotta, D., Heitlinger, K., Schneider, M., Missimer, J., Simons, L.M., Elsener, K.: Phys. Rev. **A38**,4395(1988)

[7] Leon, M., Bethe, H.A.: Phys. Rev. **127**,636(1962)

III

EXOTIC ATOMS AT HIGHER Z

"When you say 'hill'," the Queen interrupted, "I could show you hills, in comparison with which you'd call that a valley."
"No, I shouldn't," said Alice..., "a hill *can't* be a valley, you know. That would be nonsense -"
The Red Queen shook her head. "You may call it 'nonsense' if you like," she said, "but I've heard nonsense, compared with which that would be as sensible as a dictionary."

Lewis Carroll: Through the Looking-Glass

EXOTIC ATOM CASCADE PROCESSES IN ATOMS WITH Z>2

F.J. Hartmann
Physik-Department, E18, Technische Universität München
Garching, F.R.G.

1 Introduction

The topic of my talk is the cascade in exotic atoms. Let me start with a definition: The exotic-atom cascade is the deexcitation of an exotic atom from the - usually highly excited - first bound atomic state to the ground state (for muonic atoms) or to the state from which nuclear capture takes place (for hadronic atoms). We will assume for the following that the system is rotationally symmetric. So the first bound state may be characterized by principal quantum number n, angular momentum quantum number ℓ and magnetic quantum number m. As we know [1], atomic capture of exotic particles proceeds via Auger electron emission. If we accept the argument that preferentially states with maximum overlap of exotic particle and electron wave-function are populated we expect the first bound state in an exotic atom to have a principal quantum number n_{ex}

$$n_{ex} = \sqrt{\frac{m}{m_e}} * n_e \tag{1}$$

(m: exotic-particle mass, m_e, n_e: mass and principal quantum number of the bound electron, respectively). As there are good reasons to believe that Coulomb capture proceeds via interaction with the outermost electrons the first bound state should be high already for medium-Z atoms, e.g. $n \approx 28$ for muon capture by L-electron emission.

For atomic numbers Z>2 the exotic atom deexcites by two processes:

- radiative transitions,
- internal Auger effect, i.e. the ejection of an atomic electron (if electrons of sufficiently small binding energy are available).

Those processes which play a decisive role in the exotic atom cascade in hydrogen (and helium) [2], external Auger effect, Coulomb deexcitation , chemical deexcitation and Stark mixing, may be neglected for Z>2.

In principle, both types of emitted radiation (x rays *and* Auger electrons) can be observed. Quantum radiation is much easier to detect, however. It is the competition between the two deexcitation processes which determines the x-ray intensities.

The exotic atom cascade may be divided into two parts:

- the cascade through the electron cloud,
- the cascade under the cloud.

2 Cascade through the electron cloud

In the upper part of the cascade the exotic particle may be treated as moving on a classical trajectory loosing energy by interaction with the atomic electrons and by emitting bremsstrahlung. To get more quantitative results, Rooks [3] solved this problem for an 1/r potential (which, however, is only a poor approximation for the outer part of the atom). As in the early work by Fermi and Teller [4] continuous energy loss by interaction with a Fermi gas of electrons (Fermi velocity v_0) was assumed. The resulting energy loss per unit time of a particle with energy -E is

$$dE/dt = \mu * \frac{2E}{m} \tag{2}$$

with

$$\mu = \frac{2}{3} * \frac{m_e^2 * e^4}{\pi * \hbar^3} * ln(\frac{v_0}{\alpha c}).$$

The angular momentum loss per unit time can be written as

$$dL/dt = -\mu L/m, \tag{3}$$

with $L = \hbar(\ell + \frac{1}{2})$. From these two equations one easily derives

$$(\ell + \frac{1}{2})/n = const. \tag{4}$$

Thus the shape of the angular momentum distribution is *not* changed by Auger transitions.

Integration of bremsstrahlung losses over one period of the movement of the exotic particle around the nucleus leads to corresponding expressions for energy and angular momentum loss by x-ray emission. Their derivation is described by S.S. Gershtein [5]. Angular momentum and principal quantum numbers are related by

$$\frac{1}{\ell + \frac{1}{2}}[\frac{1}{(\ell + 1/2)^2} - \frac{1}{n^2}] = const. \tag{5}$$

From this equation it is easily derived that circular orbits (ℓ = n-1) are reached very quickly. Comparing the energy losses for Auger effect and bremsstrahlung, Rooks found out that for high n the exotic-atom deexcitation is completely governed by Auger effect. Hence the shape of the angular momentum remains unchanged during the cascade through the electron cloud.

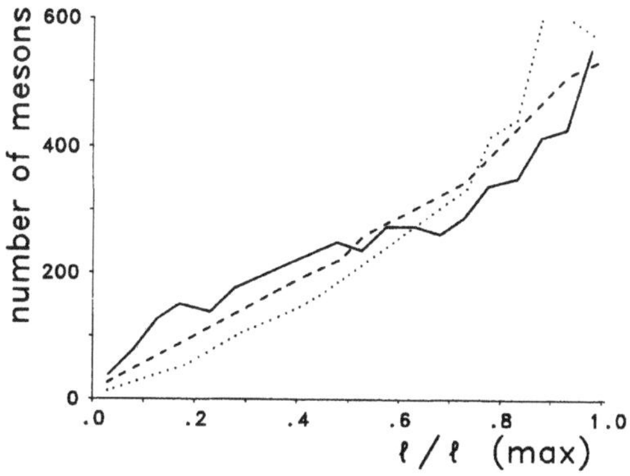

Figure 1. F(E,η) as a function of the (negative) muon energy in muonic Zr (Z=40). Dotted line: ℓ distribution immediately after capture. Dashed line: ℓ distribution at energy -270 eV. Solid line: ℓ distribution at n $\approx$ 16.

A calculation with a more realistic potential performed by Leon and Seki [6] succeeded in establishing a continuity equation for the angular momentum distribution F(E,$\eta = \ell/\ell_{max}$). After numerical integration the authors predicted a development of F(E,η) as shown in Fig. 1. The shape of the angular momentum distribution remains essentially unchanged during the way of the muon from energy zero to an energy corresponding to n $\approx$ 16.

3 Under the electron cloud: The quantal cascade

At principal quantum numbers smaller than $\sqrt{m/m_e}$ the radius of the Bohr orbits of the exotic particle is smaller than the radius of the K electron orbit. This has two consequences:

- The calculation of the exotic-particle cascade down the ladder of atomic levels has to be calculated quantum-mechanically,
- the calculation is comparatively simple as a hydrogen-like problem has to be solved.

Again deexcitation by Auger effect competes with the emission of quantum radiation. Calculations which started with the work of Burbidge and deBorde in 1953 [7] and have become more sophisticated ever since [8,9,10,11] have now firmly established the rates for both processes if the electron configuration of the host atom is known. The width for electric multipole radiation (magnetic multipole radiation usually is an order of magnitude slower than electric radiation of the same multipolarity) is, for transition energy E, given by

$$\Gamma_{rad} \propto |\langle f|\vec{r}^{\lambda}|i\rangle|^2 * E^{2*\lambda+1}. \qquad (6)$$

Here λ denotes the multipolarity of the radiation and the expression between bars is the transition matrix element. It can be shown that [11]

$$\frac{\Gamma_R^{\lambda+1}}{\Gamma_R^{\lambda}} \leq \frac{(Z\alpha)^2}{n^2}. \qquad (7)$$

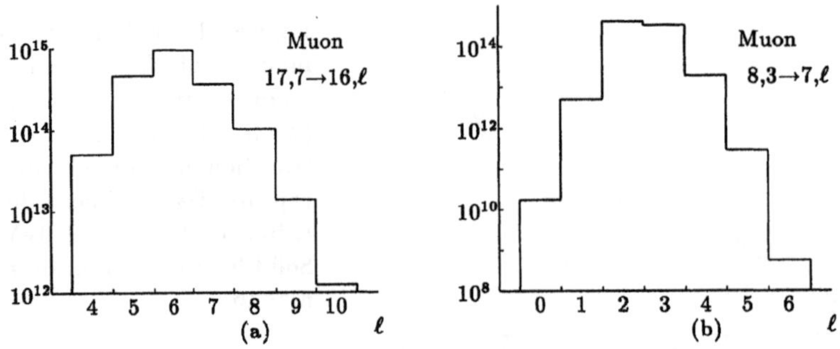

Figure 2. Rates for transitions from different levels in muonic Fe: (a) Transitions from n= 17, ℓ= 7 to different ℓ levels in n= 16. (b) Transitions from n= 8, ℓ= 3 to different ℓ levels in n= 7.

So radiative quadrupole transitions play a role only for the last transitions in the cascade in high Z muonic atoms (as corresponding levels are never reached in hadronic atoms).
For Auger transitions all multipoles from monopoles to octupoles have to be taken into consideration [11]. The most important transitions, however, are still the dipole transitions (selection rule $\Delta\ell = \pm 1$). Figure 2 shows rates for transitions in muonic Fe to different ℓ levels at n'= n-1 when starting at medium ℓ in n= 17 and n= 8, respectively [11].

The width for Auger transitions

$$\Gamma_{Auger} \propto |\langle f|\vec{r}^{\,-1}|i\rangle|^2 \tag{8}$$

decreases slightly with increasing energy. From this it becomes clear that also in the quantum cascade Auger effect dominates for high n and small Δn (which means small energies), whereas radiation is emitted preferentially in the final stage of the cascade. For the latter process Δn is as high as possible.

A formula very useful for a fast evaluation of the competition between Auger effect and radiative transitions has been derived by Ferrell [12]. It states that

$$\Gamma_{Auger}/\Gamma_{rad} = \sigma_{\gamma e}^{(Z-1)}(E)/[(Z-1)^2 * \sigma_T] \tag{9}$$

where $\sigma_{\gamma e}^{(Z-1)}(E)$ denotes the photoelectric cross section at energy E for the atom with Z-1 and $\sigma_T = \frac{8}{3}\pi r_0^2$ is the Thomson cross section. Only terms up to dipole are taken into account and a completely filled electron shell is assumed. Figure 3 shows $\Gamma_{Auger}/\Gamma_{rad}$ for Mg (Z=12) and Au (Z=79) as a function of energy. Whereas Auger and radiative rates for Mg become equal only for the muonic (4→3) transition (E≈ 20 keV), this is true for Au already for the (10→9) transition (E≈ 41 keV). Simple formulae exist for $\sigma_{\gamma e}^{(Z-1)}$(E) [13].

Which processes dominate the cascade at the different stages is displayed in Fig. 4 for transitions between circular orbits (n, ℓ = n-1) in muonic Fe. Most of the time only one process is predominant.

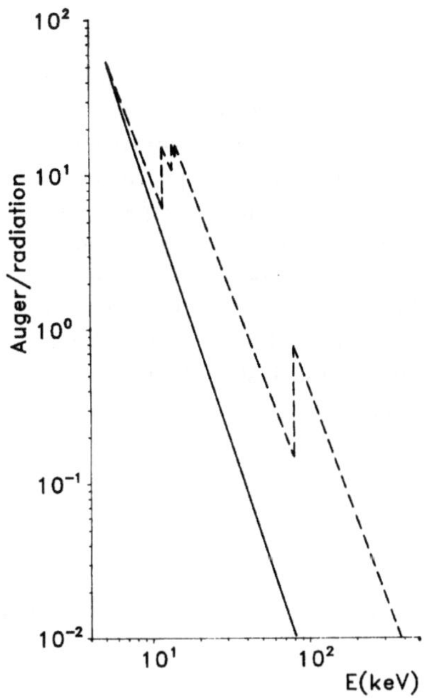

Figure 3. $\Gamma_{Auger}/\Gamma_{rad}$ for Mg (Z=12) and Au (Z=79) as a function of transition energy E. Solid line: Mg. Dashed line: Au.

For all calculations described until now it has been assumed that all electrons are present and therefore Auger effect is possible at full rate. It, however, depends strongly on the atomic structure and on the chemical and physical state of the target whether electrons ejected by Auger effect are refilled fast enough. Techniques for the calculation of electron depletion and refilling were devised. In the simplest case one takes into account only K-shell Auger effect and K-shell refilling and assumes the following:

At time t=0 the exotic atom is in the upper state with probability $\chi_0 = 1$. The probability that the level has been depopulated by radiative transitions develops as

$$d\chi_R = R * \chi_0 * dt, \tag{10}$$

with R the radiative transition rate. The probability for depopulation by Auger effect reads

$$d\chi_A = f_k * A * \chi_0 * dt, \tag{11}$$

with f_k the (continuous) K-shell filling and A the Auger transition rate for a completely filled electron shell. f_k itself changes as

$$df_k = (1 - f_k) * W_k * dt, \tag{12}$$

with W_k the K-electron refilling rate. As the initial level is depopulated only by K-Auger effect and radiation, its time evolution follows the equation

$$d\chi_0 = -d\chi_A - d\chi_R. \tag{13}$$

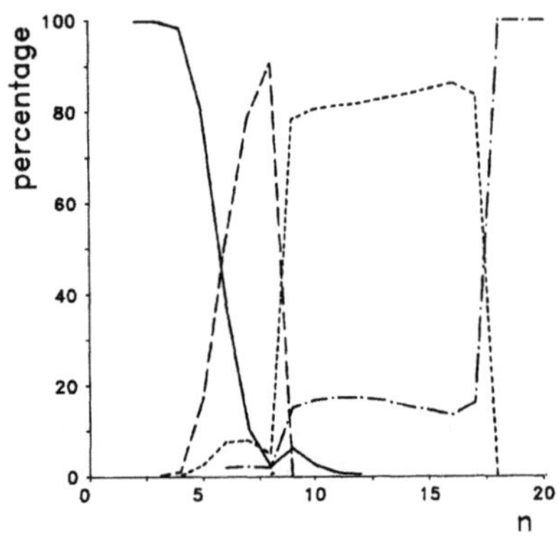

Figure 4. Dominant transition modes during the muonic cascade in Fe. The contribution of the different processes is plotted vs. the principal quantum number of the depopulated level. Solid line: Radiation. Long-dashed line: K-Auger effect. Short-dashed line: L-Auger effect. Dot-dashed line: M-Auger effect.

Integrating eq. (12) we obtain

$$f_k = 1 - (1 - f_k^0) * exp(-W_k * t). \tag{14}$$

From this and eqs. (10), (11) and (13) we derive

$$\chi_0 = exp(-\Gamma t) * exp[\beta(1 - e^{-W_K t})]. \tag{15}$$

Here f_k^0 is the K-shell filling at t=0, Γ = R+A and $\beta = (A/W_k)*(1-f_k^0)$. The overall probability $\chi_R(\infty)$ for the occurence of a radiative transition may now be obtained by integrating (15) from t=0 to t$\rightarrow\infty$. The result is

$$\chi_R = (R/\Gamma) *_1 F_1(1, 1 + \mu, \beta). \tag{16}$$

Here ${}_1F_1$ is the confluent hypergeometric function and $\mu = \Gamma/W_k$. Only ratios of transition rates enter the final result. We define a new Auger transition rate A' (corrected for electron depletion) so that the overall probability for Auger transitions χ_A = 1-χ_R is given by the branching ratio A'/(A'+R). A' is plotted in Fig. 5 as a function of the refilling rate for R=A= 1 with the initial K-shell filling as a parameter.

The simple model may be improved with respect to the following:

- L-shell refilling is taken into account, roughly in the same way as described above for the K shell,
- L-shell depletion by K refilling is considered. A number α of L holes are assumed to be generated for one refilled K hole,
- The K-shell refilling rate changes during the cascade according to the depletion of the upper electron shells (in a first approach only the L-shell depletion is taken into account).

A rather complicated system of equation results [14].

One word of caution has to be added. The approach of a continuous shell filling may become very poor for extreme conditions. Imagine the following: With a very small K-shell filling f_k= 0.01 and a ratio A/R= 100 one would expect only 50% radiation as f_k*A = R. For such a small f_k, however, the electronic K shell must be completely empty in at least 99% of the cases, resulting in a share of radiation exceeding 99%. Calculations to solve this problem have been performed [10,15].

Up to now we restricted ourselves to electromagnetic processes. In hadronic atoms nuclear absorption processes play an important role already at high n, at least for low ℓ. The absorption width of a level may be calculated by solving the Dirac or Klein-Gordon equation for a reasonably chosen strong-interaction potential. To shorten this tedious task calculations are performed only for circular orbits and scaling prescriptions are applied to extrapolate to levels with the same ℓ but different n. The simplest formula is derived from the perturbation expression for the absorption width

$$\Gamma(n,\ell) = \int W(r)|\Psi(r)|^2 d^3r, \tag{17}$$

where W(r) is the imaginary part of the optical potential and Ψ(r) is the exotic particle wave function. If the potential can be approximated by a square well and hydrogen-like wave functions are assumed for the exotic particle, this extrapolation gives [16]

$$\frac{\Gamma_{abs}(n,\ell)}{\Gamma_{abs}(m,\ell)} = (\frac{m}{n})^{2\ell+4} * \frac{(n+\ell)!(m-\ell-1)!}{(m+\ell)!(n-\ell-1)!} \tag{18}$$

and for scaling from circular orbits (m=ℓ +1)

$$\frac{\Gamma_{abs}(n,\ell)}{\Gamma_{abs}(\ell+1,\ell)} = (\frac{\ell+1}{n})^{2\ell+4} * \frac{(n+\ell)!}{(2\ell+1)!(n-\ell-1)!} \tag{19}$$

Turner and Jackson [17] pointed out that this simple scaling law gives completely wrong results for s and p levels in pionic atoms and proposed to correct eq. (19) by introducing a factor to fit absorption widths derived from exact calculations. Scaling to higher n reveals that already for low Z and for s and p states in hadronic atoms the absorption width is much larger than the combined width for Auger and radiative transitions. Some representative values are displayed in Table 1 for pionic Mg (Z=12).

Strong interaction effects have to be taken into account already in the initial stage of the quantum cascade. One even can say that the low-ℓ levels are emptied by nuclear absorption while the exotic particle is still in the electron cloud.

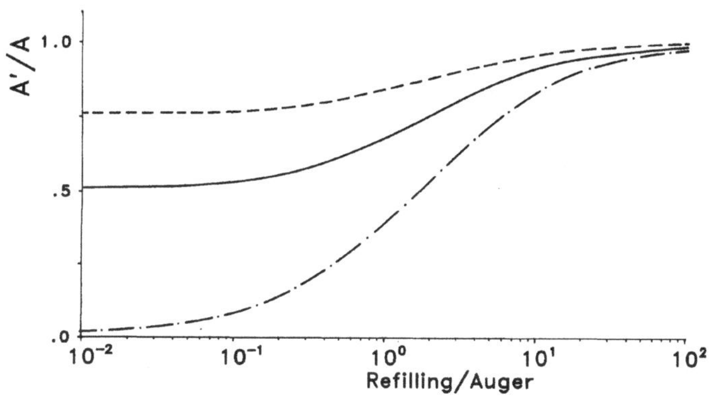

Figure 5. The effective Auger rate as a function of the K-refilling rate for different initial K-shell fillings. Dot-dashed line: K shell empty. Solid line: K shell half filled. Dashed line: K shell filled to three quarters.

Table 1. Widths for nuclear absorption and electromagnetic transitions in pionic Mg (all electron shells completely filled).

Level	n	5		10		15		20	
	ℓ	0	1	0	1	0	1	0	1
Γ_{abs} (eV)		118	5.83	14.7	0.75	4.35	0.22	1.83	0.09
$\Gamma_{el.-mag}$ (eV)		0.027	0.143	0.466	0.510	0.271	0.286	1.021	0.865

4 Experiments on the atomic cascade

What are the experimental tools we have at hand to check the picture of the exotic-atom cascade just described? The most important one is the measurement of exotic-atom x rays. The intensity pattern of these x rays has been measured for a huge number of elements, alloys, mixtures and compounds in gaseous, liquid and solid form. So I have to restrict my talk to a short overview of experiments on simple systems which nevertheless should allow to test the cascade models:

- Noble gases at low pressure (in the kPa region): The exotic atoms are well isolated systems, where electron refilling takes place at a very low rate.
- Noble gases at medium and high pressure (in the MPa region): Here electron refilling begins to compete with electron depletion.
- Metals: The ion cores of metals still are simple systems imbedded in a sea of conduction electrons. Refilling of electrons into the exotic atom depleted from electrons should be very fast. The refilling of K and L shells from higher shells plays a decisive role.

How do the tests normally proceed? The cascade is simulated and parameters are adjusted which characterize electron refilling and the population of different levels of the exotic atom at the begin of the calculation. Best agreement between calculated and experimental x-ray intensities is aimed for. Such an optimization does not only allow to perform tests on the calculations but at the same time gives hints on the status of the exotic atom in the upper part of the cascade. Direct experimental evidence on this part of the cascade essentially does not exist. This is not surprising. The Auger electrons emitted during deexcitation of muonic silver (Z=47) around n=20, for instance, have energies of only a few keV. To the best of my knowledge Auger electrons were measured only once directly in a counter experiment [18] but the energy range had to be restricted to above 20 keV due to the detector used and the thickness of the target employed.

4.1 Noble gases at low pressure

If we lower the pressure in a gas target to values below roughly 50 kPa, the rate of electron refilling into the depleting exotic atom by charge transfer from other gas atoms becomes so slow that the atom becomes heavily depleted. The refilling rate

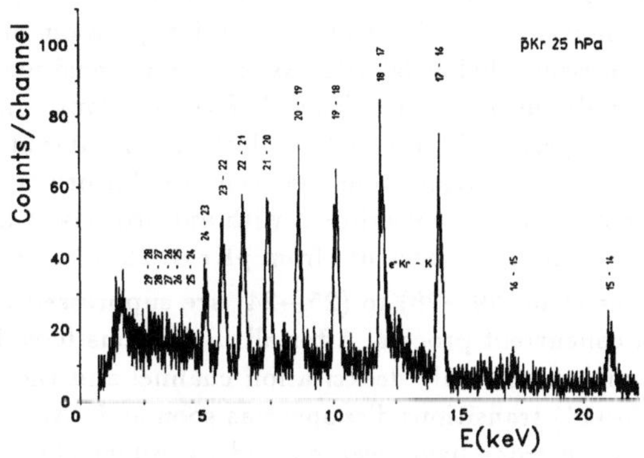

Figure 6. Antiprotonic x rays from Kr at 25 hPa.

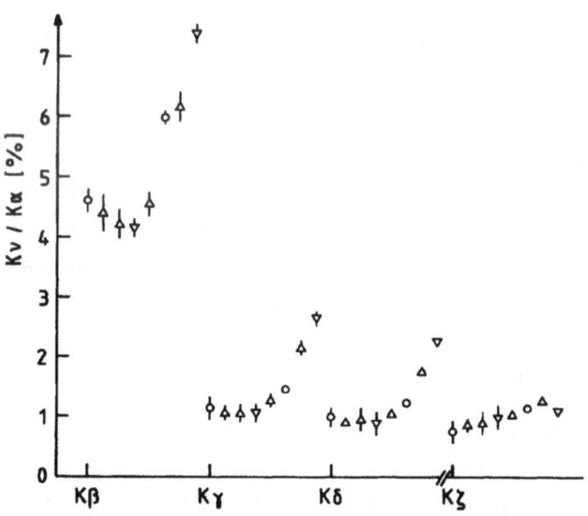

Figure 7. Muonic Lyman intensities in Ar at different densities. Experimental results are grouped for each line (K_β, K_γ, etc.) according to increasing density. Open circles: Ref.[20]. Upright triangles: Ref. [22]. Down-pointing triangles: Ref. [21].

λ_{refill} is given by

$$\lambda_{refill} = n * \sigma * v, \tag{20}$$

where n is the atom density, σ the cross section for charge transfer ($\approx 10^{-15}$ to 10^{-14} cm^2) and v the velocity of the exotic atom relative to the other atoms in the gas. Even if we assume that the total energy gained during capture ($\approx$ 30 eV) goes into the kinetic energy of the exotic atom, v stays below some 10^5 cm/s. Therefore the refilling rate for a completely stripped antiprotonic noble gas atom of medium Z does not exceed 10^9 s^{-1} at pressures below 50 kPa. As radiative transition rates are larger than 10^{10} s^{-1} the cascade normally is finished before electrons can be refilled. As Auger transitions are suppressed by the lack of electrons circular states are reached very quickly and very soon only transitions between circular orbits take place. This was beautifully demonstrated in an experiment with antiprotons at LEAR [19]. Only circular transitions are found in a spectrum from $\bar{p}$Kr at 25 hPa (see Fig. 6).

Radiative transitions from (29→28) to (25→24) are suppressed because L Auger effect is possible as a concurrent process. After all L electrons have been ejected radiative transitions again are the only deexcitation channel and their intensity rises. The (16→15) and (15→14) transitions disappear as soon as K-Auger effect becomes possible. After both K electrons have been ejected by Auger effect this concurrent process is no longer possible and radiative transitions are prominent again.

4.2 Noble gases at pressures in the MPa region

For pressures in the MPa region the rate for electron refilling into the heavily ionized exotic atom becomes comparable with the radiative transition rates. An example: A muonic Ar atom ionized tenfold and moving with thermal velocities collects electrons at 5 MPa with a rate as high as $5*10^{12}$ s^{-1} which is comparable to x-ray emission rates at levels around n $\approx$ 10. Auger transitions in the completely filled atom are much faster. Therefore the possibility of radiationless transitions is governed by refilling and hence by the target presssure. This is demonstrated by measurements of the x-ray cascade in muonic Ar at different gas pressures and as a liquid [20,21,22]. The results for the Lyman-series intensity ratios K_ν/ K_α are shown in Fig. 7. The intensity of higher transitions in the series increases with increasing pressure and is highest for liquid Ar: The higher the atom density, the higher the probability for Auger transitions which prevent the μ^- from reaching circular orbits. Noncircular transitions are stronger at higher density.

The Fribourg group has used their results on muonic Ar [22] to study depletion and refilling more quantitatively. Assuming a statistical angular momentum distribution $p(\ell) \propto 2\ell+1$ at $n_{init} = 14$ they were able to fit the Lyman series intensities quite well by adjusting as the only parameter one governing the K-shell refilling. They assumed the K-shell refilling rate Γ_K to depend linearly on the L-shell filling P_L^0 at cascade start, $\Gamma_K = \Gamma_0 * P_L^0$, with Γ_0 the refilling rate from a neutral atom. The results are shown in Table 2.

The pressure dependence of Lyman series intensities is even more pronounced in muonic nitrogen [20]. This has been explained in detail [21] by a "Coulomb explosion" of the nitrogen molecule during the muonic cascade. An electromagnetic potential of

Table 2. Experimental and calculated Lyman series intensities from muonic Ar. The experimental K_ϵ intensities are given without errors, as this line is covered by a background line.

Transition	Expt. 1.5 bar	Calc. P_L^0=2%	Expt. 22 bar	Calc. P_L^0=3.5%	Expt. 170 bar	Calc. P_L^0=15%
K_α	0.914(26)	0.909	0.905(25)	0.899	0.875(24)	0.870
K_β	0.0400(22)	0.0395	0.0410(14)	0.0427	0.0537(16)	0.0509
K_γ	0.0097(11)	0.0112	0.0116(8)	0.0128	0.0187(8)	0.0185
K_δ	0.0083(8)	0.0066	0.0095(6)	0.0085	0.0154(6)	0.0161
K_ϵ	0.0080	0.0065	0.0094	0.0086	0.0133	0.0155
K_ζ	0.0077(8)	0.0075	0.0093(6)	0.0090	0.0111(5)	0.0119

the form

$$V(r) = \frac{Q_1 * Q_2 * e^2}{r} - (\alpha_1 * Q_2 + \alpha_2 * Q_1) * \frac{exp(-r/r_p)}{(r^2 + r_p^2)^2} \tag{21}$$

between the two atoms with ionization states Q_1 and Q_2 and polarizabilities α_1 and α_2, respectively, was assumed (r_p is a parameter). A collision rate

$$\lambda_{coll}/s^{-1} = 1.6 * 10^8 * p/hPa. \tag{22}$$

resulted at pressure p, leading to refilling rates much larger than the radiation width of muonic nitrogen at medium n.

4.3 Exotic-atom cascade in metals

Conditions are completely different in metals. Refilling of electrons from the conduction band into the heavily ionized exotic atom should be extremely fast. If we (quite crudely) approximate the exotic atom as a point charge Z∗e in a degenerate gas of conduction electrons we can, following Drude [23], calculate the change of the electron momentum $\vec{p}$ as

$$d\vec{p}/dt = \vec{p}/\tau + grad[V(r)] \tag{23}$$

with the screened spherical potential $V(r) = (Z{*}e^2/r){*}\exp(-\lambda_{scr} r)$, τ the relaxation time and λ_{scr} the screening parameter. We can integrate this equation assuming $r = r_s$ at $t = 0$, with r_s the radius of a sphere which contains just one conduction electron. A refilling rate of around $5{*}10^{16}$ s^{-1} results, depending on the conduction electron density in the metal. So holes in the outer electron shells should be refilled faster than emptied by Auger effect. The bottleneck in this case is the refilling of K and L holes from higher shells. Table 3 shows calculated values for typical muonic Auger and electron refilling rates (for the Z-1 atom and one hole in the K shell).

For Mg the M holes are refilled from outside; the M shell is not expected to be empty when L Auger effect becomes predominant. The L shell, however, is strongly depleted when K Auger effect becomes possible. Hence K refilling (mostly from the L shell) should be smaller than in an atom with completely filled shells. For Fe the M shell should be strongly depleted (no refilling from outside) and the L shell nearly completely empty. For higher Z elements one may expect the L and M shells

Table 3. Typical muonic Auger rates and electronic refilling rates.

Element	Muonic K-Auger rate (s^{-1})	K-refilling rate (s^{-1})	Muonic L-Auger rate (s^{-1})	L_I-refilling rate (s^{-1})
Mg	$1.2*10^{15}$(9-8)	$3.9*10^{14}$	$3.0*10^{14}$(10-9)	$1.0*10^{13}$
Fe	$1.6*10^{15}$(8-7)	$1.6*10^{15}$	$1.4*10^{14}$(9-8)	$4.8*10^{14}$
In	$1.6*10^{15}$(8-7)	$1.2*10^{16}$	$3.6*10^{14}$(9-8)	$2.4*10^{15}$
Au	$8.6*10^{14}$(8-7)	$7.3*10^{16}$	$1.7*10^{14}$(9-8)	$6.4*10^{15}$

Table 4. Values for K-shell refilling rates in muonic atoms as fraction of the full K-shell refilling rate, as derived from experiment.

Element	Mg	Al	Fe	In	Ho	Au
W_K/W_K^0	0.33	0.30	0.16	1.0	1.0	1.0

still to be nearly full when K-Auger effect becomes possible and the K-refilling rate not smaller than for a completely filled atom. Experimental findings [24] correspond to these expectations. From a comparison of experimental results and calculations best parameter values for the initial ℓ distribution *and* the K-refilling rate could be *unambiguously* determined. The best values of W_k are shown in Table 4. The slope of the ℓ distribution changed from steeper than statistical (at Z= 12, 13) to statistical (for Z>40) with a remarkable exception: A nearly flat initial distribution was found for Fe with the electronic 3d shell filled only half . This was attributed to the large number of electronic intra-shell transitions possible in Fe [25] and the resulting large ℓ loss.

5 Conclusions

Let me summarize the results of this short description of the exotic-atom cascade.

- Widths for electromagnetic transitions and for nuclear absorption can be reliably calculated.
- The problem of electron depletion and refilling during the cascade is understood only for isolated systems like exotic noble gas atoms at low pressure. In the other cases at best a qualitative understanding has been achieved.
- Combining theoretical and experimental findings it may be concluded that the ℓ distribution of exotic particles immediately after atomic capture in normal cases is close to statistical. Exceptions have, however, been observed.

References

[1] F.J. Hartmann, Invited talk at this school.

[2] V.I. Markushin, Invited talk at this school.

[3] J.R. Rook, Nucl. Phys. B20, 14 (1970).

[4] E. Fermi, E. Teller, Phys. Rev. 72, 399 (1947).

[5] S.S. Gershtein, Invited talk at this school.

[6] M. Leon, J.H. Miller, Nucl. Phys. A282, 461 (1977).

[7] G.R. Burbidge, A.H. deBorde, Phys. Rev. 89, 189 (1953).

[8] J. Hüfner, Z. Phys. 195, 365 (1966).

[9] Y. Eisenberg, D. Kessler, Nuovo Cimento 19, 1195 (1961).

[10] V.R. Akylas, P. Vogel, Comp. Phys. Commun. 15, 291 (1978).

[11] V.R. Akylas, Ph. D. thesis, preprint CALT-63-300, California Institute of Technology, Pasadena, USA, 1978.

[12] R.A. Ferrell, Phys. Rev. Letters 4, 425 (1960).

[13] H. Bethe, E.E. Salpeter, Quantum mechanics of one- and two-electron systems, New York, 1977.

[14] F.J. Hartmann, to be published.

[15] P. Vogel, Phys. Rev. A7, 63 (1973).

[16] D. West, Repts. Progr. Phys. 21, 271 (1958).

[17] M.J. Turner, D.F. Jackson, Phys. Letters 130B, 362 (1983).

[18] R. Callies, H. Daniel, F.J. Hartmann, W. Neumann, Phys. Lett. 91A, 441 (1982).

[19] R. Bacher, P. Blüm, D. Gotta, K. Heitlinger, M. Schneider, J. Missimer, L.M. Simons, K. Elsener, Phys. Rev. A38, 4395 (1988).

[20] P. Ehrhart, F.J. Hartmann, E. Köhler, H. Daniel, Phys. Rev. A27, 575 (1983).

[21] J.D. Knight, C.J. Orth, M.E. Schillaci, R.A. Naumann, F.J. Hartmann, H. Schneuwly, Phys. Rev. A27, 2936 (1983).

[22] R. Jacot-Guillarmod, F. Bienz, M. Boschung, C. Piller, L.A. Schaller, L. Schellenberg, H. Schneuwly, D. Siradovic, Phys. Rev. A37, 3795 (1988).

[23] N.W. Ashcroft, N.D. Mermin, Solid State Physics, New York, 1976.

[24] F.J. Hartmann, R. Bergmann, H. Daniel, H.-J. Pfeiffer, T.von Egidy, W. Wilhelm, Z. Phys. A305, 189 (1982).

[25] R.A. Naumann, H. Daniel, P. Ehrhart, F.J. Hartmann, T. von Egidy, Phys. Rev. A31, 727 (1985).

[illegible]

[illegible]

[illegible]

[illegible]

[illegible]

[illegible]

[illegible]

[illegible]

[illegible]

[illegible]

[illegible]

[illegible]

[illegible] R. Hofstadter [illegible]

[illegible]

[illegible]

INTERPLAY OF EXOTIC ATOMS AND HYPERNUCLEI – CASCADE TRAPPING AND DEEPLY-BOUND EXOTIC ATOMS

Ryugo S. Hayano

Department of Physics, University of Tokyo
7-3-1 Hongo, Bunkyo-ku
Tokyo 113, Japan

INTRODUCTION

Two very recent topics on exotic atoms being studied by the University of Tokyo group are presented.

In the first part, we take up the problem of negative-meson cascade in helium. Our recent experiment has shown that a few per cent of negative kaons are trapped by metastable states (having an overall lifetime of about 40 ns) during the atomic cascade in liquid helium. The anomalous behavior of negative mesons in liquid helium was noticed many years ago when a significant fraction of pions and kaons were found to decay at rest in helium bubble chambers, and two classes of models have been proposed to explain the at-rest decay fraction of mesons in helium. One is to consider mechanisms to make the average meson cascade time in helium as slow as $\sim 10^{-10}$ s, and the other is to assume that a few % of mesons are trapped in metastable states. Our observation provides a direct evidence for the trapping hypothesis.

In the second part, structure and formation of deeply-bound exotic atoms, hitherto inaccessible to x-ray spectroscopy, are discussed. We first introduce an idea to form a Coulomb-assisted narrow Σ^- hypernuclear bound state, and generalize this idea to the cases of other negatively-charged hadrons, such as pions. It will be shown that the standard pion-nucleus optical potential predicts narrow widths for deeply-bound pionic states, even for the 1s state in ^{208}Pb. Possibilities of forming such interesting states by (n,p) or other reactions will be considered.

These two topics, which may appear unrelated to each other, stemed from our work on hypernuclear spectroscopy by using the (stopped $K^-, \pi^{\pm}$) reaction, as described briefly in the next section.

Hypernuclear Study by Using Stopped Kaons

One of the current topics in hypernuclear study is to understand the behavior of Σ hyperon in nuclei. Since a Σ hyperon embedded in nucleus can immediately decay via the strong $\Sigma N \rightarrow \Lambda N$ reaction, (widths of Σ states in nuclear matter are expected to be of the order of 20 MeV), it was quite surprising that the CERN ^{9}Be(K^-, π^-) experiment observed narrow ($\Gamma \approx 8$ MeV) peaks in the Σ unbound region [1]. Narrow Σ-hypernuclear "peaks" were also reported in the Σ unbound region by experiments done subsequently on *p*-shell nuclei at BNL, CERN and KEK [2, 3, 4, 5]. Although many theoretical papers have been written to explain why such narrow states can be

Electromagnetic Cascade and Chemistry of Exotic Atoms
Edited by L. M. Simons *et al.*, Plenum Press, New York, 1990

formed, the problem is yet to be settled down, due, largely, to the lack of conclusive data.

One of the keys to clarifying the situation is to observe Σ hypernuclear bound states. Since kaon absorption at rest is expected to be an effective tool to study Σ hypernuclei [6] because of efficient Σ production and favorable momentum transfer, the University of Tokyo group has been using this reaction to look for Σ bound states since 1984 at the National Laboratory for High Energy Physics (KEK).

The principle of measurement is quite simple: We stop negative kaons in various targets, and measure emerging pion momenta by using a high-resolution magnetic spectrometer. The mass (and thus the binding energy) of the hypernucleus, recoiling against the pion, can be precisely determined from the pion momentum.

The first topic, cascade of mesons in helium, is a byproduct of our recent experiment to look for Σ hypernuclear bound states using the ^{4}He(stopped K^-,π) reaction [7].

The second topic, deeply-bound exotic atoms, is also related to the study of hypernuclei in the following context. While looking into various possibilities of forming narrow Σ hypernuclear states, we noticed that Σ^- hypernuclei produced by the (K^-,π^+) reaction on medium-heavy nuclei have interesting aspects. Since the Σ^--nucleus potential in this case consists of both strong and Coulomb part, bound states having atomic and nuclear hybrid character may be formed; due to the reduced Σ-nucleus overlap, such states may be observable as narrow peaks in the Σ bound region by using the (K^-,π^+) reaction.

The idea of Coulomb-assisted narrow hybrid bound state can be generalized to other negatively charged particles, and interesting possibilities of studying deeply-bound pionic atoms will be discussed.

TRAPPING OF NEGATIVE MESONS DURING CASCADE IN HELIUM

At-Rest Decay of Negative Mesons in Helium

When a negatively charged unstable hadron (hereafter referred to as a "meson" for brevity) such as π^- or K^- is trapped in an atomic orbit, it either de-excites electromagnetically and gets captured by the nucleus, or undergoes weak decay. If the cascade is fast enough compared with the lifetime of the meson τ_{free} ($\tau(\pi) \approx 26$ ns, $\tau(K) \approx 12$ ns), the at-rest free decay fraction

$$f \equiv \frac{N_{decay}}{N_{stop}} ,$$

should be negligible. If, on the other hand, a significant fraction of mesons are seen to decay at rest, it is natural to assume that the cascade is rather slow, as characterized by the "average cascade time"

$$T_{av} = \frac{f}{1-f} \times \tau_{free}.$$

The average cascade time was first measured in the case of π^- stopped in hydrogen bubble chamber [8,9]. At rest $\pi \rightarrow \mu\nu$ decay events were identified based on the unique range of the μ^-. In more detail, the $\pi-\mu$ decay angle and the μ^- range were measured for each event, and the number of events satisfying the kinematical condition $v_\pi < 0.01c$ was counted in the backward hemisphere. The at-rest decay

Table 1. Fraction of at rest decays, and deduced average cascade time of π^-, K^- and Σ^- in helium

	Target	Decay fraction (%)	$T(10^{-10}$ sec)	Ref.
π^-	^{3}He Gas (17.5 atm)	0.5 ± 0.3	1.4 ± 0.7	[12]
	Liq. Helium	1.0 ± 0.3	2.5 ± 1.0	[13]
		1.2 ± 0.1	3.2 ± 0.2	[14]
K^-	Liq. Helium	2.0 ± 0.3	2.5 ± 0.4	[15]
		2.6 ± 0.4	3.2 ± 0.5	[16]
		2.5 ± 0.4	3.1 ± 0.4	[18]
Σ^-	Liq. Helium		< 0.36	[18]
		0.19 ± 0.06	0.28 ± 0.08	[19]

fraction was found to be $f = 3.8 \times 10^{-5}$, and the average cascade time was determined to be $T_{av} = 2.3 \pm 0.6 \times 10^{-12}$ s [9].

Chemically, a π^- bound to a proton should look very much like a neutron, which can easily penetrate electron clouds of nearby H_2 molecules, and can feel a strong electric field. This will cause Stark oscillation between levels of opposite parities (between nP and nS, for example), resulting in fast meson capture from high nS states. The importance of Stark mixing was first pointed out by Day *et al.* [10], and comprehensive calculations of meson cascade times, taking account of radiative, Auger and Stark transitions, were done by Leon and Bethe [11], who obtained $T = 3.5 \times 10^{-12}$s ($v = 0.01c$ to capture was estimated to be 1.2×10^{-12} s, and the subsequent cascade time was estimated to be 2.3×10^{-12} s), compatible with the experimental value.

When helium bubble chambers were exposed to meson beams, it was found that a significant fraction of mesons decay at rest in helium (see Table 1). As in the hydrogen case, average cascade times were deduced from the measured decay fractions. As shown in Table 1, cascade of mesons are quite slow in helium, of the order of 10^{-10} s. The π^- decay fractions were measured by using the same technique as in the hydrogen case. In the case of K^-, decay at rest events could be unambiguously selected by requiring that the momenta of three pions emitted in the decay $K^- \rightarrow \pi^-\pi^-\pi^+$ add up to ≈ 0 MeV/c. The procedure for obtaining the Σ^- cascade times, also listed in the table, was much more involved.

The mechanism of de-excitation of mesonic helium is expected to be different (and more complex) from the hydrogen case, because a meson bound to helium would chemically look like a heavy "proton". The average meson cascade time in helium was calculated by Day [20] by using a model as described in the next section, whose result turned out to be orders of magnitude faster than the values indicated from the free-decay fraction.

De-excitation model of mesonic helium

In Day's scenario [20], the fate of mesons captured in helium is described as follows. A slowed down meson is captured by a helium atom by an Auger process, ejecting one of the two electrons. In this process, the ionization energy (24.6 eV) is released, and the meson is brought to $n \sim \sqrt{M/m_e}$ and $r \sim 0.6a_0$, where M is the meson reduced mass and m_e is the electron mass. The principal quantum number is about 30 for K^- and 16 for π^-. Note that the radius of the meson orbit considerably overlaps with that of the electron orbit. The second electron is then

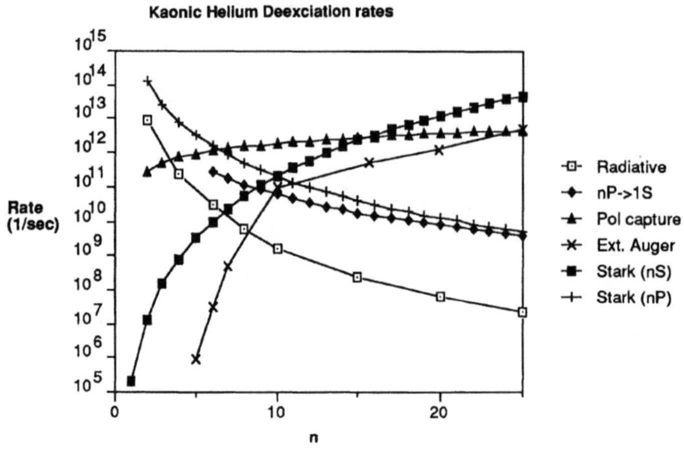

Figure 1. Day's theoretical predictions for various de-excitation processes vs principal quantum number.

ejected by another Auger process within 10^{-15} s. Since the meson has to release 24.6 eV, it should jump to $n \sim 25$ in the case of K^-, or to $n \sim 13$ in the case of π^-.

The ejection of the second electron leaves the system in a highly excited positive ion, chemically resembling a heavy proton. In Day's model, the de-excitation rates of the following processes are considered; external Auger effect, polarization capture, molecular Stark effect, radiative transition and direct nuclear capture.

The external Auger effect designates a process in which the mesonic atom ionizes a colliding helium atom. The polarization capture is a process in which a meson in the nP state is captured from the nS state by first polarizing a nearby helium atom.

Molecular Stark effect is caused by the weak molecular electric field of a metastable "proton"-helium molecule. The Stark rate can be calculated by differentiating the known Helium-proton molecular potential $\phi(R)$ as:

$$\omega_s \sim < nP|d|nS > \frac{d\phi(R)}{dR}$$

where d is the dipole matrix element and R is the inter-atomic distance. The total rate is then calculated by taking an average over R.

In Fig.1, Day's theoretical predictions for various de-excitation processes vs n are shown for the case of K^- meson. It can be seen that Stark mixing followed by nuclear capture from high nS states, whose rate was approximately given as $\Gamma_{stark}(nS) \sim 2 \times 10^5 n^6 s^{-1}$ for K^- and $\Gamma_{stark}(nS) \sim 8 \times 10^7 n^6 s^{-1}$ for π^- [13], is the most dominant process in early stages of cascade. Mesons are therefore expected to be captured dominantly from high nS states. The rate of the nearest competing process, the external Auger, was estimated to be about 10^{12} s^{-1} around $n \approx 20$.

Based on this model, the average cascade time was expected to be shorter than $\sim 10^{-12}$, very much shorter than experimental values [15].

Condo's trapping hypothesis

In order to resolve the large discrepancy between Day's estimation and measured average cascade times, Fetkovich and Pewitt [13] conjectured that Stark mixing effect is somehow inoperative in helium since a combination of radiative and external Auger rates yields approximately the experimental result.

If this model is correct, it will drastically change the fate of kaons: A large fraction of kaons in this case can cascade down without being captured from high nS states, and may be captured from P states instead (note that for π^- the S capture process is expected to be anyway dominant because the P to S radiation rate dominates P-state capture [15]). Since the knowledge of the angular-momentum states of K^- capture is essential in the detailed study of the kaon absorption, it is important to fully understand the kaon cascade process.

At that time, Condo proposed an entirely different mechanism to explain the large free-decay fraction in helium, i.e., mesons are sometimes trapped in meta-stable states. His points are as follows: When a meson is initially captured at high n, it was assumed in Day's model that the meson quickly ejects the remaining electron by a rapid Auger process. However, in order to eject this electron, the meson must release 24.6 eV which, for the π^- case, requires a $\Delta n = 3$ transition and $\Delta n = 5$ for K^-. If the meson is in a circular orbit, this is only possible via $\Delta \ell = 3(5)$ Auger transition, which is expected to be much slower than radiation.

If the Auger process cannot eject the remaining electron, the system stays neutral, and thus none of the ionic processes considered in Day's model are available to de-excite the meson. The meson therefore has to be de-excited by circular radiative transitions alone. Due to the ΔE^3 dependence, such radiative transitions are rather slow, on the order of 10^{-8} s. The mesons initially captured in high n circular orbits are "trapped", and will all decay before they are observed. Mesons trapped in nearly circular orbits, where quadrupole and high Auger transitions are required for ionization, are also expected to contribute to the free-decay fraction.

Since this argument only applies to the mesons captured in most circular orbits, the majority of mesons were expected to cascade and be captured as fast as Day's estimate.

Condo's argument was later elaborated by Russell [22] who pointed out that the trapping fraction should increase as meson mass because the number of levels which cannot de-excite by $\Delta \ell = 1$ rapid Auger process increases as the meson mass is increased, and Fetkovich *et al.* [17] performed a detailed study of the comparison of the trapping model with measured π^-, K^- and Σ^- decay fractions. They concluded that the available data prefer the trapping model over the slow cascade model, and that if the trapping model is proved to be correct, it is not possible to draw any conclusion about mesonic cascade in helium, the importance of Stark mixing etc., from the data on free-decay fraction.

Measurement of decay time spectra

In our recent experiment aiming at the search for Σ hypernuclear bound state, we obtained a direct evidence that negative kaons are indeed trapped by a long-lived metastable states.

The experiment was done at the K3 beam line of the KEK 12 GeV proton synchrotron by stopping negative kaons in a liquid He target. The incoming K^- beam trajectory as well as the trajectory of each charged particle event in the magnetic spectrometer was recorded to obtain information about the reaction/decay vertex and the particle momentum (see Fig.2). The particle identification was made by combining time of flight (TOF) and range data. After particle identification, the time difference between kaon stopping and π/μ ejection was deduced by combining

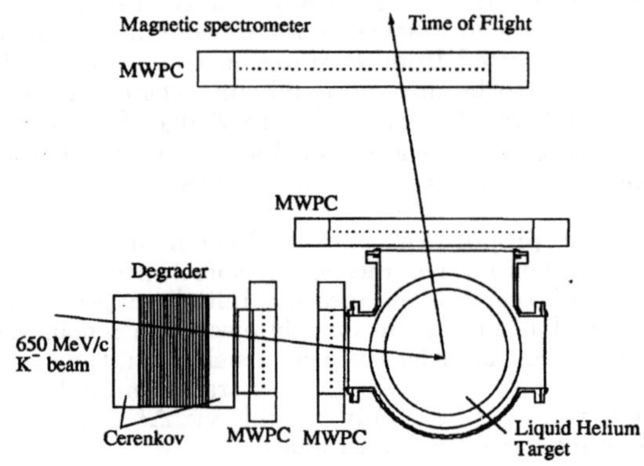

Figure 2. The target section of the setup used in the ^{4}He(stopped K^-, π) experiment.

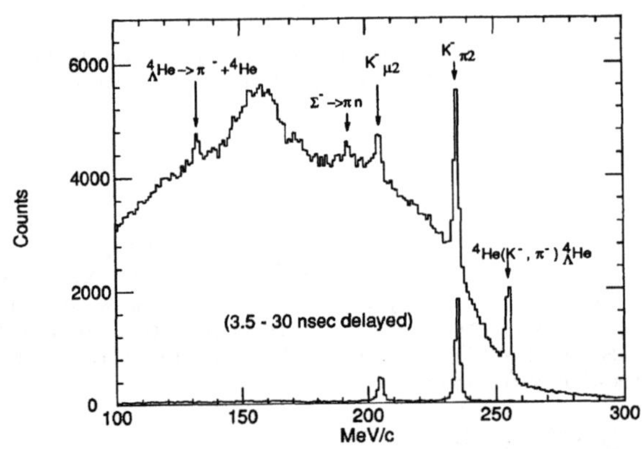

Figure 3. Momentum spectra of all the negative particles emitted from K^- mesons stopped in liquid helium. Distinct peaks due to $K^- \rightarrow \mu^-\overline{\nu}$ and $K^- \rightarrow \pi^-\pi^0$ at rest are seen; a) without timing selection and b) delayed events in the time range $3.5 < t < 30$ nsec.

the following information: kaon arrival time, kaon velocity determined by time of flight at the entrance of the helium target, position of the kaon decay vertex, π/μ path length in the spectrometer and π/μ time of flight data.

Fig.3 shows a gross momentum spectrum of all negative particles emitted from K^- stopped in liquid He. Surprisingly, the most dominant peaks are the 235 MeV/c μ^- and the 205 MeV/c π^- which are the unique signatures of $K^-_{\mu 2}$ and $K^-_{\pi 2}$ decays at rest, respectively. The total fraction of K^- decay at rest was obtained in two ways; 1) from comparison of the decay peak in the stopped-K^- spectrum with that in the stopped-K^+ spectrum, and 2) from comparison of the decay peak with the total $\pi^- + \mu^-$ in the same spectrum. The final value thus obtained is $f_K = 0.019 \pm 0.003$ which is in good agreement with the earlier values of f_K deduced from the 3π decay in He bubble chambers (see Table I).

Fig.3 shows also a delayed spectrum (3.5< t <30 nsec), which clearly indicates that the $K^-_{\mu 2}$ and $K^-_{\pi 2}$ peaks are indeed delayed. Now let us look at the time distribution of these at-rest decay peaks, as shown in Fig.4. It shows nearly an exponential decay following an effective lifetime of 9.5±0.3 nsec. This means that the average trapping time is $\tau_{\rm trap} = 41 \pm 6$ nsec. The distribution shows a slight deviation from the single exponential shape. This seems to be reasonable, since presumably many states of different lifetimes are involved in the cascade.

The spectra also show unambiguously that the "prompt" fraction is negligibly small ($f^{\rm prompt} \leq 1.2 \times 10^{-3}$). If the average cascade time were around 10^{-10} s, then a significant fraction of "prompt" component would be seen, as schematically depicted in Fig.5. The present experiment thus excludes the "slow cascade time" conjecture and indicates that the average cascade time for the majority of K^- mesons (not through trapping) is

$$T_{\rm av} = \frac{f^{\rm prompt}}{1 - f^{\rm prompt}} \times \tau^{\rm free} \leq 1.4 \times 10^{-11} {\rm sec} \qquad (1)$$

consistent with the theoretical estimates of Day including the molecular Stark effect.

The present experimental results have thus clearly shown that kaons are trapped by long-lived states, and that the average cascade time of the majority of kaons is short. It may be concluded that kaons are dominantly captured from nS states. This conclusion appears consistent with the existing K^-He x-ray data [24] in which only 16% of the kaons were found to make radiative transitions to $n = 2$. Anyway, in view of the newly obtained evidence, quantitative theoretical investigations of the early stage of meson capture and subsequent cascade processes would need to be done.

In order to further investigate the fate of mesons in helium, it would be interesting to obtain similar data for π^-. In the π^- case, the major decay mode $\pi^- \rightarrow \mu^-\overline{\nu}$ cannot be used to measure time spectra; because the μ^- momentum is only 30 MeV/c the μ^- is quickly stopped in the target, and dominantly decay via the $\mu^- \rightarrow e^-\overline{\nu_e}\nu_\mu$ mode (2.2 μsec) from the μ^-He 1s orbit. By using a high intensity π^- beam available at meson factories, it appears feasible to obtain time spectra of the $\pi^- \rightarrow e^-\overline{\nu}$ decay, whose branching ratio is known to be 1.2×10^{-4}. Such a measurement is being prepared at TRIUMF. It is interesting to perform such measurement by using other noble-gas elements having high ionization energy. Also interesting would be to use gas targets, and measure prompt decay fraction (from which we can determine the average cascade time) as a function of gas pressure. In this way, the effect of Stark mixing in Helium cascade would be directly demonstrated.

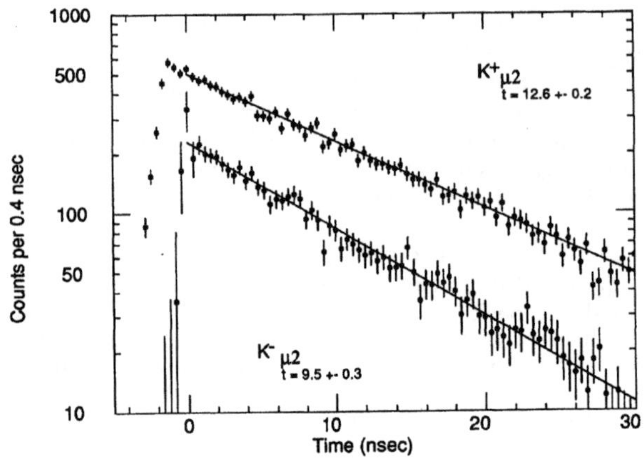

Figure 4. Time distribution of the 235 MeV/c $K^- \rightarrow \mu^- \overline{\nu}$ decay peak. For comparison, the K^+ decay time spectrum is also shown.

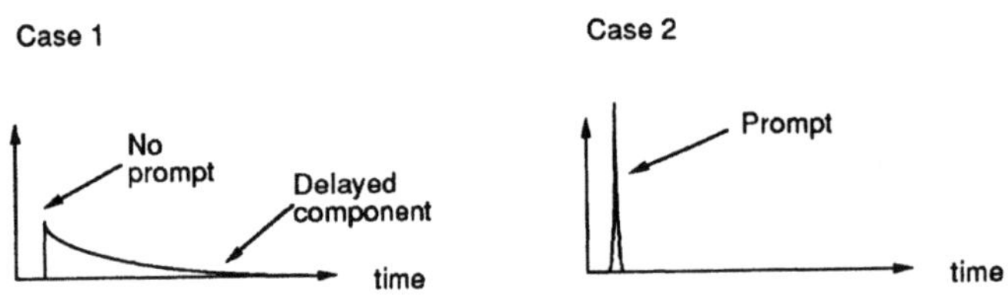

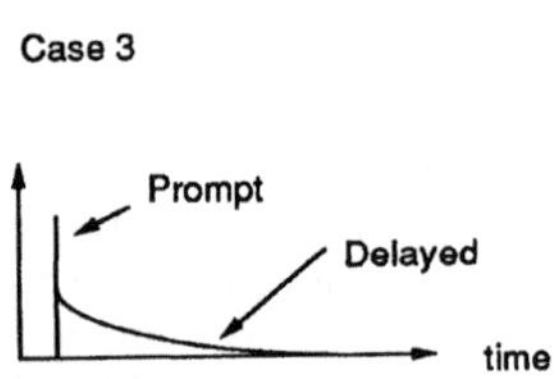

	Hypothesis		Expected Time Spectrum	
	Trapping?	Cascade	Prompt	Delayed
Case 1	Trapped	Fast	No	Yes
Case 2	Not trapped	Slow	Yes	No
Case 3	Trapped	Slow	Yes	Yes

Figure 5. Expected decay time spectra for various cascade/trapping models are shown schematically.

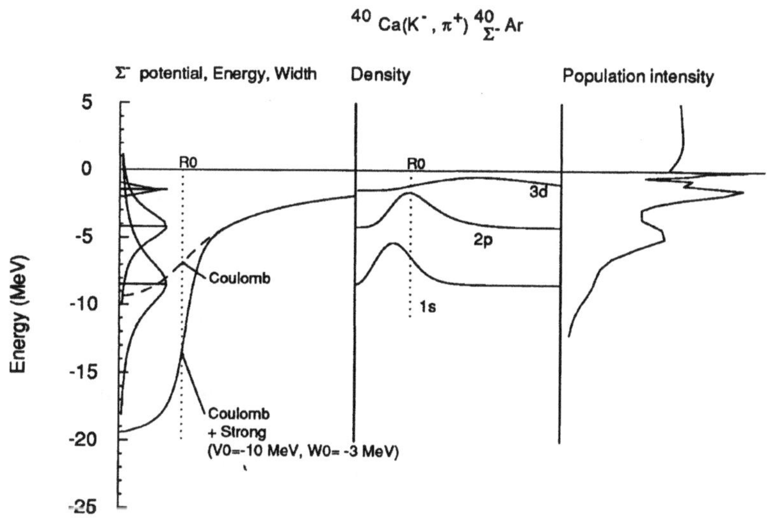

Figure 6. Coulomb-assisted hybrid bound states of Σ^- which may be populated by the ^{40}Ca(stopped K^-, π^+) reaction. The hybrid (Coulomb+strong) potential, bound-state energy levels and widths are shown to the left. The radial part of the Σ^- wave functions are shown in the middle, and the expected (K^-, π^+) strength is shown to the right.

DEEPLY-BOUND EXOTIC ATOMS

Coulomb-assisted Σ^- Bound States

As was briefly described in the introduction, the idea of producing narrow Σ^- bound states having strong-Coulomb hybrid character was theoretically proposed by Yamazaki *et al.* [25] based on the following consideration.

Although the Σ-nucleus potential depth is vaguely known at present, Σ^- can always form a bound state because of Coulomb attraction regardless of the strong potential depth. Just below the Σ binding threshold, there are ordinary Σ^- atomic levels, and x-ray measurements have yielded valuable information on Σ-nucleus interaction.

As Σ^- cascades down atomic levels, the Σ^- reaches a certain orbit where the strong width exceeds the radiative width. Beyond this orbit, so called "last orbital", no information on level energy and width can be obtained from x-ray measurement.

However, this does not mean that no Σ^- levels exist beyond this point; so far, we simply did not have any means of producing such states. Actually, if we calculate Σ^- binding energies and widths in a strong+Coulomb potential, we can see that there should be many levels beyond the "last orbital". As shown in Fig.6, a medium-heavy Σ^--nucleus system has a hybrid character. While shallow bound states are "atomic", the system becomes progressively more "hypernuclear" for more deeply bound levels and their widths become larger due to strong $\Sigma - \Lambda$ conversion.

Somewhere between these two extremes, there may exist an interesting region in which the Σ^- resides partially inside and partially outside the nucleus so that the strong-interaction widths are somewhat reduced. They can be populated in strangeness exchange reactions, and the expected (K^-, π^+) strength is shown to the left of Fig.6. This figure clearly shows that we should be able to identify those states, as long as their widths are less than level spacing.

The idea presented above may well be applied to cases of other exotic atoms. Toki *et al.* [26] has recently studied whether or not there exist quasi-stable deeply-bound pionic atoms, and, surprisingly enough, even the $1s$ of ^{208}Pb was predicted to be quasi-stable.

The properties of deeply-bound pionic atoms can be studied by solving the Klein-Gordon equation,

$$-\nabla^2 + \mu^2 + 2\mu V_{opt}(r)]\phi(r) = [E - V_{Coul}(r)]^2\phi(r),$$

where μ is the pion-nucleus reduced mass, V_{Coul} is the Coulomb potential with a finite nuclear size, and V_{opt} is the pion-nucleus optical potential. We use the optical potential of the Ericson-Ericson type [27] and take the parameter set of Seki and Masutani [28] determined by analyzing pionic atom data. The explicit form of the optical potential can be written as:

$$2\mu V_{opt}(r) = -4\pi[b(r) + \epsilon_2 B_0\rho^2(r)] + 4\pi\vec{\nabla}[c(r) + \epsilon_2^{-1}C_0\rho^2(r)]L(r)\vec{\nabla}$$

with

$$b(r) = \epsilon_1 \{b_0\rho(r) + b_1[\rho_n(r) - \rho_p(r)]\}$$
$$c(r) = \epsilon_1^{-1} \{c_0\rho(r) + c_1[\rho_n(r) - \rho_p(r)]\}$$
$$L(r) = \left\{1 + \frac{4}{3}\pi\lambda[c(r) + \epsilon_2^{-1}C_0\rho^2(r)]\right\}^{-1}$$

and

$$\epsilon_1 \equiv 1 + \mu/M, \quad \epsilon_2 = 1 + \mu/2M.$$

We use the same Woods-Saxon form for all the density distributions, ρ, ρ_p and ρ_n. The parameters b's are coming from the S-wave πN interaction and c's are from the P-wave interaction.

In Fig.7, we show energy levels and width of π^- bound states of ^{90}Zr and ^{208}Pb. Finite-Coulomb results (strong interaction turned off) are shown by dashed lines, and the results with full optical potential parameters are in solid lines. The hatches indicate the widths of the levels, which, for the case of π^{-208}Pb $1s$ state is 0.63 MeV, much smaller than the $2p - 1s$ level spacing of 1.8 MeV. Thus, such deeply-bound pionic levels are narrow enough to be observed if proper production methods are employed. A similar result was reported by Friedman and Soff related to superheavy elements [29].

Why do deeply-bound pionic states are quasi-stable in spite of the usual expectation that such states should be broad due to the strong pion absorption? In a nonrelativistic point-nucleus estimate, the radius of π^- Bohr radius comes out to be smaller than the nuclear size, i.e., the pion is expected to stay mostly inside nuclear volume. If this is so, the width of $1s$ level should be as broad as $\Gamma \sim 20$ MeV due to the imaginary part of the π-nucleus potential.

The use of a more realistic finite-Coulomb wave function reduces the $1s$ pion density $|r\phi(r)|^2$ in the nucleus to about 50%, as shown by the dashed lines in Fig.8. The 50% overlap between the pion density and the absorptive potential would still make the $1s$ width as broad as ~ 10 MeV.

Turning on the optical potential has a drastic effect on the pion density, as shown by the full lines in Fig.8. The pion now stays mostly outside of the nucleus, and the overlap between the pion density and the nucleus is very much reduced. It can be readily seen that this small overlap between the pion density and the nucleus is the source of the small widths of deeply-bound states.

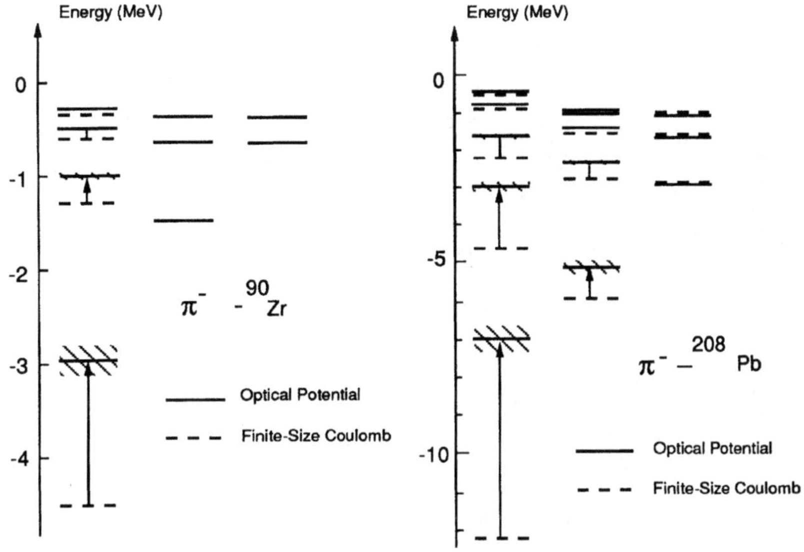

Figure 7. Energy levels and widths of π^- bounds states of ^{90}Zr and ^{208}Pb. Finite-Coulomb results are shown by dashed lines, and the results with full optical potential parameters are in solid lines.

The reason why pion density is very much reduced inside the nucleus may be graphically understood by examining the upper part of Fig.8, in which a combined strong+Coulomb potential felt by the $1s$ pion in ^{208}Pb is shown. The local part of the pion nucleus potential (coming from the S-wave πN interaction) is strongly repulsive, so that the pion is expelled from the nucleus. Outside the nucleus, there exists an attractive pocket made by Coulomb attraction, where pion stays most of the time.

In a more detailed study, the above results have been found to be unchanged as long as the optical potential parameters are varied within the ambiguities already found from the analysis of pionic x-ray data [30].

Formation of deeply-bound pionic atoms

The standard method of populating pionic atoms levels via cascade is inapplicable to the formation of deeply-bound pionic atoms. So far, the following methods to produce pionic atoms have been discussed; resonance capture of proton [31], resonance capture of photon [32], $(e, e'\pi^-)$ [33].

As a suitable method to produce heavy pionic atoms, the (n,p) reaction was proposed by Toki *et al.* [26]. In this reaction, a π^- is transferred from the neutron to the target, and is "soft-landed" to form a pionic atom. Such an experiment would be possible by using a monochromatic neutron beam and by measuring precisely the energy of the outgoing proton, provided the cross section is large enough.

In a plane-wave approximation, the formation cross section can be written as [26]

$$(\frac{d\sigma}{d\Omega})_{PW} = \frac{M^2}{2\pi^2}\frac{k_f}{k_i}(\frac{f}{\mu})^2\frac{q_L^2}{\mu - E_{n\ell}}\sum_m |\phi_{n\ell m}(\vec{q})|^2,$$

where $\phi(\vec{q})$ is the Fourier transform of the pionic wave function, M and μ respectively are the nucleon and the pion masses, k_i and k_f are the incoming and outgoing nucleon

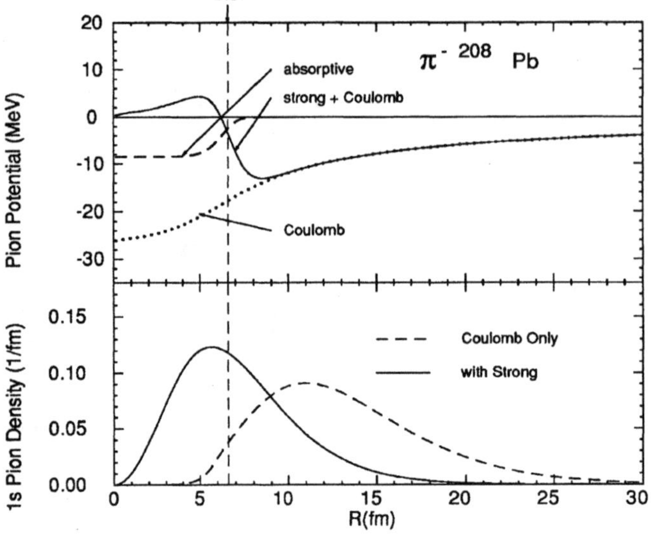

Figure 8. The π^- optical potential for ^{208}Pb (upper) and the pion densities $|r\phi(r)|^2$ of the 1s state (bottom).

momenta, $\vec{q} = \vec{k_i} - \vec{k_f}$ is the momentum transfer, and q_L is the Lorentz invariant momentum transfer defined as $q_L^2 = q^2 - (E_i - E_f)^2$.

A very rough estimate of the cross section was made by using a non-relativistic point-Coulomb wavefunction and taking $q \sim 200$ MeV/c ~ 1 fm^{-1} (an appropriate choice for the neutron incident energy of about 400 MeV), to be

$$\frac{d\sigma}{d\Omega} \sim 0.5\text{mb/sr} \times (Z/40)^5$$

where Z is the atomic number of the target nucleus.

Although the cross section is reduced when a relativistic atomic wave function is used, it is still appreciably large, around 0.5 mb/sr for the $1s$ state of π^-208Pb as long as the plane wave approximation is retained. At this level, observation of discrete peaks of deeply bound pionic states appears to be possible by using existing neutron beams and spectrometers, such as the MRS spectrometer at TRIUMF [35], and we have recently performed (n,p) measurement on Pb by using the MRS.

However, recent calculations which take account of neutron and proton distortions have indicated that the distortion effect is quite appreciable, which reduces the forward-angle cross sections by roughly an order of magnitude [36]. Although the data obtained at TRIUMF is still being analyzed, the signal to noise ratio currently achieved does not seem to allow us to identify deeply-bound π^{-208}Pb states yet. Possibilities of further improving the (n,p) method, or to use alternative methods such as the (d,2p) reaction, are being pursued.

ACKNOWLEDGMENTS

The author would like to thank Professor T. Yamazaki and other members of the University of Tokyo group, Drs. T. Ishikawa, M. Iwasaki, E. Takada, A. Sakaguchi and Mrs. H. Outa and M. Aoki for daily collaboration and discussions. The author is also grateful to Professor D.H. Davis and H. Toki for stimulating for stimulating discussions. This work is supported in part by the Grant-in-Aid for Special Project Research on Meson Science of the Japan Ministry of Education, Science and Culture.

REFERENCES

1. R. Bertini *et al.*, Phys. Lett. 90B (1980) 375.
2. H. Piekarz *et al.*, Phys. Lett. 110B (1982) 428.
3. R. Bertini *et al.*, Phys. Lett. 136B (1984) 29.
4. R. Bertini *et al.*, Phys. Lett. 158B (1985) 19.
5. T. Yamazaki *et al.*, Phys. Rev. Lett. 54 (1985) 102.
6. T. Yamazaki *et al.*, Phys. Lett. 144B (1984) 177.
7. R.S. Hayano, T. Ishikawa, M. Iwasaki, H. Outa, E. Takada, A. Sakaguchi, M. Aoki and T. Yamazaki, 1988 International Symposium on Hypernuclear and Low Energy Kaon Physics, Nuovo Cimento, in press.
8. T.H. Fields, G.B. Yodh, M. Derrick and J.G. Fetkovich, Phys. Rev. Lett. 5 (1960) 69.
9. J.H. Doede, R.H. Hildebrand, M.H. Israel and M.R. Pyka, Phys. Rev. 129 (1963) 2808.
10. T.B. Day, G.A. Snow, J. Sucher, Phys. Rev. Lett. 3 (1959) 61.
11. M. Leon and H. Bethe, Phys. Rev. 127 (1962) 636.
12. O.A. Zaĭmidoroga, R.M. Sulyaev and V.M. Tsupko-Sitnikov, JETP 52 (1967) 97.
13. J.G. Fetkovich and E.G. Pewitt, Phys. Rev. Lett. 11 (1963) 290.
14. M.M. Block, T. Kikuchi, D. Koetke, J. Kopelman, C.R. Sun R. Walker, G. Culligan, V.L. Telegdi and R. Winston, Phys. Rev. Lett. 11 (1963) 301.
15. M.M. Block, J.B. Kopelman and C.R. Sun, Phys. Rev. 140 (1965) 143.
16. J.G. Fetkovich, J. McKenzie, B.R. Riley, I.-T. Wang, K. Bunnell, M. Derrick, T. Fields, L.G. Hyman and G. Keyes, Phys. Rev. D2 (1970) 1803.
17. J.G. Fetkovich, B.R. Riley and I.-T. Wang, Phys. Lett. 35B (1971) 178.
18. C. Comber, D.H. Davis, J. Gordon, D.N. Tovee, R. Roosen, C. Vander Velde-Wilquet and J.H. Wickens Nuovo Cimento 24A (1974) 294.
19. J.G. Fetkovich, J. McKenzie, B.R. Riley and I.-T. Wang, Nucl. Phys. A240 (1975) 485.
20. T.B. Day, Nuovo Cimento, 18 (1960) 381.
21. G.T. Condo, Phys. Lett. 9 (1964) 65.
22. J.E. Russell, Phys. Rev. Lett. 23 (1969) 63.
23. J.E. Russell, Phys. Rev. 188 (1969) 187.
24. C. E. Wiegand and R.H. Pehl, Phys. Rev. Lett. 27 (1971) 1410: S. Baird, C.J. Batty, F.M. Russel, P. Sharman, P.M. Bird, A.S. Clough, K.R. Parker, G.J. Pyle and G.T.A. Squier, Nucl. Phys. A392 (1983) 297.

25. T. Yamazaki, R.S. Hayano, O. Morimatsu and K. Yazaki, Phys. Lett. 207B (1988) 393.
26. H. Toki and T. Yamazaki, Phys. Lett. 213B (1988) 129.
27. M. Ericson and T.E.O. Ericson, Ann. Phys. 36 (1966) 496.
28. R. Seki and K. Masutani, Phys. Rev. 27C (1983) 2799.
29. E. Friedman and G. Soff, J. Phys. 11G (1985) L37.
30. H. Toki, S. Hirenzaki, T. Yamazaki and R.S. Hayano, Nucl. Phys., in press.
31. G.T. Emery, Phys. Lett. 60B (1976) 351.
32. C. Tzara, Nucl. Phys. 18B (1970) 246.
33. V.F. Dmitriev, Sov. J. Nucl. Phys. 17 (1973) 417.
34. J.H. Koch, Phys. Lett. 59B (1975) 45.
35. K.P. Jackson et al., Phys. Lett. 201B (1988) 25.
36. H. Toki, private communication.

AN EXPERIMENTAL STUDY OF THE $(\mu^{-}He)^{+}$ LIFETIME IN THE METASTABLE 2S-STATE USING NaI-CRYSTALS

H. Orth

GSI Darmstadt, Postfach 11 05 52, D-6100 Darmstadt

1. Introduction

The level structure of the first excited states of muonic helium isotopes has attracted physics interest since more than twenty years. From the viewpoint of physics theory the fine structure, hyperfine structure and in particular the Lambshift of this two body bound state are excellent testing grounds for quantum electrodynamics. Experimentally, laser spectroscopy, a precision tool in atomic physics, can here be transfered to a fundamental structure in a muonic atom, hence providing accurate measurements for theory to compare with.

Application of laser spectroscopy relies on some metastability of the 2S-state of the muonic helium ion within the buffer gas, where its formation conventionally takes place. For this reason the 2S-lifetime was the subject of many experimental[1-6]and theoretical[7-13]investigations. First measurements carried out at CERN[1-4]indicate that the influence of the surrounding helium atoms is small even at pressures up to 50 atmospheres and that the 2S-lifetime at such high densities differs only slightly from the free, unperturbed ion. The 2S-2P energy differences determined there after by laser resonance excitation[14-15]are in good agreement with theory, a result which requires the 2S-state to be metastable around 500 ns. Nevertheless, the experimentally established long lifetime of the 2S-state is in contradiction to theoretical calculations as well as recent experimental results. Since Lambshift measurements in the muonic helium(3) ion have not yet been performed, a better understanding of the 2S-lifetime especially at high pressure is quiet important. The fact that the CERN data seem not to reconcile with the body of new data from SIN, strengthened our assessment to redo the experiment with an apparatus which resembles the one from CERN in particular, by using NaI crystals.

2. Experimental

A complete account of the experimental method has been given elsewhere[16]. In brief, negative muons are stopped in gaseous helium, and the photons from the radiative transitions of the cascade are monitored. This "light" from the muonic cascade splits into a prompt and delayed component. The fast part is used to compute the formation fraction of the 2S-state, the slow part consists of a continuous spectrum from retarded one-and two-photon decays of the metastable 2S-state. The goal is to measure the time distribution of these photons relative to the incoming muon.

Fig. 1 shows the experimental set-up. Incoming negative muons are detected by the plastic szintillators S_1 and S_2, actively confined by the veto counter SA to a spotsize of 18 mm. After passing a 0.1 mm thick beryllium-copper foil and an energy loss surface barrier detector (205 μm) the muons stop in the target gas of about 1000 cm^3 volume. Twelve NaI crystals (size $40 \times 40 \times 1$ mm^3) cover $0.45 \times 4\pi$ solid angle of the gas volume. They monitor the prompt cascade photons, the delayed ones from radiative 2s-1s transitions and the decay electrons from the subsequent muon decay. The target vessel was built of stainless steel and coated with a thin layer of silver to suppress low energy characteristic x-rays from the wall materials. Together with all target components it was tested to resist a pressure of 80 atmospheres, and after been pumped to a rest gas pressure of 5×10^{-6} mbar, was filled with 40 atmospheres of helium (4) at room temperature under normal running conditions. Gas impurities at a very low level might have a great influence on the lifetime of the 2s-state, therefore special care was taken to prevent gas contaminations caused by outgasing materials. For this reason, the light guides of the NaI crystals consist of optical glas. The crystals themselves were only covered with an 0.1 μm thick vapour deposited aluminium layer to avoid optical cross talk and a 0.05 μm thick SiO_2 protection layer to avoid oxidation of the aluminium. The helium gas was continuously circulated through a purification system consisting of a titanium getter heated to 800 °C and a coldtrap at liquid nitrogen temperature. A quadrupol mass spectrometer was used to analyse the gas during the measurements. The main contamination was found to be 3 ppm neon, a component of the used helium gas which could not be removed by the purification system. All other contaminations, especially water vapour or nitrogen, were less than 1 ppm.

The measurements were carried out at the μE4 channel of SIN. The incoming muons at 33 MeV/c momentum was limited to a rate of 100 s^{-1}. At any time, only one single muon was allowed to be in the apparatus and was observed individually. A triggered event consisted of a muon stopped, followed by at least two hits in NaI crystals. Then

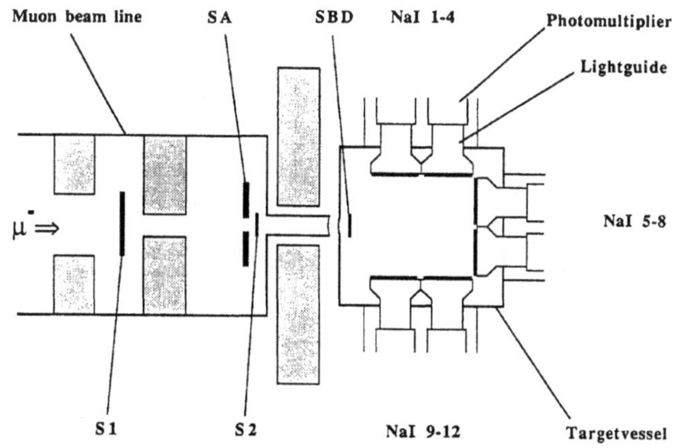

Fig. 1 Experimental setup: (S1,S2,SA) beam defining plastic szintillators, (SBD) surface barrier detector to detect the incoming muons and measure their energy loss, (NaI) sodiumiodide crystals to detect low energy photons emitted during the deexcitation of the muon as well as the decay electrons from subsequent muon decay.

ADC and TDC information from all the detectors was stored on tape. Depending on some further run conditions data spectra were generated for an line inspection.

The 12 NaI-crystals were calibrated by muonic X-rays of various target gases. The efficiency for photons in the energy regime of the K-series in helium was determined to be 40 %. There is good agreement between the measured response of the NaI crystals and Monte Carlo simulations based on the overall photon statistics and the special geometry of the glas light guides. From the time distribution of the prompt photons from helium with energy between 4 and 10.4 KeV a time resolution of 110 (20) ns was extracted. The time distribution of decay electrons detected in addition to the prompt photons fits well the exponential muon decay with a lifetime of $\tau = 2.24(3)$ μs. We conclude at this point that the influence of wallstops has been minimized and we satisfactorily understand the response of the detector set-up.

3. Results

Measurements have been performed in pure helium at 14 and 40 atm pressure and with added oxygen and xenon contaminations. Tab. 1 summarizes the results including data runs with methane at like target thickness and with and empty target vessel.

Tab. 1 The number of delayed and prompt photons in the energy interval of 4 to 10.4 keV, observed after negative muons had stopped in various target gases and gas mixtures. The decay electron of the muon had also been detected.

	μ_{IN} [10^{6}]	N_p (0-400 ns)	N_p/μ_{IN} [10^{-3}]	N_d (650-4400 ns)	N_d/μ_{IN} [10^{-6}]	N_d/N_p [%]
40 atm helium	7.80	38.959	4.99	441	56.2(2.7)	1.13(5)
14 atm helium	4.06	12.615	3.11	155	38(3)	1.2(1)
empty target	1.16	9	8×10^{-3}	0	---	---
38 atm (97 4 % helium + 2.6 % oxygen)	2.13	8.709	4.09	95	45(5)	1.1(1)
36 atm (99 % helium + 1 % xenon)	0.39	475	1.22	21	54(12)	4.4(1.0)
26 atm (97.7 % helium + 2.3 % xenon)	2.06	1.143	0.55	61	30(4)	5.3(7)
9 atm methan	2.33	3.897	1.67	60	26(4)	1.5(2)

In the off-line analysis data were accepted only if a low energy (4-10.4 KeV) photon signal as well as a decay electron were observed in two different NaI detectors after a μ-stop. The electron would be identified easily by its high energy loss in a crystal. To rule out crosstalk or bremsstrahlung events during the 500 ns long integration time of the first NaI puls, a minimum delay of 550 ns was required to the second pulse.

Fig. 2 shows the time distribution of the prompt and delayed events in 40 atm. pure helium. The exponential fit to the data results in a lifetime of 2.0(1) μs. This can be interpreted as due to two photon radiative 2s-1s transition, where only one of the two photons is observed. It would lead to the conclusion, that the metastability of 2S-state muonic helium is little effected by the interaction with other target atoms at 40 atm pressure. This is supported by the data at 14 atm He. The run with the empty target clearly shows that wall stops are efficiently eliminated. However, several test measurements with other targets throw serious doubts on this interpretation, which can be seen from the ratio of the number of delayed to prompt photons in Tab. 1. Gas contaminations such as oxygen are supposed to play an important role in destroying the 2S-state, reducing the delayed-to-prompt photon ratio. The data cannot indicate this effect. Similarly, adding xenon to helium (1 % or more) will neutralize the muonic helium ion, which in turn quenches the 2S-state by Auger electron emission. We observe the absorption of low energy photons in the helium/xenon gas mixture to be stronger for the prompt fraction than it is for the delayed fraction, which is the opposite of what one does expect for NaI signals coming from the μ-cascade in helium. Finally, we used also methane, in this context a more esoteric choice of target

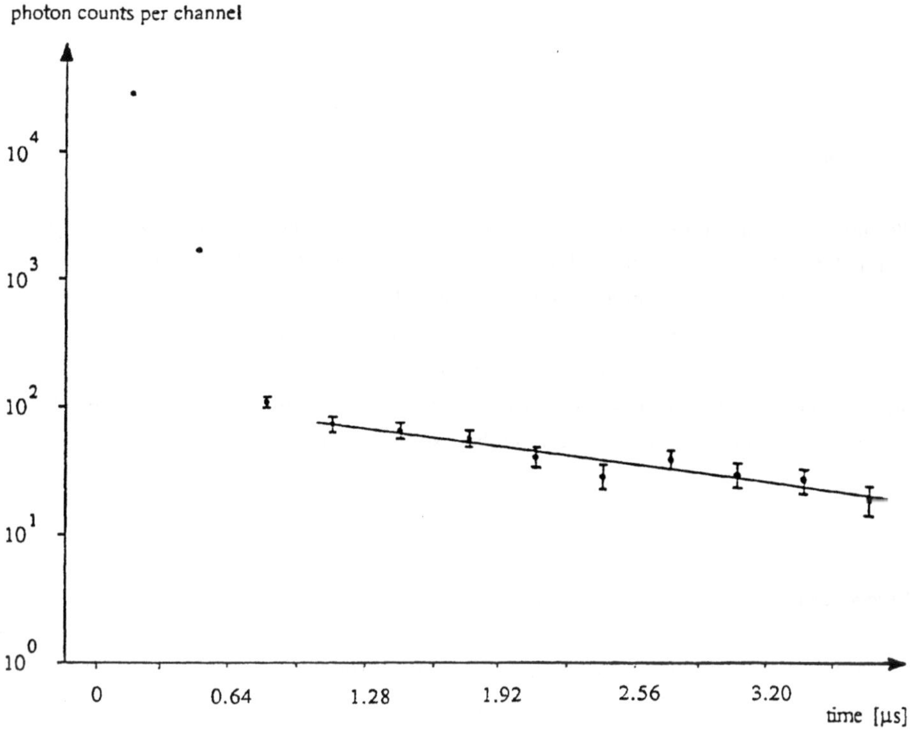

Fig. 2 Time spectrum of the delayed photons after negative muons had stopped in 40 atmosheres of pure helium(4). The decay electron had to be detected in the time interval of 550 ns $\leq t_{electron} \leq$ 1550 ns with respect to detected photon. The slope of the curve corresponds to a lifetime of 2.0(1) μs. For the unperturbed 2S-state, a lifetime of 1.75 μs is expected.

filling. Still, delayed photons are observed as frequent as it was in helium. We therefore reject the hypothesis, that the lifetime of 2 μs shown in Fig. 2 can be attributed to the radiative decay of a long lived 2S-state in muonic helium.

The high NaI counting rate of 8 kHz especially at low pulse heights enforces a more detailed study of the influence of the background sources on the results. Due to the requirement in the analysis of a decay electron a flat background would exhibit an exponential decrease with the muon lifetime of 2,2 μs. For comparison, the data shown in Fig. 2 could be well fit by an exponential function with fixed muon decay rate, which demonstrates that the delayed component could also be explained by noise of the NaI-crystals. We remind the reader her, that only 91 % of excitation energy is converted into prompt scintillation light. The rest is distributed over rather slow components which are likely to be responsible for the delayed events. We favour this hypothesis and note that all the results in Tab. 1 can now be consistently explained.

4. Conclusion

We have investigated the radiative transitions in muonic helium at high gas pressure using NaI-detectors in anticipation to elaborate the delayed photons from a metastable 2S-state. The detector did sense signals in accordance with the expected energy, lifetime and yield of the 2S-1S two-photon deexcitation. However, these events cannot be attributed to delayed photons from a muon cascade. Rather surprisingly they can be attributed to random coincidences in which detector noise and muon decay conspire in order to delude the experimenter.

References

1. A. Placci, E. Polacco, E. Zavattini, K. Ziock, G. Carboni, U. Gastaldi, G. Gorini, G. Neri and G. Torelli, Nuovo Cimento 1A, 445 (1971)

2. G. Carboni, A. Placci, E. Zavattini, U. Gastaldi, G. Gorini, O. Pitzurra, G. Neri, E. Polacco, G. Torelli, J. Duclos, J. Picard and A. Vitale, Lett. Nuovo Cimento 6, 233 (1973)

3. A. Bertin, C. Carboni, G. Gorini, O. Pitzurra, E. Polacco, G. Torelli, A. Vitale and E. Zavattini, Phys. Rev. Lett. 33, 253 (1974)

4. A. Bertin, G. Carboni, A. Placci, E. Zavattini, U. Gastaldi, G. Gorini, G. Neri, O. Pitzurra, E. Polacco, G. Torelli, A. Vitale, J. Duclos and J. Picard, Nuovo Cimento 26B, 433 (1975)

5. H.P. von Arb, F. Dittus, H. Heeb, H. Hofer, F. Kottmann, S. Niggli, R. Schaeren, D. Taqqu, J. Unternährer and P. Egelhof, Phys. Lett. 136B, 232 (1984)

6. M. Eckhause, P. Guss, D. Joyce, J.R. Kane, R.T.Siegel, W. Vulcan, R.E. Welsh, R. Whyley, R. Dietlicher and A. Zehnder, Phys. Rev. A33, 1743 (1986)

7. G. Carboni and O. Pitzurra, Nuovo Cimento 25B, 367 (1975)

8. R.O. Mueller, V.W. Hughes and H. Rosenthal, Phys. Rev. A11, 1175 (1975)

9. J.S. Cohen and J.N. Bardsley, Phys. Rev. A23, 46 (1981)

10. J.S. Cohen, Phys. Rev. A25, 1791 (1982)

11. J.E. Russell, Phys. Rev. A34, 3865 (1986)

12. D.P. Grechukhin, S.V. Romanov and A.A. Soldatov, Sov. Phys. JETP 64, 942 (1986)

13. L.I. Menshikov, L.I. Ponomarev and L.P. Sukhanov, Z. Phys. D7, 203 (1987)

14. G. Carboni, U. Gastaldi, G. Neri, O. Pitzurra, E. Polacco, G. Torelli, A. Bertin, G. Gorini, A. Placci, E. Zavattini, A. Vitale, J. Duclos and J. Picard, Nuovo Cimento 34A, 493 (1976)

15. G. Carboni, G. Gorini, G. Torelli, L. Palffy, F. Palmonari and E. Zavattini, Nucl. Phys. A278, 381 (1977)

16. J. Rosenkranz, PhD-thesis University of Heidelberg 1987, to be published

LIFETIME OF METASTABLE MUONIC HELIUM

L. Bracci[(a)] and E. Zavattini[(b)]

a) Dipart. di Fisica dell'Univ. and Sez. I.N.F.N., Pisa, Italy

b) CERN, Geneva, Switzerland and Universita' di Trieste, Italy

The lifetime of $(He\mu)_{2S}$ in an atmosphere of helium has been the subject of several investigations. There seems to be disagreement between different experiments, the problem being whether Stark mixing and Auger effect, in a range of pressures from a few to some tens atmospheres, allow the survival of these states till times of the order of some hundreds ns. Evidence that this is the case is given by experiments performed at CERN[1,2], whereas claims that no long-lived $(He\mu)_{2S}$ states exist were supported by the SIN experiments[3,4]. In the CERN experiments delayed X-rays in the range 0-8.2 keV (8.2 keV being the separation between the ground state and the n=2 state in muonic helium) were observed. The estimated yield ϵ_{2S} and lifetime τ_{2S} of muonic helium were found to be[1] $\epsilon_{2S} = 3.4\%$, $\tau_{2S} = 1.3\ 10^{-6}$s. These values were both derived from the data as independent parameters. Later experiments[2] substantially confirmed this finding. This in turn implied that, contrary to any expectation, the mechanisms responsible for 2S-quenching were depressed, and invoked an explanation. In the experiments performed at SIN, on the other hand, the detection of long-lived 2S states was first based[3] on the detection of delayed X-rays of 8.2 keV, and subsequently[4] on the search of delayed non-monochromatic photons in the range 0-8.2 keV, but with the underlying assumption that the yield of long-lived 2S states was the same as the total yield of 2S states, which can be determined by inspection of the intensities of the K-lines[2]. This is questionable, as will be made clear below.

An explanation of the discrepancy between the two groups of experiments can be found by assuming that during the cascade some of the $(He\mu)^+$ ions can be bound to another helium atom to form an analog of the hydride molecular ion $(H^+He)^+$, where the proton is substituted by the $(He\mu)^+$ ion via multi-body reactions of the

Electromagnetic Cascade and Chemistry of Exotic Atoms
Edited by L. M. Simons *et al.*, Plenum Press, New York, 1990

type

$$(He\mu)_n^+ + He + He \longrightarrow [(He\mu)_m^+ He] + He \tag{1}$$

n, m being the principal quantum numbers. Henceforth, we will loosely denote as "clusters" these states. Also the formation of more complex clusters is possible, but for the sake of simplicity we will consider only the case of $((He\mu)^+He)^+$.

In order to understand why the presence of these clusters is relevant, some data about their properties are in order. The spectrum of these clusters can be derived from the spectrum of the helium hydride molecular ion, since, already at $n \leq 10$, the $(He\mu)^+$ system has dimensions smaller than the Bohr radius. There are several excited states[5], the excitation energy being in the range 10-20 eV. Thus, excited clusters can be formed in transitions with $\Delta n = 1$ starting from $n \leq 9$. The rate for cluster formation can be estimated using the law[6] $\lambda_m = \sigma_m v n^m V^{m-1}$ for a $X + mY$ reaction, where σ_m is the effective cross section of a two-body reaction, v the relative velocity, V the interaction volume and n the density. It turns out that, using the value of V found in ref. 6, the multi-body reactions are dominant starting from a pressure of $p \approx 2$atm.

The formation of excited clusters, which in the pressure conditions of the aforementioned experiments certainly occurs, can heavily affect the lifetime of $(He\mu)_{2S}^+$. Because of their peculiarities (see table I) they are responsible for a heavy depression of the quenching mechanisms considered so far. The internal Auger rate λ_A^{int} is proportional to the electron density at the muon site ($\lambda_A^{int} = \rho_e 8.1\ 10^{-17}s^{-1}$, with ρ_e in cm^{-3}). As a consequence of the big dimensions of the excited cluster, λ_A^{int} is now of the order of the muon decay rate. Stark mixing due to collisions with the neighbouring atoms is also now reduced, because of its dependence on ω_{vib}^3 and the tiny value of ω_{vib} for the excited clusters: $\lambda_{St} \leq 10^5$ for the $(He\mu)_{2S}^+$ ions bound in the excited clusters. Finally, the external Auger effect ($\lambda_A^{ext} = 3.6\ 10^3 ps^{-1}$, p in atm) is negligible up to $p \approx 200$ atm.

Table I

State	D_e (eV)	r_e (Bohr)	ω_{vib} (cm^{-1})	λ_{St} ($10^4 s^{-1}$)	E(R=∞) (Hartree)
X^1S^+	1.850	1.444	2128	5800	-2.8757
A^1S^+	0.0353	5.68	132	1.4	-2.5000
B^1S^+	1.051	8.486	240	7.4	-2.1428
C^1P^+	0.2228	7.905	138	1.5	-2.1250
a^3S^+	0.0814	4.472	234	6.8	-2.5000
b^3S^+	0.7324	7.654	249	8.5	-2.1742
c^3P^+	0.1988	7.720	141	1.6	-2.1307

In the excited clusters, on the other hand, it is possible to have transitions to the ground state, where the internal quenching mechanisms regain their efficiency. In order to estimate whether excited clusters are long-lived, we distinguish between triplet (S=1) clusters and singlet (S=0) clusters. Since the ground state is a singlet state, transition by dipole radiation can occur only from the singlet states. Using the dipole matrix elements (properly rescaled) of ref. 5, and harmonic oscillators wave functions, we estimate[7] that, depending on the vibrational state, transitions from the singlet excited levels to the ground state can occur in a time as short as $\approx$ 10 ns. In a similar way, transitions from the triplet excited states to the lowest one ($a^3\Sigma^+$) can occur quite rapidly. But triplet-to-singlet transitions due to collisions are inhibited, since, due to total spin conservation in the electric dipole interaction, such transitions would require the exchange of two electrons. A rough calculation performed within the impact parameter approximation yields a rate of order 100 s^{-1} in standard conditions of temperature and pressure for the triplet-to-singlet transitions. In conclusion, we can safely state that those $(He\mu)^+$ ions which during the cascade are trapped in excited triplet clusters, if they cascade down to 2S state, there they are not efficiently quenched by any mechanism.

Within this framework, we propose the following picture. The $(He\mu)^+$ excited ions can form, at pressure $p \geq 2$ atm, excited clusters, where they cascade to $(He\mu)^+_{2S}$ with probability $\epsilon \approx 3.5\%$. These states are not appreciably quenched by any of the mechanisms which are responsible for the decay of $(He\mu)^+_{2S}$. In this sense, the situation for these ions is quite similar to what happens at very low pressure. On the other hand, for those ions which are bound in the ground state $X^1\Sigma^+$, or cascade down to it from excited singlet clusters, the Auger effect is still important, and dissociation of the cluster ensues quickly, while the $(He\mu)^+$ ion is still in an excited state and consequently is available for formation of an excited cluster. We recall that the cluster formation time estimated in ref. 6 is τ=4.2 10^7p^2 (p in atm), which, as soon as $p \geq 5$ atm, is of the order of (or less than) the cascade time. Thus, a population of long-lived $(He\mu)^+_{2S}$ exists (those which are bound in an excited triplet states), whose amount cannot be decided on the basis of the observation of the K-lines intensity, but must be determined by the experiment.

With this scheme in mind, the results of refs. 1,2 are easily explained by assuming that approximately 60% of muonic helium ions are bound in the ground state cluster ($X^1\Sigma^+$) at late times, since the total yield of 2S muonic helium (long-lived and short-lived) can be derived from observation of K-lines to be $\epsilon_{2S}^{tot} \approx 7\%$. As for the results of refs. 3,4, the estimate of the $(He\mu)_{2S}$ lifetime was obtained with the assumption that the yield of long-lived $(He\mu)_{2S}$ ions was equal to the total yield. Because of the mechanisms outlined above, this assumption is not tenable.

The existence of excited clusters $((He\mu)^{+}He)^{+}$ can also yield a possible explanation of the results on the residual polarizations of muons in helium[8]. The basic facts are that in a wide range of pressures the residual muon polarization is found to be lower than theoretically estimated, but there is a rise up when contamination of Xenon is added and formation of $((\alpha\mu)^{+}e^{-})$ is made possible. We argue that depolarization occurs quickly for those $(He\mu)^{+}$ ions which are bound in a cluster in the ground state, because of the random orientation of the field B due to the rotation of the cluster, this field being of order 100 G. This mechanism is ineffective in excited clusters, because of their dimension. Due to energy conservation, formation of $((\alpha\mu)^{+}e^{-})$ is possible only in the excited clusters, where depolarization is smaller. This channel of formation is in competition with the direct formation, the transfer of an electron from Xenon to the $(He\mu)^{+}$ ion being energetically allowed. The rate should be rather large, however, since at the pressures of the experiment most likely the ion is trapped in a cluster. On the other hand, a clean revelation of these excited clusters could be allowed by resorting to a contamination with some atom, like Argon, for which direct formation of $((\alpha\mu)^{+}e^{-})$ atoms by transfer of an electron to the ion is forbidden by energy conservation. In this case, the detection of $((\alpha\mu)^{+}e^{-})$ atoms, as well as the measurement of the residual polarization, could yield a clear proof of the existence of the excited clusters.

REFERENCES

[1] A. Placci et al., Nuovo Cimento **1A**, 445 (1971).

[2] A. Bertin et al., Nuovo Cimento **26B**, 433 (1975).

[3] H.P. von Arb et al., Phys. Lett. **136B**,232 (1984)

[4] M. Eckause et al., Phys. Rev. **A33**, 1743 (1986)

[5] H.H. Michels, J. Chem. Phys. **44**, 3834 (1966)

[6] C.P. de Vries and H.J. Oskam, Phys. Rev. **A22**, 1429 (1980)

[7] L. Bracci and E. Zavattini, CERN preprint, to be published

[8] P.A. Souder et al., Phys. Rev. **A22**, 33 (1980)

A STUDY OF THE STRONG INTERACTION EFFECTS OF DEEPLY BOUND PIONIC LEVELS IN PIONIC ATOMS

J. Konijn[1], C.T.A.M. de Laat[2], A. Taal[1], P.David[3], H.Hänscheid[3], F.Risse[3], Ch.F.G.Rösel[3], W. Schrieder[3],W.Lourens[2], C.Petitjean[4], A. van der Schaaf[5]

[1]NIKHEF-K, P.O.Box 4395, NL-1009 AJ Amsterdam, The Netherlands

[2]Physics Laboratory, University of Utrecht,P.O.Box 800000, NL-3508 TA Utrecht

[3]Institut für Strahlen- und Kernphysik der Universität Bonn, D-5300 Bonn

[4]Paul Scherrer Institute (PSI), formerly SIN, CH-5234 Villigen

[5]Physik-Institut der Universität Zürich, CH-8001 Zürich

ABSTRACT

The pionic X-ray spectra of ^{181}Ta, ^{nat}Re, ^{nat}Pt, ^{197}Au, ^{208}Pb, ^{209}Bi and ^{237}Np have been investigated to obtain data which provide for a systematical investigation of the pionic 3d and 4f levels. Special attention was paid to experimental methods for obtaining clearly improved results as compared to earlier experiments, by using an array of Compton suppression BGO-shields and by using neutron time-of-flight discrimination to reduce background. Also to explore X-rays populating deeply bound pionic 1s-levels in ^{24}Mg, ^{27}Al, ^{28}Si and 2p-orbits in ^{93}Nb, ^{nat}Ru, respectively, coincidence techniques in combination with these high resolution Ge-BGO Compton suppression spectrometers were applied. Only deviations in the width of the pionic 3d-levels are observed. They are narrower by a factor 1.5 as compared to theoretical predictions. The same deviation as for Γ_0(3d) is found for the strong interaction quadrupole shift, ε_2, for both the 4f- and 3d-orbits. An explanation for the difference in Γ_0(3d) can be found by adding an isovector absorption term to the optical potential with a negative value for the parameters ImB_1 or ImC_1.

INTRODUCTION

The pionic atom experiments were performed at NIKHEF in Amsterdam and at the Paul Scherrer Institute (formerly SIN) at Villigen in Switzerland. At NIKHEF, measurements of pionic ^{24}Mg and ^{27}Al were started to determine the properties of the deeply bound pionic 2p→1s X-ray transition. Our experimental method, though including modern spectroscopic techniques such as the use of Compton suppression BGO-crystals and time-of-flight discrimination to reduce neutron induced background, was found to be insufficient to measure these low-intensity transitions which have appreciable widths. Moreover, in pionic ^{24}Mg and ^{27}Al, and also in ^{28}Si, the pionic 2p→1s X-ray transition is obscured by strong nuclear γ-ray transitions. To reduce the effect of this γ-interference additional coincidence requirements had to be applied. Therefore, these experiments were repeated at the PSI with much better beam properties (100% duty factor versus 1% at NIKHEF and a higher beam intensity). The result was an order of magnitude improvement of both the statistics and the peak-to-background ratio in these X-ray spectra. For the measurement of the pionic 3d→2p X-ray transition in ^{93}Nb and ^{nat}Ru and of the 4f→3d transition in ^{93}Nb, ^{nat}Ru, ^{nat}Ag, ^{nat}Cd, ^{181}Ta, ^{nat}Re, ^{nat}Pt, ^{197}Au, ^{208}Pb and ^{209}Bi a conventional setup was used in combination with neutron time-of-flight discrimination but without the application of additional coincidence techniques was found to be adequate.

EXPERIMENTAL

The experimental setup and analysis methods are extensively described by de Laat[1] and Taal[2]. In fig.1 we show the influence of the neutron time-of-flight discrimination. By selecting suitable windows on the time spectrum one effectively can separate the prompt spectra from the neutron induced γ-rays. These neutrons are emitted after pion absorption in the target nuclei.

1. Pionic ^{24}Mg and ^{27}Al measurements

Some years ago the maximum Z-values for which strong interaction shifts and widths of pionic levels had been investigated were Z=10 and Z=33 for the 1s and 2p-orbits, respectively. Beyond these Z-values, the relevant X-ray transitions become increasingly wider and weaker and therefore more difficult to separate from the background, especially if these transitions are also obscured by nuclear γ-ray transitions. In the case of ^{27}Al the pionic 2p→1s X-ray transition at about 363 keV is obscured by the muonic 2p→1s transition at 345 keV and by a nuclear γ-ray at 350 keV, as is illustrated in fig.2. This nuclear γ-ray at 350 keV is mainly produced by the reaction $^{27}Al(\pi^-,2p4n\gamma)^{21}Ne$ which occurs after the pion ends its atomic cascade in the nucleus. The spectra of fig.2 were obtained at the NIKHEF facility with a rather thick Al target of 3.97 g/cm^2, using a Compton-suppressed Ge-detector, of which the suppression shield consisted of NaI scintillation material only. To reduce the strong muon

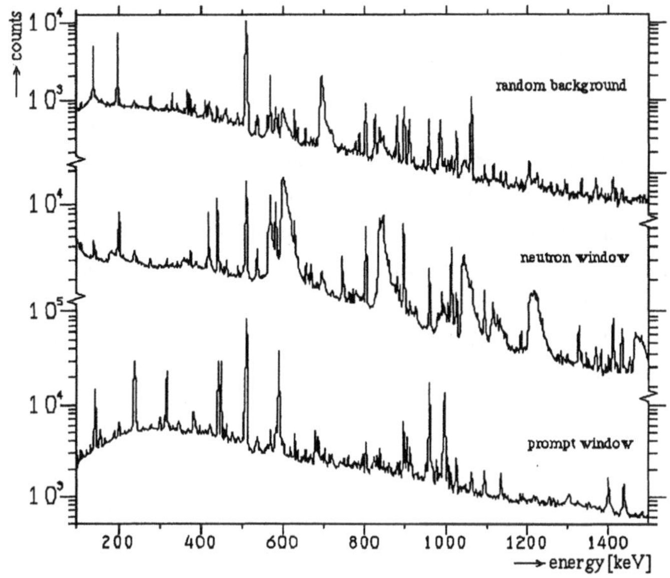

Fig.1 Prompt, delayed and random spectrum of pionic ^{209}Bi. In the delayed spectrum the neutron induced γ-rays in Ge and Al (detector cap) are recognized.

contamination in the spectrum a clean pion trigger had to be installed. As is shown in the righthand side of fig.2 a significant improvement was obtained; the μX-transition strength is reduced by a factor of about 3.5. This, however, is still not good enough for a meaningful analysis of the pionic 2p→1s X-ray transition.

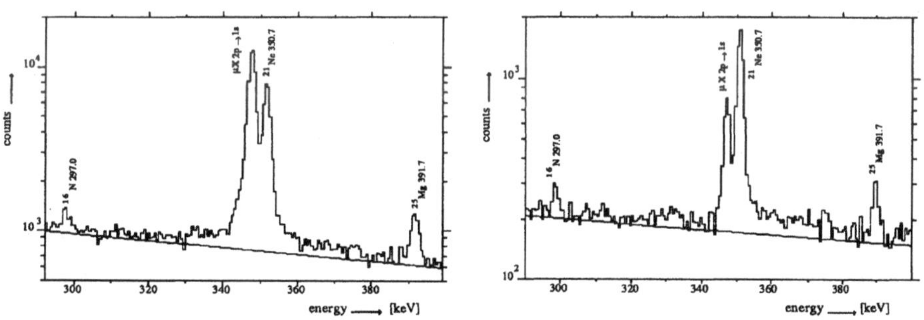

Fig. 2 Pionic ^{27}Al measured with a target of 3.97 g/cm^2 thickness. The left spectrum was recorded during a test run at NIKHEF. The right-hand spectrum was measured at PSI with a thin target and a well tuned pion telescope to suppress the influence of the muonic 2p→1s X-ray transition.

Besides the use of Compton suppression an additional method has been applied to reduce the interference of possible nuclear γ-rays. In the ^{27}Al measurement the aim was to reduce the occurrence of the 350 keV nuclear transition in ^{21}Ne in the energy spectra of the Ge-detectors. There are several possible strong nuclear γ-rays besides the 350 keV transition in ^{21}Ne, e.g. the 440 keV in ^{23}Na. By measuring the Ge-energy spectra in coincidence with events detected in an array of NaI-crystals viewing the target, it may be possible to reduce this disturbing γ-ray background

However, the pion-absorption reactions in ^{27}Al which end with a ^{21}Ne nucleus and a ^{23}Na nucleus in an excited state, producing the 350 keV and 440 keV γ-rays, respectively, are of the same probability and by far the most important ones. An inconvenient side-effect of these two absorption reactions, $^{27}Al(\pi^-,2p4n\gamma)^{21}Ne$ and $^{27}Al(\pi^-,p3n\gamma)^{23}Na$, is the emission of neutrons (of roughly 25 MeV). Off-line discrimination against the (n,n') reactions of these prompt neutrons in the NaI-crystals by means of the time-of-flight method, would require to measure with a rather large solid angle.

A total of eight 12.7 cm (diameter) × 12.7 cm (long) NaI(Tl)-crystals around the target at a rather short distance of about 12.5 cm, (see fig. 3) were used to measure the coincident X-ray

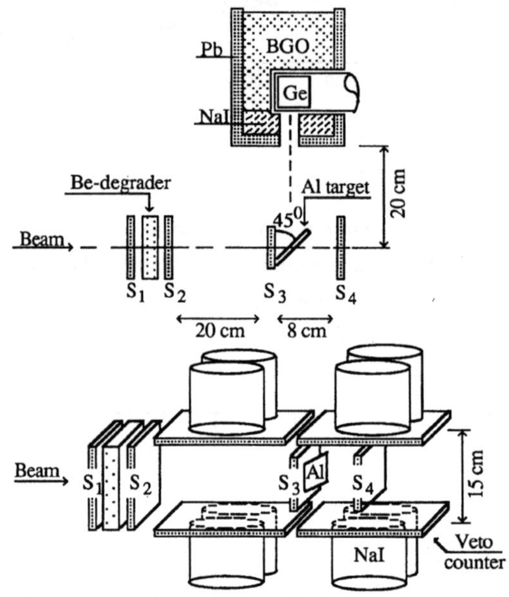

Fig. 3 Schematic view of the setup for the measurement of pionic ^{27}Al. An array of eight 12.7 cm (diameter) × 12.7 cm (long) NaI(Tl)-crystals was used to detect nuclear γ-rays after pion absorption. Plastic veto counters were placed in front of the NaI(Tl)-crystals to discriminate against charged particles. Compton-suppressed Ge-detectors were placed in the horizontal plane; the Na(Tl)-detectors viewed the target from above and below.

spectra without applying the time-of-flight method to the NaI-counters. The incorporation of the coincidence with the NaI-crystals in the event trigger improved the quality of the energy spectra of the Ge-detectors, which is illustrated by the two spectra of fig.4. This ^{27}Al spectrum was recorded at the PSI-facility, using a 0.54 g/cm^2 thick Al-target. The energy spectrum is

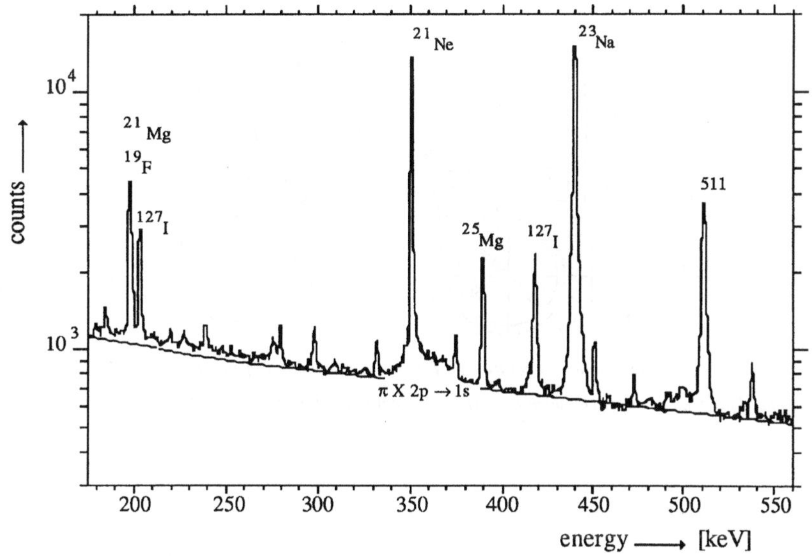

Fig. 4 Pionic X and γ-ray spectra of ^{27}Al for one of the Ge-detectors. The spectrum results from the additional requirement of a coincidence signal in any of eight NaI(Tl)-crystals viewing the target.

Compton suppressed and recorded in coincidence with a valid pion stop in the target, and in addition coincident with a signal from any of the NaI-crystals. As a result of this additional requirement the peak-to-background ratio of the pionic 2p→1s transition improved even more. Furthermore, the 2p→1s μX-ray transition has almost entirely disappeared in the spectrum due to the much cleaner pion trigger accomplished with the additional NaI coincidence requirements.

A similar improvement was found for the two spectra of pionic ^{24}Mg. In this measurement also a remarkable reduction of the 2p→1s μX-ray transition was achieved. The target thickness of ^{nat}Mg was 1.74 g/cm^2.

2. *Pionic ^{28}Si measurement*

In the case of ^{28}Si the muonic 2p→1s X-ray transition at 400 keV (see fig.6) and the nuclear γ-rays from the reactions $^{28}Si(\pi^-,p2n\gamma)^{25}Mg$, $^{28}Si(\pi^-,2n\gamma)^{26}Al$ and $^{28}Si(\pi^-,2p3n\gamma)^{23}Na$ at 392, 417 and 440 keV, respectively, strongly hinder a relevant analysis of the pionic transition. To improve the peak-to-background ratio and to reduce the influence of the muonic 2p→1s X-ray transition, this measurement was performed in coincidence with charged particles emitted after pion absorption. The target, composed of circular plates of ^{nat}Si with a diameter of 76 mm and a total thickness of 0.8 mm, was mounted in a vertically placed cylindrical wire chamber. The advantage of this method, compared to the former one with NaI(Tl)-counters, is that a wire chamber is not sensitive to neutrons. Thus, there is no need to record time-spectra for an off-line analysis to do time-of-flight discrimination.

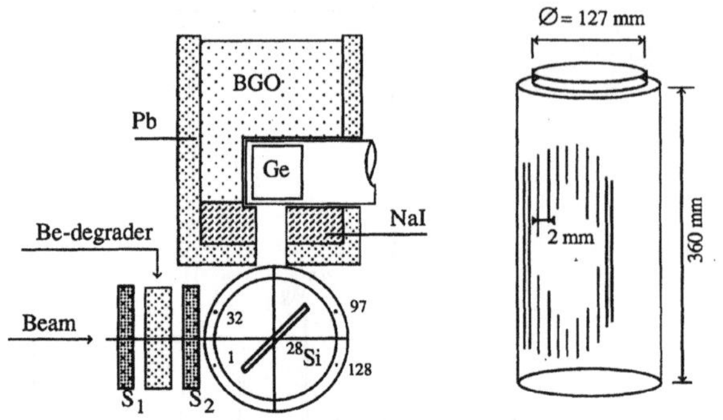

Fig. 5 Schematic view of the experimental setup used for the measurement of pionic ^{28}Si. A cylindrical wire chamber was used for charged particle detection, which are emitted after pion absorption, is vertically placed in the beam.

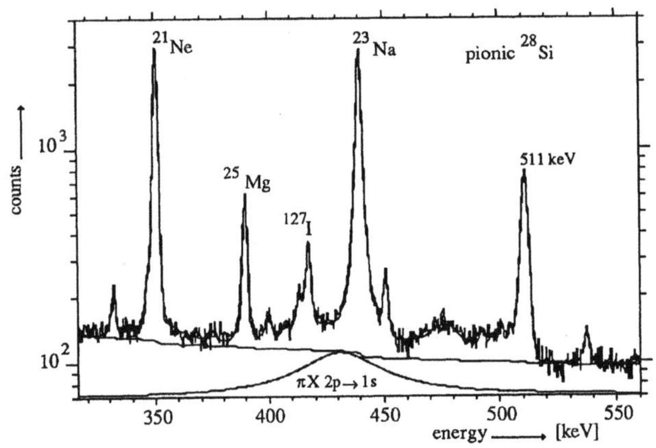

Fig.6 The pionic 2p→1s X-ray transition in ^{28}Si at 424.1± 2.0 keV.

The experiment was performed using one of the low mass cylindrical multiwire proportional chambers which are part of the large-solid-angle magnetic spectrometer SINDRUM. The wire chamber, as depicted in fig.5, consisted of 192 wires and had a length of 360 mm and an inner diameter of 127 mm. It was divided into 6 sections of 32 wires, each with its own electronics. The front section of 32 wires facing the incident beam was used as a trigger on the incident pions and was as such a part of the telescope together with the rear section of the chamber (wires 97-128), providing a stop veto for every traversing particle.

As a result of the clean pion trigger, accomplished by the coincidence with charged particle emission following the absorption, the recorded spectrum, fig.6, is almost free from the muonic 2p$\rightarrow$1s X-ray transition as well as to a large extent free from the 417 keV nuclear γ-ray transition from the reaction $^{28}Si(\pi^-,2n\gamma)^{26}Al$.

3. *Pionic ^{93}Nb, ^{nat}Ru, ^{nat}Ag, ^{nat}Cd, ^{181}Ta, ^{nat}Re, ^{nat}Pt, ^{197}Au, ^{208}Pb, ^{209}Bi and ^{237}Np measurements*

The pionic X-ray spectra of ^{93}Nb, ^{nat}Ru, ^{nat}Ag, ^{nat}Cd, ^{181}Ta, ^{nat}Re, ^{nat}Pt, ^{197}Au, ^{208}Pb, ^{209}Bi and ^{237}Np were measured by using an array of Compton suppressed (BGO-shields) Ge-spectrometers only. As is reflected in fig.1 the suppression of neutron induced background is of importance for the measurement of these isotopes. To discriminate against neutrons emitted after pion absorption by means of time-of-flight, the Ge-detectors were placed at distances of about 40 cm from the target. Only for pionic ^{93}Nb and ^{nat}Ru did the measured spectra permit an analysis of the 3d$\rightarrow$2p pionic X-ray transition and hence for a detemination of the width and shift of the 2p-level. In contrast to the ^{93}Nb and ^{nat}Ru spectra the pionic spectra of ^{nat}Ag and ^{nat}Cd did not show a 3d$\rightarrow$2p transition and therefore it was only possible to determine the shift and width of the pionic 3d–level in these cases. The results of our analysis for the strong interaction shifts, ε_0, and widths, Γ_0, are presented in table 1. For the pionic 1s-, 2p- and 4f-orbits theoretical calculations by Batty et al.[3] reproduces the data reasonably well. We observe, however, that the data establish a clear discrepancy between theory and experiment for the deeply bound pionic states.

Olivier et al.[4] have made an attempt to relate these anomalously small widths to an increase in s-wave repulsion in the optical potential. The parameter set of Seki and Masutani[5] essentially describes the same s-wave repulsion as suggested by Olivier et al.[4]. The result of their calculations come closest to the experimental widths for the 3d-levels. Recently Wright et al.[6] at the PANIC 1987 conference in Kyoto, Japan, reported on low-energy elastic scattering experiments of $\pi^{\pm}$ on ^{58}Ni and ^{40}Ca at 30 and 20 MeV at LAMPF. They obtain a better agreement between the π^--scattering data and theory if the parameter set [5] used in the optical potential is adjusted so that it confirms the data presented in this work. They conclude that the evidence of pionic-atom anomalies agrees with that of in the π^--scattering data. We must remark, however, that this set destroys the agreement for the pionic 4f-data on the shifts.

In contrast to the 4f-levels, the deeply bound 3d-levels in e.g. ^{nat}Pt, ^{197}Au, ^{208}Pb and ^{209}Bi are influenced not only by the p-wave interaction but also to a large extent by the local s-wave interaction. This can best be illustrated as follows. We calculate that the contribution to ε_0 from s- and p-wave terms for the case of ^{208}Pb are ε_0(s-wave) = -71.6 keV and ε_0(p-wave) = +110.5 keV, respectively, showing a somewhat delicate balance between the two terms.

The agreement between the present experimental values for pionic ^{208}Pb and those reported by Olin et al.[4] is reasonably good.

Table 1. Strong interaction shifts and widths for the pionic 1s-, 2p-, 3d- and 4f-levels obtained in the present work. All values are given in keV. The theoretical values are calculated with the parameter set of ref.[3].

Nucleus	ε_0	ε_0^{theory}	Γ_0^{direct}	Γ_0^{theory}
		1s orbit		
^{24}Mg	-82.7 ± 1.2	-76.1	24.3 ± 1.0	24.8
^{27}Al	-115.5 ± 1.4	-111.3	28.8 ± 1.2	29.3
^{28}Si	-131.6 ± 2.0	-128	41. ± 4	39
		2p orbit		
^{24}Mg			0.076 ± 0.007	0.070
^{27}Al			0.111 ± 0.012	0.117
^{93}Nb	-11. ± 3	-17	64. ± 8.	57
^{nat}Ru	-48. ± 7	-43	77. ± 24	70
		3d orbit		
^{93}Nb	0.74 ± 0.02	0.73	0.402 ± 0.016	0.459
^{nat}Ru	1.39 ± 0.08	1.16	0.75 ± 0.08	0.81
^{181}Ta	16.0 ± 1.3	13.8	24.6 ± 1.5	35.7
^{nat}Re	16.5 ± 1.3	16.2	29.7 ± 2.7	41.9
^{196}Pt	22.8 ± 1.9	19.9	37. ± 5	51.6
^{197}Au	20.6 ± 1.9	19.9	34. ± 4	56.2
^{208}Pb	22.7 ± 2.4	18.9	47. ± 4	68.2
^{209}Bi	20.1 ± 3.0	19.3	52. ± 4	73.5
		4f orbit		
^{181}Ta	0.56 ± 0.04	0.53	0.31 ± 0.05	0.33
^{nat}Re	0.76 ± 0.04	0.69	0.41 ± 0.05	0.43
^{196}Pt	1.09 ± 0.04	1.04	0.59 ± 0.05	0.65
^{197}Au	1.25 ± 0.07	1.15	0.77 ± 0.04	0.71
^{208}Pb	1.68 ± 0.04	1.61	0.98 ± 0.05	1.06
^{209}Bi	1.78 ± 0.06	1.80	1.24 ± 0.09	1.19
^{237}Np	5.26 ± 0.19	4.32	3.88 ± 0.26	3.85

EXTENSION OF OPTICAL POTENTIAL TERMS

In the theoretical considerations, the parameter c_p for the proton and c_n for the neutron Fermi distributions in the nucleus have until now been taken equal. For light nuclei, for which the semi-empirical optical potential has been used to fit the strong interaction shifts and widths, the differences between proton and neutron density distributions are indeed small and hardly influence the parameters in these nuclei. We will here investigate whether taking these quantities into account for our heavier elements may explain the "discrepancy". In the

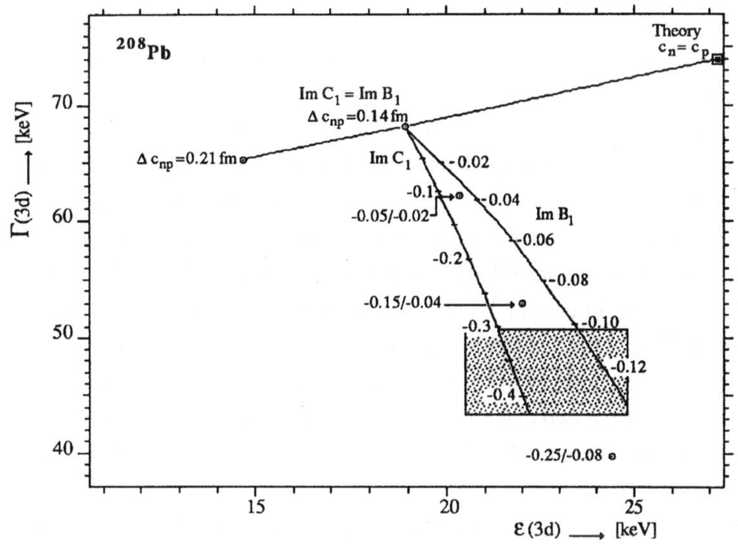

Fig. 7 Experimental and theoretical data on pionic ^{208}Pb, in a plot of widths versus shifts[3,4]. The double square point gives the theoretical value for $c_n=c_p$ (using the parameter set of Batty et al.[2]). The almost horizontal line shows the linear behaviour for non-zero c_n-c_p but zero isovector absorption terms. Starting from the value for c_n-c_p= 0.14 fm, the influence of non-zero ImB_1 and ImC_1 are shown (lines for only one of the two differing from zero, points for three combinations of non-zero terms).

formalism given below the expressions for the radial neutron and proton distributions $\rho_n(\mathbf{r})$ and $\rho_p(\mathbf{r})$ are therefore kept distinct. In general $\rho_p(\mathbf{r})$ is deduced from experimental charge distributions obtained in muonic atom and electron-scattering experiments by unfolding the nucleon charge form factors.

The difference of the neutron and proton matter distribution has been investigated by Hoffman et al.[7] using the results from scattering of 1 GeV protons. The values for the half-density radii in the Fermi distributions of the protons and neutrons, c_p and c_n respectively, result in a difference of rms radii $\Delta r_{np} = 0.14 \pm 0.04$ fm. Hartree–Fock calculations by Angeli et al.[8] gave a somewhat larger value for the difference of about 0.20 fm. The calculations by Negele and Vautherin[9] also result in a difference of about 0.20 fm. Comparing the strong interaction monopole shifts and widths with respect to the standard optical potential calculations for pionic 4f- and 3d-levels for the different parameter sets, using Fermi distributions with $c_p = c_n$ and with $c_p - c_n = 0.14$ fm, respectively. One notices that shifts and widths are indeed reduced, resulting in a better agreement with experiment. In total, a better agreement of the calculated values with the experimental results by almost a factor of two for $c_p - c_n = 0.14$ fm. For a special case, ^{208}Pb, we show in fig. 7 the influence of increasing values of Δc_{np} in a plot of shifts versus widths. Though indeed a somewhat better agreement is reached for $\Delta c_{np} \neq 0$, the situation is still not satisfactory.

Making use of the non-zero difference in these radii, a further improvement can be reached by adding isovector absorption terms for heavier nuclei in the standard optical potentials, being the terms with the parameters B_1 and C_1 in eq. (1) and (2); both constants being imaginary if we only consider absorption by them. Kunselman et al.[10] in an attempt to determine neutron radii for nuclei with $20 \le Z \le 26$ already used non-zero (positive) terms of the type $\mathrm{Im}B_1$ and $\mathrm{Im}C_1$ in the optical potential. Their calculations did not show any improvement. Seki and Masutani[5] point out that the isovector density distribution is quite sensitive to the choice of the radial dependences of $\rho_n(\mathbf{r})$ and $\rho_p(\mathbf{r})$. The values of the isovector parameters thus varies among the different analysis and their determination requires extra care. We have tried to explain our results with additional isovector absorption terms $\mathrm{Im}B_1$ and $\mathrm{Im}C_1$. The result is shown in fig. 7 for the 3d-case of ^{208}Pb. The situation is definitely improved for negative values of both $\mathrm{Im}B_1$ and $\mathrm{Im}C_1$. In reality, a combination of the two might occur which results in a deviation from the $\mathrm{Im}B_1 = \mathrm{Im}C_1 = 0$ point which is essentially the vector sum of the deviations for each of the two quantities separately. The influence of introducing isovector terms in the calculated values for the other cases investigated is presented in a separate paper at this workshop.

The Ericson-Ericson optical potential in second order is of the form given in eq. (1) with the s- and p-wave parts of the optical potential given by eq. (2a) and (2b), respectively.

$$V_S(\mathbf{r}) = -\frac{2\pi}{\mu}\left[q(\mathbf{r}) - \nabla\cdot\frac{\alpha(\mathbf{r})}{1+\frac{4}{3}\pi\xi\alpha(\mathbf{r})}\nabla\right] \tag{1}$$

with

$$q(\mathbf{r}) = \left(1+\frac{m_\pi}{m_N}\right)\left(b_0\rho(\mathbf{r}) + b_1\delta\rho(\mathbf{r})\right) + \left(1+\frac{m_\pi}{2m_N}\right)\left(B_0\rho^2(\mathbf{r}) + B_1\rho(\mathbf{r})\delta\rho(\mathbf{r})\right) \tag{2a}$$

and

$$\alpha(\mathbf{r}) = \left(1+\frac{m_\pi}{m_N}\right)^{-1}\left(c_0\rho(\mathbf{r}) + c_1\delta\rho(\mathbf{r})\right) + \left(1+\frac{m_\pi}{2m_N}\right)^{-1}\left(C_0\rho^2(\mathbf{r}) + C_1\rho(\mathbf{r})\delta\rho(\mathbf{r})\right) \tag{2b}$$

where we define the isoscalar and isovector densities as

$$\rho(\mathbf{r}) = \rho_n(\mathbf{r}) + \rho_p(\mathbf{r})\,, \qquad \delta\rho(\mathbf{r}) = \rho_n(\mathbf{r}) - \rho_p(\mathbf{r}) \tag{3}$$

SUMMARY

New or improved values have been found for widths and energies of strongly bound pionic atom levels. To explain the values found, the theory of these levels has been extended in two ways. On one hand, the known difference between the distributions of neutrons and protons in the nucleus has been taken into account. This gave a noticeable but far from sufficient improvement in the relation between theory and experiment. A further and almost completely satisfactory agreement could be obtained by adding some isovector absorption terms to the optical potential. In another contribution to this workshop we will show that a fit to 140 carefully selected pionic atom data of the pionic optical potential parameters gives a satisfactory agreement between theory and experiment.

REFERENCES

1. C.T.A.M.de Laat; Ph.D.Thesis, University of Technology Delft, March 1988
2. A.Taal; Ph.D.Thesis, University of Technology Delft, March 1989
3. C.J.Batty, S.F.Biagi, E.Friedman, S.Hoath, J.D.Davies, G.J.Pyle and G.T.A.Squier; Phys.Rev.Lett. **40** (1978) 931
4. J.G.J. Olivier, M.Thies and J.H.Koch, Nuclear Physics **A429** (1984) 477
5. R. Seki and K. Masutani; Phys. Rev. **C27** (1983) 2799 and R. Seki, K. Masutani and K. Yazaki; Phys.Rev. **C27** (1983) 2817
6. D.H.Wright, M.Blecher, K.Masutani, R.Seki, R.L.Boudrie, R.L.Burman, M.J.Leitch, M.Alsolami, G.Blanpied, J.A.Escalante, C.S.Mishra, G.Pignault, B.M.Preedom, C.S.Whisnant and B.G.Ritchie; Phys.Rev. **C35** (1987) 2258
7. G.W. Hoffman, L.Ray, M.Bartlett, J.McGill, G.S.Adams, G.J.Igo, F.Irom, A.T.M.Wang, C.A.Whitten Jr, R.L.Boudrie, J.F.Amann, C.Glashauser, N.M.Hintz, G.S.Kyle and G.S.Blanpied; Phys. Rev. **C21** (1980) 1488
8. I. Angeli, M.Beiner, R.J.Lombard and D.Mas; J. of Phys. **G6** (1980) 303
9. J.W. Negele and D. Vautherin.; Phys. Rev. **C5** (1972) 1472
10. R. Kunselman, R.J.Powers, M.V.Hoehn and E.B.Shera; Nucl. Phys. **A405** (1983) 627

AN IMPROVED PARAMETRIZATION OF THE OPTICAL POTENTIAL FOR PIONIC ATOMS

J. Konijn[1], C.T.A.M. de Laat[2], A. Taal[1] and J.H. Koch[1]

[1]NIKHEF-K, P.O.Box 4395, NL-1009 AJ Amsterdam, The Netherlands

[2]Physics Laboratory, University of Utrecht,P.O.Box 800000, NL-3508 TA Utrecht

ABSTRACT

We extended the parametrization of the phenomenological optical potential for pionic atoms by allowing for an isospin-dependence of the absorption parameters. Least squares fits to the 140 selected pionic atom data are performed. The new parameter set indicates a significant isospin dependence and is able to provide a simultaneous description of shifts and widths for the previously problematic deeply bound 3d-orbits in elements such as ^{181}Ta, ^{nat}Re, ^{nat}Pt, ^{197}Au, ^{208}Pb and ^{209}Bi.

INTRODUCTION

There are two main sources of information on the low-energy π^--nucleus interaction: π-nucleus scattering and pionic atoms. In pion scattering, only a few angular momentum states of the π-nucleus system contribute to the scattering wave function. For light targets and at the lowest energies one presently can measure ($T_\pi \approx 20$ MeV) usually only the central s-wave and the more peripheral p-wave contribute significantly. To change the relative importance of these partial waves or to bring in higher l -states one can increase the scattering energy.

The situation is different in pionic atoms. After its capture, the pion finally cascades through circular orbits, each having a well defined angular momentum state $l = n - 1$. As the orbits come close enough to the nucleus, the pion feels the strong interaction. This interaction shifts the energy of the pionic level and - due to the possibility of pion absorption - leads to an increasing broadening of the level. In heavier targets, in fact, the absorption process becomes so strong that the lowest possible level (1s and even 2p) is not reached by the pion. By measuring the position and width of pionic atom states, we can, therefore, get information on the π-nucleus interaction for a fixed, discrete angular momentum state at essentially constant energy (on the scale relevant for the π-nucleus interaction).

Electromagnetic Cascade and Chemistry of Exotic Atoms
Edited by L. M. Simons *et al.*, Plenum Press, New York, 1990

In a first generation of measurements, the shifts and widths of the more peripheral states for a variety of nuclei across the periodic system were measured. For the heaviest targets, such as ^{208}Pb, no state lower than the 4f-state could be observed. All lower lying states were too broad to be extracted from the emitted γ-ray spectrum. These data could be described quite satisfactorily by an optical potential with a few fitted parameters. It was also possible to relate this semi-phenomenological potential to the one needed to describe low-energy pion-nucleus scattering.

This situation changed when it became possible to also extract the shift and width of the previously unobserved deeply bound and much broadened levels for many targets[1-7], e.g. the 3d-level in ^{208}Pb. It turned out that these states could not be described by the optical potentials obtained earlier. Particularly disturbing was the fact that the width of these states was smaller than expected. One possible explanation was that an additional nuclear s-wave repulsion, which is also indicated by scattering data, may expel the pion from the nuclear interior, thereby reducing the pion absorption probability. However, the data base was too small to test this and other hypothesis.

The use of high-resolution, Compton-suppressed solid state γ-ray detectors[1-7] and the availability of intense π^--beams at pion factories have yielded much more accurate data on pionic atoms than previously achieved, both for the peripheral and low lying states. Given this extensive and accurate data base, we have performed a new fit of the optical potential parameters for pionic atoms. As the nuclei range from N=Z nuclei to nuclei with a large neutron excess, we have extended the standard parametrization to allow for the isospin dependence of the pion absorption parameters. Of course, extending the number of parameters always will lead to an improved fit. As we will show below, by carefully choosing the fitting strategies, a rather reliable new parameter set can be obtained that yields an isospin dependent absorption for s- and p-wave.

Below, we briefly outline the previous parametrization of the optical potential for pionic atoms and describe our extended model. The results of various fits to a total of 140 pionic atom data points will be discussed (see also Taal[1]) and we give our 'best' new set of phenomenological parameters. At the end we briefly summarize our results and put them into perspective.

THE OPTICAL POTENTIAL

The optical potential, used to describe pionic atoms and low-energy pion-nucleus scattering has two components: one part is derived from multiple scattering theory, the other is purely phenomenological and includes π-absorption. By assuming that only the elementary s- and p-wave πN-scattering is important, one can sum the repeated scattering of the pion from the target nucleons, yielding a real potential of the form

$$U(\mathbf{r}) = -\frac{4\pi}{2\overline{m}}\left\{ q(\mathbf{r}) - \nabla\cdot\frac{\alpha(\mathbf{r})}{1 + 4/3\pi\xi\alpha(\mathbf{r})}\nabla\right\} \quad , \qquad (1)$$

Here m is the reduced pion-nucleus mass and

$$q(\mathbf{r}) = (1 + \frac{m_\pi}{m_N})\{ b_0\rho(\mathbf{r}) + b_1\, \delta\, \rho(\mathbf{r}) \} \quad , \tag{2a}$$

$$\alpha(\mathbf{r}) = (1 + \frac{m_\pi}{m_N})^{-1}\{ c_0\rho(\mathbf{r}) + c_1\, \delta\, \rho(\mathbf{r}) \} \quad , \tag{2b}$$

where we have defined

$$\rho(\mathbf{r}) = \rho_n(\mathbf{r}) + \rho_p(\mathbf{r}) \qquad \text{and} \qquad \delta\rho(\mathbf{r}) = \rho_n(\mathbf{r}) - \rho_p(\mathbf{r}). \tag{3}$$

The term involving $q(\mathbf{r})$ in eq. (1) is due to the elementary π-nucleon s-wave interaction and b_0 and b_1 represent the in medium isoscalar and iso-vector scattering lengths, respectively. Analogously, the term $\alpha(\mathbf{r})$ involves the elementary p-wave scattering volumes, c_0 and c_1, respectively. However, this term enters non-linearly into the potential as shown by Ericson and Ericson[8]. The parameter ξ depends crucially on the assumed range $r_{\pi N}$ of the π-N interaction and r_c the nucleon-nucleon correlation length of the nuclear ground state. Assuming both to be of zero range, as was done in the original derivation of this non-linear term, yields $\xi=1$, but $r_c \approx r_{\pi N}$ yields $\xi << 1$.

In the optical potential of eq. (1) an important dynamical aspect of the pion-nucleus interaction is still absent: π-absorption. To make the optical potential suitable to describe pion absorption, parameters must be added which reflect the absorption mechanism in the nucleus. One expects that absorption on a single nucleon is highly suppressed due to energy and momentum conservation. If a stopped pion is to be absorbed on a single nucleon, the nucleon must have an initial momentum of roughly 500 MeV/c. The Fermi momentum-distribution makes this process rather unlikely to occur. One therefore includes a phenomenological term proportional to the square of the nuclear density to account for possible absorption processes such as

$$\pi^- + (pp) \rightarrow pn \qquad \text{and} \qquad \pi^- + (pn) \rightarrow nn$$

It is, however not clear that indeed only two nucleons participate in the absorption process[9-12]. One therefore extends the s- and p-wave terms in eq. (1) according to

$$q(\mathbf{r}) \rightarrow \quad q(\mathbf{r}) = (1 + \frac{m_\pi}{m_N})\{ b_0\rho(\mathbf{r}) + b_1\, \delta\rho\, (\mathbf{r})\} + (1 + \frac{m_\pi}{2m_N})\ B_0\rho^2(\mathbf{r}) \quad , \tag{4a}$$

and

$$\alpha(\mathbf{r}) \rightarrow \quad \alpha(\mathbf{r}) = (1 + \frac{m_\pi}{m_N})^{-1}\{ c_0\rho(\mathbf{r}) + c_1\, \delta\rho(\mathbf{r}) \} + (1 + \frac{m_\pi}{2m_N})^{-1}\ C_0\, \rho^2(\mathbf{r}) \quad , \tag{4b}$$

where B_0 and C_0 are two complex constants fit to the experimental information.

From experiments it is well known that there is dependence of the isospin of the two nucleons involved in the process. Experiments on multi-nucleon removal after pion absorption reveal that the main process is pion absorption on a correlated nucleon pair, although pion absorption on a cluster of more than two nucleons is not much less important. Measurements to reveal the absorption mechanism are strongly hindered by the multi-nucleon knock-out due to final state interactions. An important quantity is the ratio R of np to pp pairs that can absorb the pion. Experimentally, the value of R is not determined unambiguously. This is illustrated for instance by the different values of R for ^{12}C. Nordberg et al.[13] obtained R = 2.5±1.0 whereas Ozaki et al.[14] found a value of R = 5.0±1.5 and Lee et al.[15] R = 8.8±1.3. The value for R should, statistically speaking, be given by the ratio 2N/(Z-1), for a nucleus with Z protons and N neutrons. Since the experimental values for R are significantly larger than the statistical expectation value, the pion absorption must occur preferentially on a proton in a strongly correlated np-pair, called quasi-deuteron absorption(QDA).

In an analogous way as for the πN-scattering amplitude, a low-energy π2N-amplitude in the zero range approximation can be constructed[8]. This parametrization is incomplete since it is assumed that the two nucleons involved are in a relative s-state and, therefore, certain possible terms are absent. It is in fact a phenomenological two-nucleon amplitude wherein the constants simulate all the information on the short ranged absorption process and even short ranged nucleon pair correlations. The parameters are complex numbers with the imaginary part accounting for the width of the bound pion level.

As already mentioned, the absorption on a cluster of nucleons is also of importance. The process of pion absorption is still not fully understood. A study of ten absorption channels with stopped pions on ^{6}Li by coincidence detection of two outgoing nucleons was performed by Dörr et al.[9]. The quasi-deuteron process was found to be responsible for 72.5 % of the absorption strength while Isaak et al.[10] found a percentage of 51%. Three other channels of somewhat less importance are characterized by the outgoing particles n-p, n-d and n-t, each having a strength of about 10%. The six channels involving the coincidence of two charged particles leaving the nucleus after pion absorption are weak (< 1%). For heavy nuclei, the QDA may be the dominant process again. This is confirmed for instance by Isaak et al.[11] and Shinohara et al.[12] who could explain the experimental data after stopped π^- absorption in Ni-isotopes and ^{209}Bi, respectively, to a considerable extent, using quasi-deuteron absorption in an extremely peripheral region of the nucleus. The cluster absorption channels cannot be assumed to be governed by the parameters mentioned so far since these reactions should be represented in the optical potential by terms proportional to $\rho^3(\mathbf{r})$ for 3N-absorption and to $\rho^4(\mathbf{r})$ for α-cluster absorption and so on. A strong channel, therefore, is the 'quasi-deuteron' mechanism on a pn (T=0) pair. To allow for this isospin dependence, we therefore have extended the previous parametrizations, which were based on two absorption parameters B_0 and C_0, by using

$$q(\mathbf{r}) = (1+\frac{m_\pi}{m_N})\{b_0\rho(\mathbf{r}) + b_1\,\delta\rho(\mathbf{r})\} + (1+\frac{m_\pi}{2m_N})\{B_0\rho^2(\mathbf{r}) + B_1\rho(\mathbf{r})\,\delta\rho(\mathbf{r})\}, \tag{5a}$$

and

$$\alpha(\mathbf{r}) = (1+\frac{m_\pi}{m_N})^{-1}\{c_0\rho(\mathbf{r}) + c_1\,\delta\rho(\mathbf{r})\} + (1+\frac{m_\pi}{2m_N})^{-1}\{C_0\,\rho^2(\mathbf{r}) + C_1\rho(\mathbf{r})\,\delta\rho(\mathbf{r})\}, \tag{5b}$$

Several authors do not take the renormalization due to the Lorentz-Lorenz effect into account for the p-wave part of the 2N-interaction, since it is not clear how this phenomenon should occur in the potential. In those fits the C_0 and C_1 terms are not incorporated into the Lorentz-Lorenz term. Isovector terms in the absorption, represented by B_1 and C_1, are mostly neglected in the fits to pionic atom data.

De Laat[3] showed that an improved fit of the deviating 3d-shifts and widths could be achieved by adding absorption isovector terms ImB_1 and ImC_1. In this proceedings, see fig.7 of Konijn et al.[28], we show the result of the experiment on pionic ^{208}Pb as performed by De Laat. The theoretical value according to the parameter set by Batty et al.[16] is indicated by the double square in the upper right corner of this figure. By taking the neutron density different from the proton density as given by Hoffman et al.[17] and by Angeli et al.[18], the experimental 3d-data of ^{208}Pb can be reached by adding isovector absorption terms ImB_1 and ImC_1 different from zero to the standard optical potential[28]. The conviction that the standard optical potential needs the extension discussed, is supported by similar substantial improvements[3,4] obtained for ^{181}Ta, ^{nat}Re, ^{nat}Pt, ^{197}Au and ^{209}Bi.

Several remarks should be made about the absorption part of the optical potential and the extended form we use here.

i. First, it is of course possible to improve the fit of a given data set by introducing four new parameters (the two complex constants B_1 and C_1). Therefore, the strategy of fitting these constants should be outlined carefully, by e.g. fitting the isoscalar parameters to Z=N nuclei only. Then, in a second step, investigating the change of the parameters for the local interaction in a fit pionic 1s orbits and observing how these parameters change when introducing the non-local p-wave interaction terms with its parameters.

ii. The ImB_0, ImB_1, ImC_0 and ImC_1 terms represent the absorptive part of the potential. However, it is not quite clear what influence (in second and higher order) the ReC_0 and ReC_1 have as these occur in the velocity dependent part of the potential.

iii. In theoretical calculations the parameter c_p for the proton and c_n for the neutron Fermi distributions in the nuclei have been taken equal. For light nuclei the differences between proton and neutron density distributions are indeed small and hardly influence the parameters in these nuclei. De Laat et al.[3,4] showed that by using experimental Δc_{np}–values (or values obtained from Hartree-Fock calculations) the agreement with experiment improved, although it still was not satisfactory.

iv. To calculate the various pionic atom orbits the optical potential is used in a modified Klein–Gordon equation that entails the π-nuclear interaction to be treated in a non-relativistic way. The measurable quantities Γ_0 and ε_0 are then compared with the calculated ones by correcting for higher order terms like vacuum polarization, orbital electron screening, Lamb shift, nuclear polarization and relativistic reduced mass effects[1-4]. The strong interaction shift ε_0 always includes finite size effects as we compare the transition energies with the values from Coulomb corrected nuclear point values. The experimental shifts and widths of the pionic transitions have all been corrected for the radiative transition probability (width) and the strong interaction shift and width of the initial upper atomic level.

SHIFTS AND WIDTHS OF PIONIC ENERGY LEVELS

The energy levels of a pionic atom are obtained by the bound-state eigenvalues of the modified Klein-Gordon equation of the form ($\hbar=c=1$)

$$\left\{ \Delta + [(E-V^{C}(\mathbf{r}))^2 - \bar{m}^2] \right\} \psi(\mathbf{r}) = 2\bar{m} U(\mathbf{r}) \psi(\mathbf{r}) \quad , \tag{6}$$

where $V^C(r)$ is the Coulomb potential, E the pion binding energy including the pion rest mass m, $\bar{m}=m[1+(m/M)]^{-1}$ the reduced pion mass with M the nuclear mass and $U(\mathbf{r})$ the potential, describing the π-nuclear interaction as given e.g. by eq.(1).This modified form of the Klein-Gordon equation entails that the π-nuclear interaction can be treated in a non-relativistic way. Whether such an approach is justified depends on the values of the optical potential (see for some comments and further references Giovanetti et al.[19]).

Since the optical potential $U(\mathbf{r})$ contains complex parameters, the eigenvalues of the Klein-Gordon equation (6) for a given pionic orbit (n, l) are also complex

$$E_{nl} = \mathrm{Re}(E_{nl}) + i\,\mathrm{Im}(E_{nl}) = \mathrm{Re}(E_{nl}) - \frac{i}{2}\Gamma_{nl} \quad , \tag{7}$$

where the imaginary part of the eigenvalue is equated to the width, Γ_{nl} , for the given orbit. The level energy shift caused by the strong interaction, ε_{nl} , is defined as the difference between the energy in the field of a point charge Ze and the energy in the combined Coulomb field of the extended nuclear charge distribution $\rho_p(\mathbf{r})$ and the optical potential

$$\varepsilon_{nl} = E^{C}_{nl}\,(\text{point charge Ze}) - \mathrm{Re}(E_{nl}) \quad . \tag{8}$$

Using this convention a negative value for ε_{nl} corresponds to a repulsive shift. The shift so defined contains strong interaction effects as well as the finite size effect. In order to compare the theoretically calculated shift ε_{nl} (for a given set of optical potential parameters) with the value deduced from experiments, we must subtract the electromagnetic contribution from the measured energy value E^{exp} for the pionic transition between the levels with quantum numbers $(n+1,l+1)$ and (n,l). Since the effect of the strong interaction on an upper atomic level is usually down by at least two orders of magnitude with respect to the lower level of the transition, the shift ε_{nl} is related to the measured energy of the corresponding transition by the following relation

$$\varepsilon_{nl} = E^{exp}(n+1,l+1 \rightarrow n,l) - (E^{C} + E^{Cor}) \quad , \tag{9}$$

where E^C is the transition energy according to the point nucleus approach and E^{Cor} the other contributions to the pionic transition energy. These other contributions due to effects like vacuum polarization, orbital electron screening, nuclear polarization etc. are discussed by Taal[1] and de Laat[3].

PARAMETER FITS

The following strategy was used to fit the optical potential parameters to the experimental values for the strong interaction level shifts and widths obtained in pionic atoms. First a fit was made of the s-wave parameters only using the experimental values for the 1s-shifts and widths as input. To investigate the dependence of the p-wave interaction on the calculated s-wave data we performed a fit with all the p-wave parameters equal to zero. The data on all the Z=N nuclei were then used to determine the isoscalar parameters, while the isovector ones were fitted to the whole set of data, keeping $\xi=1$. We will then proceed by making a number of fits imposing different constraints among the various free parameters. Finally, fits will be made in which all the parameters are set free to vary, one fit keeping $\xi=1$ fixed and another where ξ is free to vary.

In order to make the comparison of the calculated shifts and widths with the experimental values meaningful, we have in all cases taken the uncertainties in the shifts to be ≥5% of the measured widths. The resulting errors are never smaller than the ones given in the corresponding original publications. The reason for doing this lies in the fact that one is never certain that the broadened lines do not contain contaminant lines that affect the actually measured energy to some extent.

Taal[2] gives in his appendix A a listing of the experimental and fitted values for all the 140 pionic atom level shifts and widths, used in the present fit. For the most recent values we also refer to the paper presented at this workshop on the study of deep lying pionic orbits. The 1s-level shifts are all negative, all higher level shifts positive except for the recently measured shifts of deeply bound pionic 2p-orbits[1,2] in ^{93}Nb and ^{nat}Ru. This means that the pion experiences a repulsive interaction in the 1s-orbit but an attractive one in higher orbits. Therefore, the parameters b_0, b_1 and ReB_0 are generally negative, as these parameters mainly determine the 1s-level shifts in the calculation. Furthermore, the shift values increase steadily with the atomic mass A and additional isotopic effects seem to be hardly present. One can expect that the addition of neutrons increases the probability for scattering on nucleons, causing an increase in the shift. The values for the width are more sensitive to isotopic effects. The widths for ^{13}C, ^{18}O and ^{22}Ne are significantly lower than those for the isotopes ^{12}C, ^{16}O and ^{20}Ne, respectively. Here the statistical ratio of pn to pp pairs (see remarks above) increases when the number of neutrons increases, which would lead to the expectation of a corresponding increase of the level width; this is in contrast to the experimentally observed decrease. Considering the pionic 2p-data the shifts for the Ca-isotopes show a clear isotopic effect as do the widths of the Cr-isotopes. As already predicted by Krell and Ericson[20] the effective potential predicts a change of the attractive p-wave interaction to a repulsion for deeply bound pionic 2p-orbits in the region of $Z\approx36$. This is illustrated by the recently measured 2p-shifts[1,2] in $^{93}_{41}$Nb and $^{nat}_{44}$Ru for which values were found of $\varepsilon(2p)= -11\pm3$ keV and -48 ± 7 keV, respectively. Considering the 3d-widths of the heavier (^{181}Ta, ^{nat}Re, ^{nat}Pt, ^{197}Au, ^{208}Pb, ^{209}Bi) elements, they are clearly overestimated by at least 35% for all the earlier

1

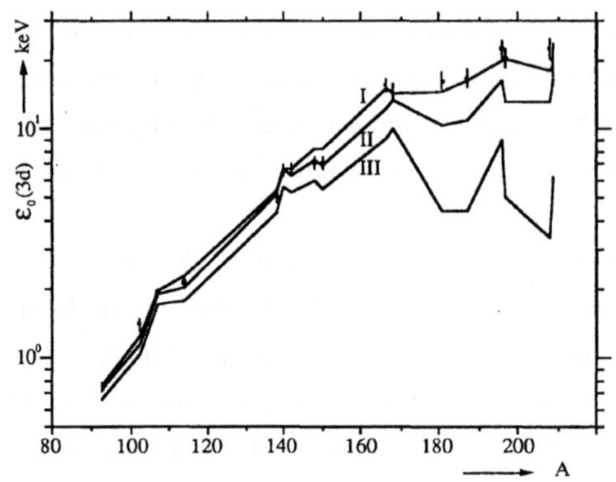

2

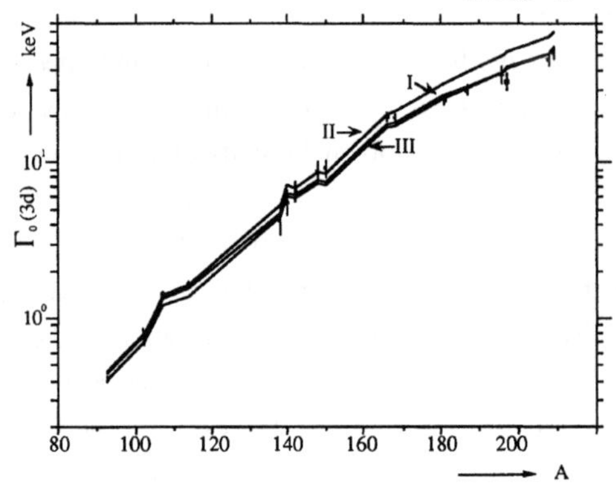

Figures 1 and 2.The strong interaction shifts and the strong interaction widths of the pionic 3d-level are plotted as a function of A, the atomic mass. The curves are the result of various fits of the strong interaction parameters in the pion-nucleus optical potential to pionic atom data. Curve I represents the new fit I to 140 pionic atom data points (see table 1). The new fit includes, contrary to earlier fits, isovector πNN-scattering terms in the strong interaction and is able to describe the 3d-level shifts and widths simultaneously in a consistent way. For comparison the curves representing the calculated values from the parameter sets otained by Batty et al.[16](curve II) and Seki and Masutani [21](curve III) are shown.

parameter sets. The 3d-shifts for the same elements are underestimated by the same parameter sets.

To illustrate the deviations between experimental and calculated values, the quadratic differences, weighted by the experimental uncertainties, $Q^2=(y^{exp}-y^{calc})^2/\sigma^2_{exp}$, for the shifts and the widths of each level are calculated according to each fitted parameter set.

Fit I given in table 1 is the final fit to the full pionic atom data base (140 data points) leaving the 11 s-wave and p-wave parameters free and is presented as our best overall fit with the parameter $\xi=1$. If we compare the quality of this fit with the performance of the set by Batty et al.[16], we observe that a significant improvement has been obtained in matching the pionic strong interaction level shifts and widths with experiment. The improvement is clearly shown in figs. 1 and 2, where we compare the experimental and calculated strong interaction 3d-shifts and widths for various parameter sets.

Of course, the pionic 1s and 2p-level shifts and widths can be somewhat better described by an optical parameter set that is fitted to 1s and 2p-data only, giving χ^2 values of 55 and 99 for the 1s- and 2p-data, respectively. The performance of this set, however, is much worse for the 3d-levels, giving a χ^2 value of 146, compared to the value of 72 for the best fit (fit I).

For the best fit case, only the s-wave parameter ImB_1 has changed significantly due to the inclusion of the 3d-levels in nuclei with large neutron densities ($\delta\rho(\mathbf{r})\neq0$). All the other s-wave parameters still have values similar to those deduced from the pionic 1s-level shifts and widths. Considering the p-wave parameters, these are all fitted with errors that are a factor of two smaller as compared to the fit on the s-p data only. The value of ReC_0 is not significantly different from zero, which is affirmed by a 10-parameter fit of the same quality with this parameter set to zero (see fit II in table 1).This indicates that the 'anomalous' 3d-level shifts are better described by adding the p-wave πNN-isovector parameter ReC_1. Since the p-wave parameter ImC_1 has not significantly changed compared to the 1s-2p-value, the conclusion arises that the deviating 3d-level widths are better matched with the experimental data due to a s-wave πNN-isovector parameter ImB_1 in the optical potential. The large negative value for this parameter indicctes an extra repulsion. Figs.1 and 2 show the pionic 3d-level shifts and widths, respectively, according to the calculations of our best fit parameter set (curve I) and the sets by Batty et al.[16] (curve II) and Seki and Masutani[21] (curve III) compared to the experimental values.

Leaving the ξ-parameter in front of the Lorentz-Lorenz term free to vary results in a fit of the same quality. In this case (fit VII in table 1) the value of ReC_0 is found to be zero. Such an 11-parameter fit gives the lowest χ^2-value of 256 for the 140 data points considered with the parameter $\xi = 2.4\pm0.4$. Having a closer look to the results of the best fits, given as fits I, II and VII, we find excellent agreement with the measured values with χ^2_{total} of 261, 266 and 256, respectively. This results in a value of $\chi^2 = 1.83$ per data point for fit VII.

FITS USING OTHER CONSTRAINTS

In a fit approach to adjust the isoscalar parameters b_0, ReB_0, ImB_0, c_0, ReC_0 and ImC_0

Table 1. Fits to all 140 pionic atom data points [28-1s, 58-2p, 34-3d and 20-4f] with and without various constraints all with ReB_1=0.

	fit I 11-par.fit ReB_1=0	fit II 10-par.fit $ReB_1=ReC_0$=0	fit III 11-par.fit,Z=N nuclei ReB_1=0	fit IV 10-par.fit $ReB_1=ImC_1$=0	fit V 9-par.fit $ReB_1=ImC_1=ReC_0$=0	fit VI 10-par.fit $ReB_0=-ImB_0$	fit VII 11-par.fit with ξ free $ReB_1=ReC_0$=0	
χ^2_{total}: =	**261**	**266**	**285**	**301**	**305**	**284**	**256**	
b_0: =	0.002 ± 0.005	0.006 ± 0.004	-0.003 ± 0.005	0.010 ± 0.005	0.013 ± 0.004	-0.0177± 0.0003	0.000 ± 0.005	m_π^{-1}
b_1: =	-0.073 ± 0.005	-0.075 ± 0.005	-0.076 ± 0.005	-0.077 ± 0.005	-0.078 ± 0.005	-0.068 ± 0.005	-0.071 ± 0.005	m_π^{-1}
ReB_0: =	-0.153 ± 0.026	-0.176 ± 0.021	-0.123 ± 0.023	-0.195 ± 0.025	-0.210 ± 0.021	-0.0533± 0.0010	-0.140 ± 0.026	m_π^{-4}
ImB_0: =	0.0569 ± 0.0014	0.0573± 0.0013	0.0546 ± 0.0014	0.0548± 0.0011	0.0546± 0.0011	0.0533	0.0566 ± 0.0015	m_π^{-4}
ReB_1: =	0.	0.	0.	0.	0.	0.	0.	m_π^{-4}
ImB_1: =	-0.215 ± 0.024	-0.207 ± 0.024	-0.244 ± 0.023	-0.115 ± 0.011	-0.109 ± 0.010	-0.239 ± 0.022	-0.211 ± 0.023	m_π^{-4}
c_0: =	0.29 ± 0.03	0.247 ± 0.003	0.35 ± 0.03	0.269 ± 0.024	0.243 ± 0.003	0.371 ± 0.020	0.35 ± 0.04	m_π^{-3}
c_1: =	0.41 ± 0.07	0.50 ± 0.06	0.32 ± 0.07	0.38 ± 0.07	0.42 ± 0.06	0.35 ± 0.07	0.42 ± 0.09	m_π^{-3}
ReC_0: =	-0.16 ± 0.10	0.	-0.34 ± 0.10	-0.09 ± 0.08	0.	-0.41 ± 0.07	0.	m_π^{-6}
ImC_0: =	0.107 ± 0.004	0.103 ± 0.004	0.108 ± 0.005	0.119 ± 0.004	0.117 ± 0.003	0.113 ± 0.004	0.26 ± 0.08	m_π^{-6}
ReC_1: =	-1.5 ± 0.3	-1.8 ± 0.3	-1.2 ± 0.3	-1.2 ± 0.3	-1.4 ± 0.3	-1.6 ± 0.3	-1.9 ± 0.6	m_π^{-6}
ImC_1: =	0.31 ± 0.06	0.32 ± 0.06	0.37 ± 0.05	0.	0.	0.33 ± 0.05	0.50 ± 0.12	m_π^{-6}
ξ: =	1.0	1.0	1.0	1.0	1.0	1.0	2.4 ± 0.4	
$\chi^2_{per\ deg.freed.}$ =	2.02	2.05	2.21	2.31	2.33	2.18	1.99	
$Q^2(1s)$ =	59	62	72	58	63	79	56	
$Q^2(2p)$ =	100	105	112	104	105	103	94	
$Q^2(3d)$ =	72	71	68	75	74	74	70	
$Q^2(4f)$ =	29	28	33	64	63	28	36	

to the 26 available experimental 1s- and 2p-data points for Z=N nuclei, we first kept the isovector parameters b_1, ImB_1, c_1, ReC_1 and ImC_1 fixed. After b_0, ReB_0, ImB_0, c_0, ReC_0 and ImC_0 have reached their ultimate values in the Z=N fit, those values are kept fixed in the next step, i.e. in fitting the isovector parameters to all the 140 data points. Iterating this procedure yields the final values of this fit for b_0, ReB_0, ImB_0, c_0, ReC_0 and ImC_0 from Z=N nuclei and values for b_1, ImB_1, c_1, ReC_1 and ImC_1 determined from all the 140 data points. The parameters from this fit are also listed in table 1 as fit III and give a much better overall agreement with experiment especially with the 3d-data then the fit performed in a similar way but using the 1s- and 2p-data only. Finally, fits were made with the constraints $ReB_1=ImC_1=0$ (fit IV) and $ReB_1=ImC_1=ReC_0=0$ (fit V) which all gave χ^2_{total}-values (see table 1) only 8-11 % worse than our best fit.

FITS USING THE CONSTRAINT $ReB_0=-ImB_0$

Stricker et al. fitted both low energy pion-nucleus scattering[22] and pionic atom data[23]. They imposed the condition $ReB_0=-ImB_0$ as a restriction in their fits to the data used by Krell and Ericson[20] in 1969. We also made a fit including such a condition as a constraint. Such a 10-parameter fit to all the 140 data points, resulted in a fit with a χ^2_{total}-value of 284, see fit VI in table 1, which is only 11 % worse than the best fit to our data set. Dover[24] obtained from a T-matrix analysis $ReB_0 \approx ImB_0$ whereas Nishimoto et al.[25] calculated for the π-deuteron interaction $ReB_0 \approx 2ImB_0$.

CONCLUSIONS

An extension of the experimental data on strong interaction pionic 1s-levels with rather precise experimental values for shifts and widths is very difficult to obtain within a reasonable measuring period. This is clear from the work by Taal et al.[1,2]. Similar arguments hold for a possible extension of the pionic 2p-level data . In the case of pionic 3d and 4f-levels, the end of the periodic system for stable isotopes has been reached. Since for these elements no deviations between experiment and optical potential calculations have been observed, further measurements on weaker-bound pionic 3d-levels seem to be of minor interest. Suggestions by Kilian[26] to create a negatively charged pion in a deeply bound pionic orbit may in the future offer a possibility to further explore the π^--nucleus interaction in even more exotic pionic orbits.

The experimental results on strong interaction effects in pionic atoms have in recent years become significantly more accurate and the presently available data throughout the periodic table enable a rather detailed study of the optical potential. The phenomenological π^--nucleus optical potential apparently can quite successfully reproduce the experimental results for these strong interaction level shifts and widths. The optical potential parameter set, presented in this paper, results from a fit to 140 pionic atom data. The main improvement achieved in comparison with other standard optical potentials is the successful reproduction of pionic 3d-

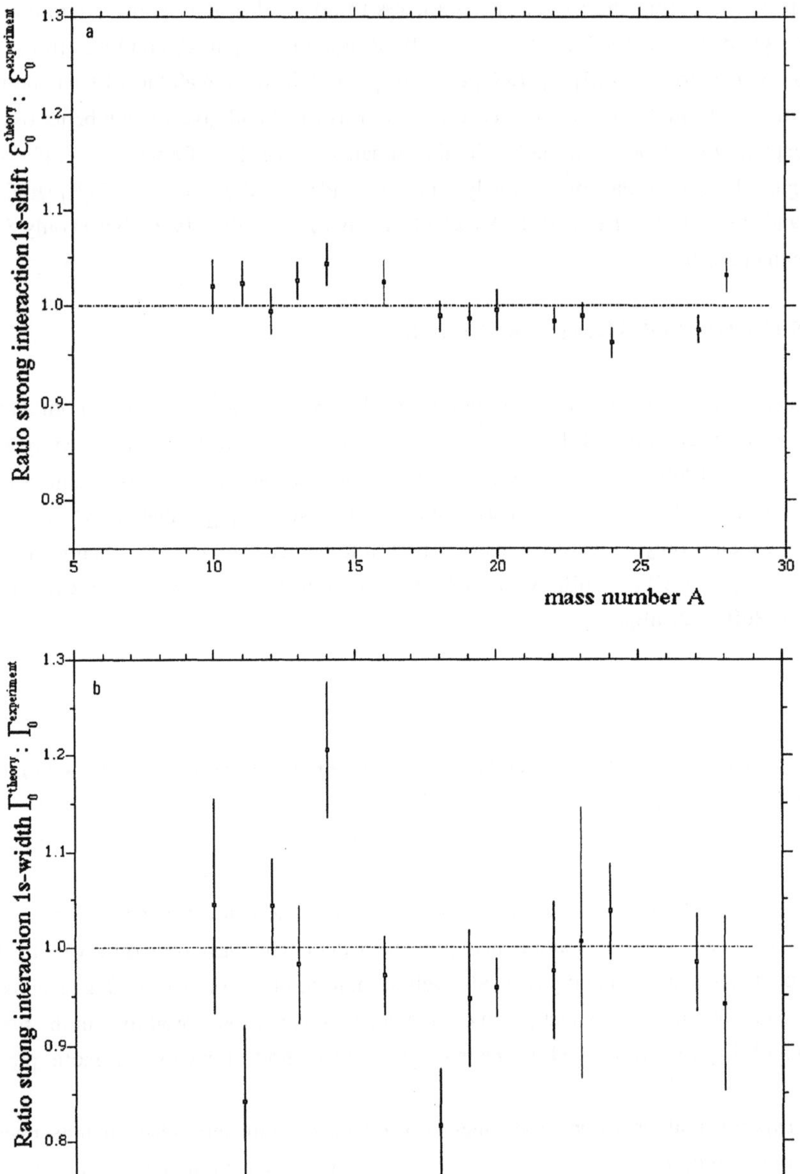

Fig.3 a and b. Pionic 1s-orbit ratios of theoretical and experimental values for the strong interaction shifts, $\varepsilon_0(1s)$, and widths, $\Gamma_0(1s)$, plotted as a function of the mass number A.

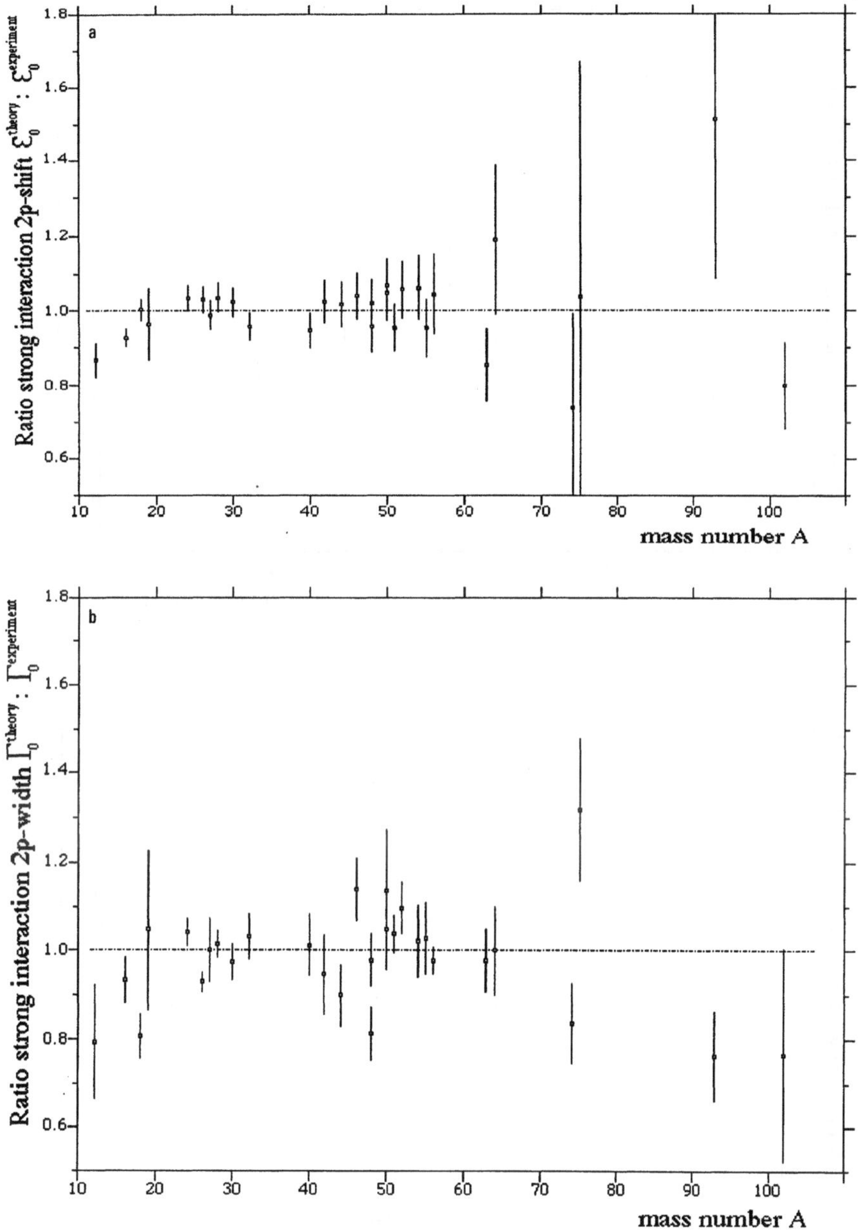

Fig.4 a and b. Pionic 2p-orbit ratios of theoretical and experimental values for the strong interaction shifts, $\varepsilon_0(2p)$, and widths, $\Gamma_0(2p)$, plotted as a function of the mass number A.

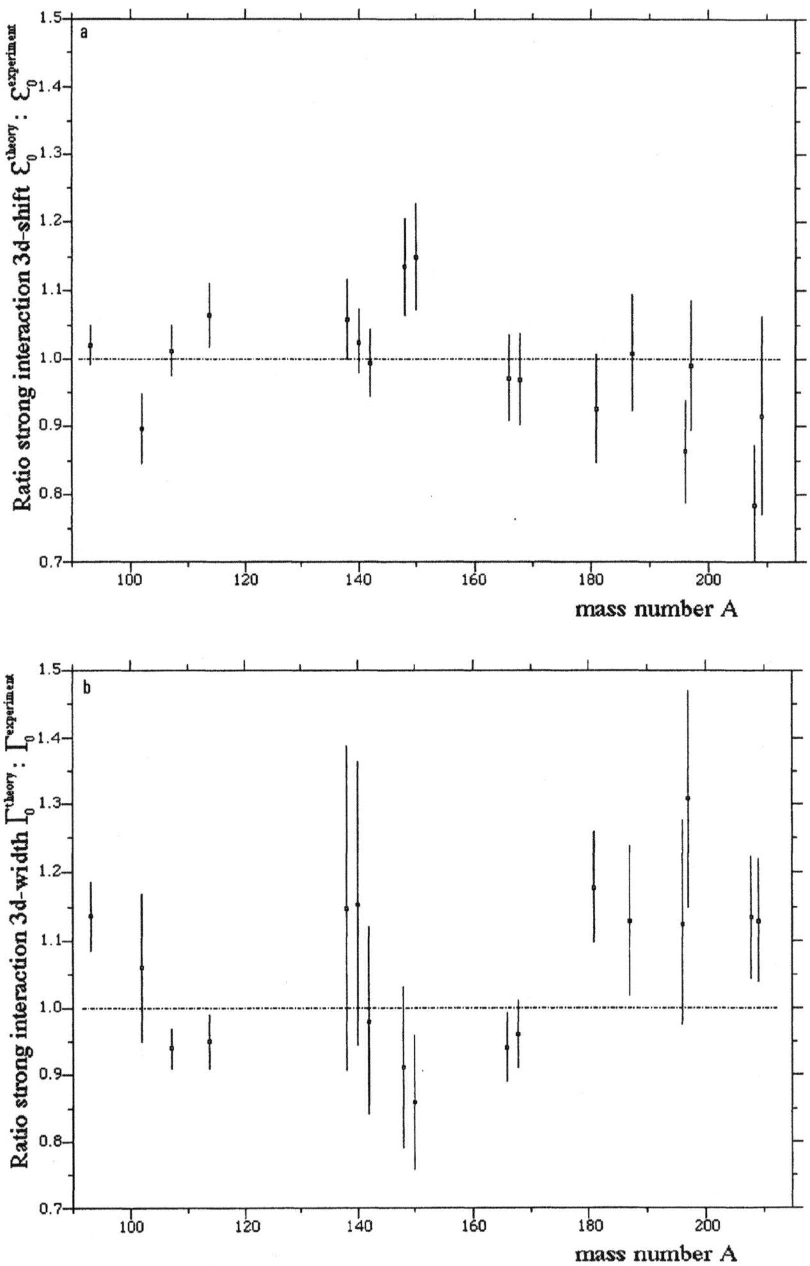

Fig.5 a and b. Pionic 3d-orbit ratios of theoretical and experimental values for the strong interaction shifts, $\varepsilon_0(3d)$, and widths, $\Gamma_0(3d)$, plotted as a function of the mass number A.

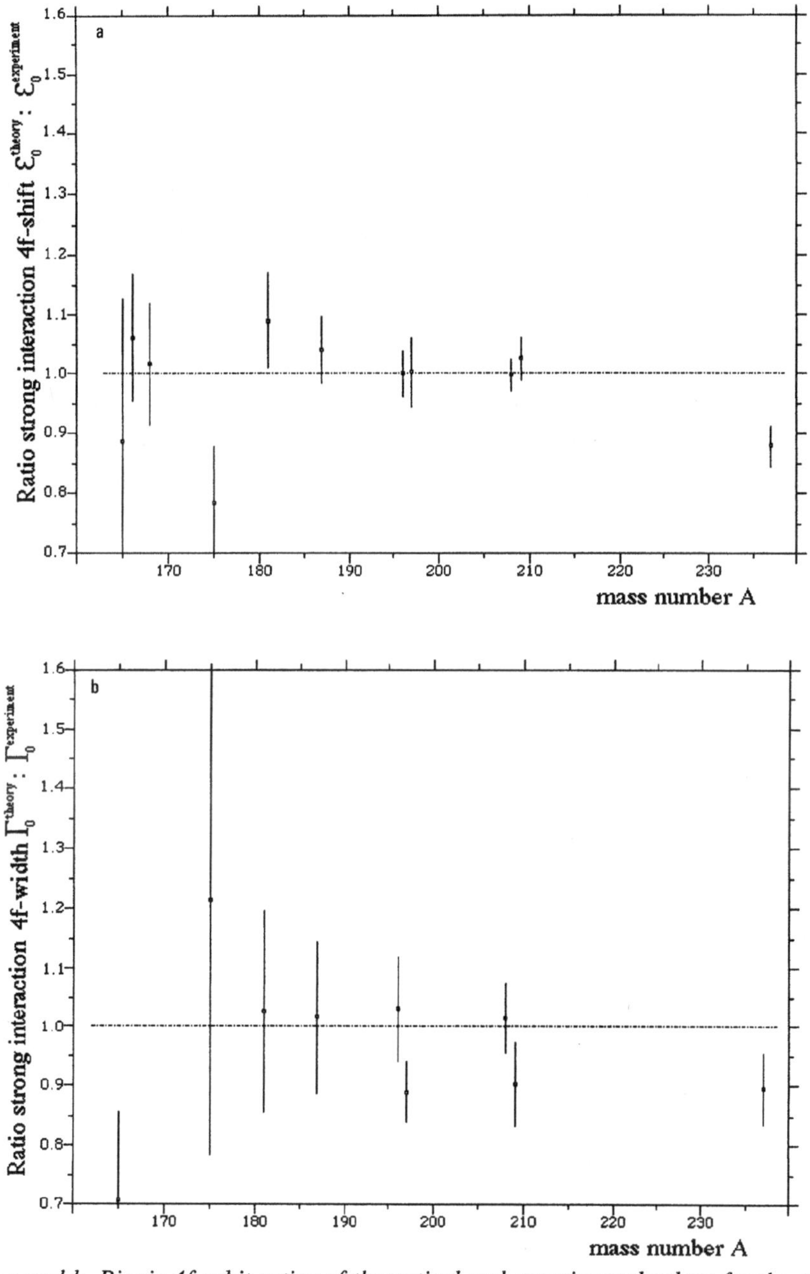

Fig.6 a and b. Pionic 4f-orbit ratios of theoretical and experimental values for the strong interaction shifts, $\varepsilon_0(4f)$, and widths, $\Gamma_0(4f)$, plotted as a function of the mass number A.

level shifts and widths simultaneously in heavy elements. So far, optical potential parameter sets could not reproduce these shifts and widths in elements like ^{181}Ta, ^{nat}Ru, ^{nat}Pt, ^{197}Au, ^{208}Pb and ^{209}Bi. The quality of our best fit with ξ=1 (fit I of table 1) is best illustrated by plotting (as a function of the mass number A) the ratios of calculated and experimental values, see figs. 3 through 6.

Summarizing the present finding is that by including the πNN-isovector terms with suitable strengths in the standard optical potential, both the 3d-level shifts and widths are much better reproduced. A large negative πNN-isovector term ImB_1 gave a satisfactory improvement of the calculated pionic 3d-level widths in these heavy elements. The negative value for ImB_1 implies an increased s-wave repulsion experienced by pions absorbed from these deeply bound 3d-levels. This result is more satisfactory than the ad hoc introduction of an extra hard core repulsion in the optical potential as proposed by Olivier et al.[27]. The pionic 3d-level shifts in the lighter elements are quite well reproduced by the extended optical potential. These observations, along with the fact that the best fit to pionic 1s and 2p-data only is not able to reproduce the 3d-data for deeply bound states, show that the extra repulsion needed to describe the 3d-data is not structural for deeply bound pionic orbits but is confined to these 3d-levels only. This is consistent with the fact that for a suitable choice of the πNN-isovector term also the 3d-data can be fairly well reproduced, since only heavy nuclei have a large neutron excess.

REFERENCES

1. A.Taal, Ph.D. Thesis, University of Technology Delft, March 1989
2. A.Taal, P.David, H.Hänscheid, C.T.A.M.de Laat, W.Lourens, C.Petitjean, F.Risse, Ch.F.G. Rösel, A. van der Schaaf, W. Schrieder and J. Konijn; to be published
3. C.T.A.M. de Laat, Ph.D. Thesis, University of Technology Delft, March 1988
4. C.T.A.M.de Laat, P.David, H.Hänscheid, W.Lourens, C.Petitjean, F.Risse, Ch.F.G. Rösel, A. van der Schaaf, W. Schrieder, A.Taal and J. Konijn; to be published
5. J.F.M. d'Achard van Enschut, J.B.R. Berkhout, W. Duinker, C.W.E. van Eijk, W.H.A. Hesselink, T. Johansson, T.J. Ketel, J.H. Koch, J. Konijn, C.T.A.M. de Laat, W. Lourens, G. Van Middelkoop and W. Poeser; Physics Letters **136B** (1984) 24
6. C.T.A.M. de Laat, A. Taal, W. Duinker, A.H. Wapstra, J. Konijn, J.F.M. d'Achard van Enschut, P. David, J. Hartfiel, H. Janszen, R. von Mutius, C. Gugler, L.A. Schaller, L. Schellenberg, T. Krogulski, C. Petitjean and H.W. Reist; Physics Letters **162B** (1985) 81
7. A. Olin, J.W.Forsman, J.A.Macdonald, G.M.Marshall, T.Numao, P.R.Poffenberger, P.van Esbroek, G.A.Beer, D.I.Britton, G.R.Mason, A.R.Kunselman and B.H.Olaniyi; Nuclear Physics **A439** (1985) 589
8. M. Ericson and T.E.O. Ericson; Ann. Phys. **36** (1966) 323
9. M. Dörr, W. Fetscher, D. Gotta, J. Reich, H. Ullrich, G. Backenstoss, W. Kowald and H.-J. Weyer; Nuclear Physics **A445** (1985) 557
10. H.P. Isaak, P. Heusi, H.S. Pruys, R. Engfer, E.A. Hermes, T. Kozlowski, U. Sennhauser and H.K. Walter; Helv.Phys. Acta **55** (1982) 477
11. H.P. Isaak, H.S. Pruys, R. Engfer, E.A. Hermes, F.W. Sclepütz, A. Zglinski, T. Kozlowski, U. Sennhauser and H.K. Walter; Nuclear Physics **A392** (1983) 385

12. A. Shinohara, A. Yokoyama, S. Moriyasu, T. Saito and H. Baba; Nuclear Physics **A456** (1986) 701
13. M.E. Nordberg, K.F. Kinsey and R.L. Burman; Phys.Rev. **165** (1968) 1096
14. S. Ozaki, R. Weinstein, G. Glass, E. Loh, L. Neimala and A. Wattenberg; Phys.Rev.Lett. **4** (1960) 533
15. D.M. Lee, R.C. Minehart, S.E. Sobottka and K.O.H. Ziock; Nuclear Physics **A182** (1972) 20
16. C.J. Batty, S.F. Biagi, E. Friedman, S.D. Hoath, J.D. Davies, G.J. Pyle and G.T.A. Squier; Phys.Rev.Lett. **40** (1978) 931
17. G.W. Hoffmann, L. Ray, M. Barlett, J. McGill, G.S. Adams, G.J. Igo, F. Irom, A.T.M. Wang, C.A. Whitten Jr., R.L. Boudrie, J.F.Amann, C. Glashausser, N.M.Hintz, G.S. Kyle and G.S. Blanpied; Phys.Rev. **C21** (1980) 1488
18. I. Angeli, M. Beiner, R.J. Lombard and D. Mas; Journ.of Physics **G6** (1980) 303
19. K.L. Giovanetti, G. de Chambrier, P.F.A. Goudsmit, H.J. Leisi, B. Jeckelmann and Th. Karapiperis; Physics Letters **186B** (1987) 9
20. M. Krell and T.E.O. Ericson; Nuclear Physics **B11** (1969) 521
21. R.Seki and K. Masutani; Phys.Rev. **C27** (1983) 2799
22. K. Stricker, H. McManus and J.A. Carr; Phys.Rev. **C19** (1979) 92
23. K. Stricker, H. McManus and J.A. Carr; Phys.Rev. **C22** (1980) 2043
24. C.B. Dover; Ann.Phys. **79** (1973) 441
25. K. Nishimoto, H. Ohtsubo and H. Narumi; Progr.Theor.Phys. **46** (1971) 135
26. K. Kilian; CERN-EP/85-17, and Proc.IUCF Workshop on Nuclear Physics "New possibilities with recoilless kinematics using high quality proton beams", Bloomington, Indiana, October 15-17, 1984
27. J.G.J. Olivier, M.Thies and J.H.Koch; Nuclear Physics **A429** (1984) 477
28. J. Konijn, C.T.A.M. de Laat, A. Taal, P. David, H. Hänscheid, F. Risse, Ch.F.G. Rösel, W. Schrieder, W. Lourens, C. Petitjean and A. van der Schaaf; "A study of the strong interaction effects of deeply bound pionic levels in pionic atoms" (This proceedings)

ADIABATIC INVARIANT OF RADIATIVE TRANSITIONS IN THE COULOMB FIELD

S.P. Alliluev and S.S. Gershtein

Institute for High Energy Physics
142284, Serpukhov
Moscow Region, USSR

I. The adiabatic invariants[1] turn out quite useful in describing nonconservative systems with slowly changing parameters. The examples of their existence are well-known. Thus, for an oscillator such adiabatic invariant is the ratio of the oscillator energy (E) its frequency (ω), which remains constant under slow changes of the latter (as compared with the period). If we take a charged particle moving in a weakly variable magnetic field (H), its adiabatic invariant is $P_\perp^2/H$ ($P_\perp$ is a particle momentum component, perpendicular to the field). In this Lecture we want to point out that such an adiabatic invariant does exist for the finite motion of a charged particle emitting electro-magnetic waves in the Coulomb field. The change of the energy (E), as well as of the momentum (L) of the particle due to electromagnetic waves radiation is described by the equations

$$-\frac{dE}{dt} = \frac{2e^2}{3c^3}(\ddot{\vec{r}})^2, \tag{1}$$

$$\frac{d\vec{L}}{dt} = \left[\vec{r},\ \vec{F}_r\right] \tag{2}$$

with $\vec{F}_r$ being the radiative friction force:

$$\vec{F}_r = \frac{2e^2}{3c^3}\dddot{\vec{r}}\ . \tag{3}$$

If we take into account the equations of motion for a particle of mass "m" and charge (-e) in the Coulomb centre field with the charge Ze

$$m\ddot{\vec{r}} = -\frac{Ze^2}{r^3}\vec{r} \tag{4}$$

equations (1), (2) will take the form

Electromagnetic Cascade and Chemistry of Exotic Atoms
Edited by L. M. Simons *et al.*, Plenum Press, New York, 1990

$$- \frac{dE}{dt} = \frac{2e^2}{3c^3}\left(\frac{Ze^2}{m}\right)^2 \frac{1}{r^4} , \tag{1'}$$

$$- \frac{d\vec{L}}{dt} = \frac{2Ze^4}{3m^2c^3}\vec{L}\frac{1}{r^3} . \tag{2'}$$

Since the changes in the energy and the momentum during the period of motion are very small,* here in eqs. (1'), (2') we can pass to the quantities averaged with respect to the period of motion (T) (i.e. the trajectory unperturbed during $t \sim T$): it is this approximation that is assumed so as to get adiabatic invariants. Calculations of the mean values of the quantities

$$\overline{\frac{1}{r^{\nu}}} = \frac{1}{T}\int_0^T \frac{dt}{r^{\nu}(t)} \tag{3}$$

reduce to the integrals

$$\overline{\left(\frac{1}{r^{\nu}}\right)} = \frac{1}{T/2}\int_{r_{min}}^{r_{max}} \frac{1}{r^{\nu} v_r(r)} dr = \frac{1}{T/2}\int_{r_{min}}^{r_{max}} \frac{dr}{r^{\nu}\sqrt{\frac{2}{m}(E+Ze^2/r-L^2/2mr^2)}} . \tag{4}$$

Here r_{min}, r_{max} are the points of stopping which are the roots of the equation:

$$E + \frac{Ze^2}{r} - \frac{L^2}{2mr^2} = 0. \tag{5}$$

The values of r_{min}, r_{max} are connected with the parameters of the ellipse along which our particle is moving (if radiative friction is neglected), as well as with the values of its semi-major (a) and semi-minor (b) axes:

$$r_{min} + r_{max} = 2a = \frac{Ze^2}{E} , \tag{6}$$

$$r_{min} \cdot r_{max} = b^2 = \frac{L^2}{2m\,E} \quad \text{(see fig.1).} \tag{6'}$$

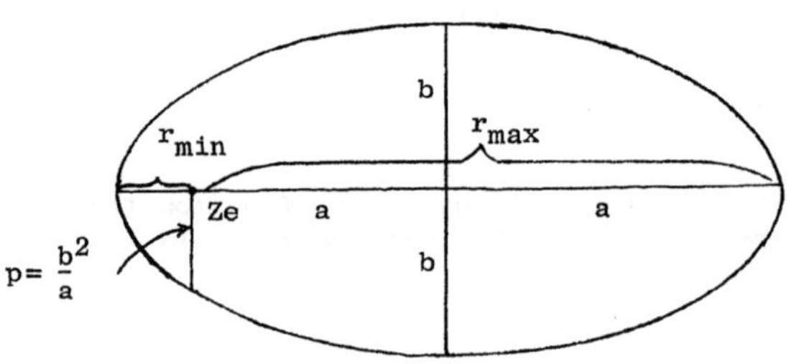

Fig. 1.

---

* The ratios $\Delta E/E$ and $\Delta L/L$ (where ΔE and ΔL are the changes of the energy and the momentum during one turnover, respectively) equal in the order of magnitude $(r_o/r)^{3/2}$ ($r_o = e^2/mc^2$ being a classical electromagnetic radius of the particle). Naturally, we imply $r \gg r_o$.

Taking into account that the period of motion is

$$T = \frac{2\pi a}{\sqrt{\frac{2}{m} E}},$$

one can express the mean value of eq. (4) as:

$$\overline{\left(\frac{1}{r^{\nu}}\right)} = \frac{1}{\pi a} \int_{r_{min}}^{r_{max}} \frac{1}{r^{\nu-1} \sqrt{(r-r_{min})(r_{max}-r)}} dr. \tag{7}$$

Writing down (7) as an integral in the complex plane along the contour enveloping the cut between the points r_{min} and r_{max}, one can reduce (7) at $\nu \gg 2$ to the residue at r=0. Finally, we obtain

$$\overline{\left(\frac{1}{r^{\nu}}\right)} = \frac{1}{ab^{\nu-1}} P_{\nu-2}\left(\frac{a}{b}\right) \tag{8}$$

with $P_{\nu-2}(x)$ being a Legendre polynomial. Hence,

$$\overline{\left(\frac{1}{r^{3}}\right)} = \frac{1}{b^{3}}, \quad \overline{\left(\frac{1}{r^{4}}\right)} = \frac{3a^{2}-b^{2}}{2ab^{5}}, \tag{9}$$

Making use of (6) and (9), we can reduce now eqs. (1'), (2') to the equations defining the changes of the parameters of the ellipse:

$$-\frac{da^{2}}{dt} = \frac{4Ze^{4}}{3m^{2}c^{3}} \frac{(3a^{2}-b^{2})a^{2}}{b^{5}},$$

$$-\frac{db^{2}}{dt} = \frac{2Ze^{4}}{3m^{2}c^{3}} \frac{(3a^{2}+b^{2})}{b^{3}}. \tag{10}$$

Dividing then one of equations (10) by the other, we come to a homogeneous equation:

$$\frac{db^{2}}{da^{2}} = \frac{(3a^{2}+b^{2})b^{2}}{2(3a^{2}-b^{2})a^{2}}, \tag{11}$$

that can be solved in an ordinary way by introducing a new variable

$$u = \frac{b^{2}}{a^{2}}. \tag{12}$$

Finally we come to

$$a = \text{const} \frac{(1-u)^{2/3}}{u}. \tag{13}$$

Thus, the adiabatic invariant turns out to be the combination

$$I = \frac{b^{2}}{a(1-b^{2}/a^{2})^{2/3}}, \tag{14}$$

which may be expressed via the eccentricity of the ellipse

$$\varepsilon = \sqrt{1-(b/a)^{2}}$$

and the parameter $p = b^{2}/a$

$$I = \frac{p}{\varepsilon^{4/3}} \tag{15}$$

Now, if we use the Runge-Lentz vector

$$\vec{A} = \frac{\vec{r}}{r} - \frac{1}{Ze^2}\left[\vec{V}, \vec{L}\right] \tag{16}$$

which conserves when a particle moves in the Coulomb field (with the modulus equal to the eccentricity of the ellipse and is directed along the principal axis) we can show that an account for emission does not change the direction of vector (16). It means that the space position of the principal axis of the ellipse will remain the same for the case when a moving particle is emitting electromagnetic waves.

II. In the Coulomb field with large quantum numbers, the quantum mechanical motion is quasi-classical, and the parameters of a classical orbit can be expressed via the orbital (ℓ) and the principal (n) quantum numbers. Assuming

$$E = -\frac{Z^2e^4m}{2h^2n^2}, \quad L^2 = h^2(\ell + 1/2)^{2*} \tag{17}$$

one obtains according to (6), (6') that

$$a = \frac{h^2n^2}{mZe^2}, \quad b = \frac{h^2}{mZe^2}n(\ell + 1/2),$$
$$\varepsilon = \sqrt{1 - \frac{(\ell + 1/2)^2}{n^2}} . \tag{18}$$

The radiative transition invariant here takes the form:

$$\frac{(\ell + 1/2)^2}{\left[1 - \frac{(\ell + 1/2)^2}{n^2}\right]^{2/3}} = \text{const.} \tag{19}$$

Through simple transformations it can be brought to a more convenient form:

$$\frac{1}{\ell + 1/2}\left[\frac{1}{(\ell + 1/2)^2} - \frac{1}{n^2}\right] = \text{const.} \tag{20}$$

As far as we know, the existence of a classical (or quasi-classical) invariant of radiative transitions (14), (15), (19), (20) in the Coulomb field has never been announced. (Although, its existence could be traced back, e.g., to the formula given in the Example of the Section "Retardation by Emission" from the book "Field Theory"[2] by L.D.Landau and E.M.Lifschitz.

Note, for the case of a three-dimensional oscillator eqs. (1), (2) lead to the presence of the invariant L/E = const. Then the eccentricity of the ellipse, along which a charged particle moves, remains constant provided there is some radiative friction (simultaneously, its axes decrease in an exponential way).

*The substitution $L^2 = h^2(\ell + 1/2)^2$ corresponds to the well-known Langer transformation, responsible for the correct behavior of a quasi-classical wave function at $r \to 0$. This guarantees, in its turn, a proper accuracy of the quasi-classical approximation up to $\ell = 0$.

III. Since motion along highly excited orbits with large quantum numbers, $n \gg 1$, $\ell \gg 1$, is quasi-classical, invariant (20) can be applied for evaluation of the general picture of the electromagnetic cascade from the viewpoint of high levels. Since in classical motion the radiative friction makes the momentum smaller, the transitions with $\Delta\ell = +1$, allowed by the quantum selection rules, must be considerably less probable than transitions with $\Delta\ell = -1$ (what one can see from the direct calculations, for instance[3]). For the transitions from the initial state with n_o, ℓ_o to the state with $\ell = \ell_o - 1$ invariant (20) permits to define the proncipal quantum number (n) of the most probable transition:

$$\frac{1}{n^2} = \frac{(\ell_o - 1/2)}{n_o^2(\ell_o + 1/2)} + \frac{3\ell_o^2 + 1/4}{(\ell_o^2 - 1/4)^2(\ell_o + 1/2)} . \tag{21}$$

For large quantum numbers eq. (21) yields

$$n = \frac{n_o}{\sqrt{1 + \frac{3n_o^2 - \ell_o^2}{\ell_o^3}}} . \tag{22}$$

Note, the accuracy of eq. (22) (which is more than 4%0 is preserved up to small $_o$ 2. If we assume now $_o = n_o - k$ (with 1 k n), it follows from (22) that

$$n = n_o - 1 - \frac{4k}{n_o} ,$$

which means that for the states luing near the circular orbit, $\ell_o = n_o - 1$ ($k < n_o/8$), most probable are transitions changing the principal quantum number by unity: $\Delta n = -1$. On the other hand for the states with the small ℓ_o from the inequality

$$\frac{1}{n^2} > \frac{3\ell_o^2 + 1/4}{(\ell_o^2 - 1/4)(\ell_o + 1/2)}, \quad n \geqslant \ell + 1 = \ell_o,$$

It follows that the most probable transition to circular orbits is performed with the maximum change in n, irrespective of the initial quantum number n_o. This is true for the states with $\ell_o = 1,2,3,4$, which turn directly (with great probability) into the 1s, 2p, 3d, 4f states. As an example, see Fig. 2 showing most probable transitions from the states with the principal quantum number $n_o = 60$ and various ℓ_o.

With the help of the invariant (20) it is easy to estimate the possibility for a particle to pass, as a result of radiative decays, from the level $n_o \gg n$ to the circular orbit n, $\ell = n-1$ if the distribution $f(\ell_o)$ is known. Obviously, all the states with $\ell_o \geqslant (\ell_o)_{min}$, where $(\ell_o)_{min}$ satisfies the conditions

$$\frac{1}{n-1/2}\left[\frac{1}{(n-1/2)^2} - \frac{1}{n^2}\right] = \frac{1}{((\ell_o)_{min}+1/2)}\left[\frac{1}{((\ell_o)_{min}+1/2)^2} - \frac{1}{n_o^2}\right]$$

must pass to this circular orbit. If n $n_o^{3/4}$ then with a good accuracy

$$(\ell_o)_{min} \simeq n^{4/3}\left(1 - \frac{n^{2/3}}{3n_o^2}\right).$$

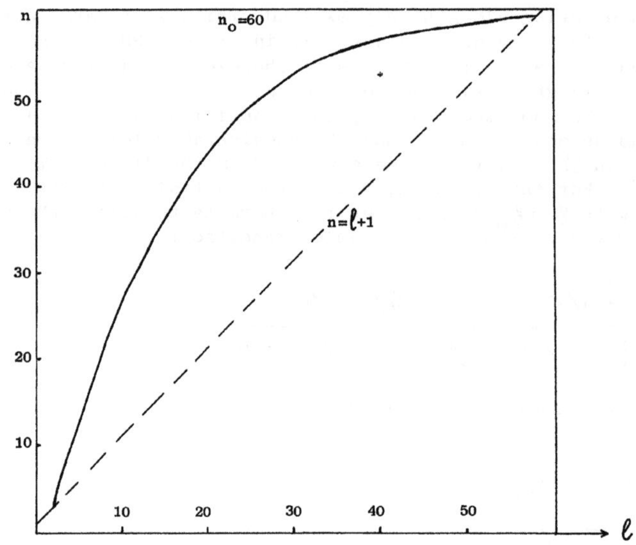

Fig. 2.

The probability of a transition to the circular orbit in the state with the principal quantum number n (the statistic distribution on the initial level being $f(\ell_o) = (2\ell_o + 1)/n_o^2$) is

$$W = 1 - \frac{((\ell_o)_{min} + 1)^2}{n_o^2} = 1 - \frac{n^{8/3}}{n_o^2} . \tag{24}$$

Then for $n_o = 60$ and $n = 7$, $(\ell_o)_{min} = 13$ and $W = 1 - 0.05 = 95\%$.

REFERENCES

1. L.D.Landau and E.M.Lifschitz, "Mechanics".
2. L.D.Landau and E.M.Lifschitz, "Field Theory".
3. H.A.Bethe and E.E.Salpeter, "Quantum mechanics of one-and two electron atoms", Springer-Verlag, 1957.

IV

MUON TRANSFER

Alice laughed. "There is no use trying," she said: "one *can't* believe impossible things."
"I daresay you haven't had much practice," said the Queen. "When I was your age I always did it for half-an-hour a day. Why, sometimes I've believed as many as six impossible things before breakfast."

Lewis Carroll: Through the Looking-Glass

MUON CAPTURE AND TRANSFER IN GAS MIXTURES WITH HYDROGEN

H. Schneuwly

Institut de Physique de l'Université
CH-1700 Fribourg (Switzerland)

INTRODUCTION

According to our present knowledge a negative muon, which is stopped in hydrogen gas of a few atmospheres, say 10 bar, forms a muonic hydrogen atom in an excited state and deexcites to the ground state or the metastable 2s state[1] in a time of about 10^{-10} s, which is shorter than the time resolution (a few nanoseconds) of current experiments dealing with atomic capture and transfer of negative muons.[2] The deexcitation takes place through dissociation of the collision partner, external Auger effect and radiative transitions. Even the metastable 2s state with a population[3] of about 8% has a short lifetime which is estimated to about 10-20 ns at a pressure of 10 bar.[1,4] Through collisions the triplet state should flip to a singlet state in a few nanoseconds.[5] Even if the capture and the thermalization processes are not very well known, one can assume that after a few tens of collisions the μp atoms are thermalized,[6] and after about 10 ns most μp atoms are in the singlet state.

The muonic hydrogen atom in its ground state can form a muonic pμp molecule with a rate $\Lambda_{pp} = 2.5\cdot10^6$ s^{-1}, can transfer its muon to deuterium ($\Lambda_d = 1.7\cdot10^{10}$ s^{-1}) or form a pμd molecule ($\Lambda_{pd} = 5.5\cdot10^6$ s^{-1}). Corresponding rates for excited μp states are not yet known. These rates, which are reduced to liquid hydrogen density, determine together with the muon decay rate $\lambda_o = 0.455\cdot10^6$ s^{-1} the lifetime of the muonic hydrogen atom μp in its ground state.

In hydrogen containing a small admixture of another gas, there is an additional disappearance channel. In collisions, the muon can be transferred from the muonic hydrogen atom to an excited state of the μZ atom.[7] In addition to this charge transfer mechanism, there exist, at least for very light elements like helium, another mechanism where, in an intermediate stage, a muonic molecule pμZ is produced in an excited state, which deexcites to the ground state μZ(1s) in very short time (10^{-12} s for helium).[8] In the case of the first mechanism, the muon-atomic states, to which the muon is transferred in the μZ atom, i.e. the initial (n,l) population in the μZ atom, can be predicted and has been found in agreement with the measured muonic x-ray intensities.[9,10]

The lifetime of the μp atom in the ground state has been recently measured in hydrogen-argon gas mixtures under different experimental conditions of pressure and relative concentration.[10,11] Lifetimes of the μp atom in the ground state as short as 40 ns have been measured. The reduced transfer rates to argon Λ_{pAr}, deduced from the exponential structure of the time spectra of the muonic argon x-rays, are in very good agreement with each other. These results give us confidence in our present understanding of the muon transfer process from the μp atom in its ground state.

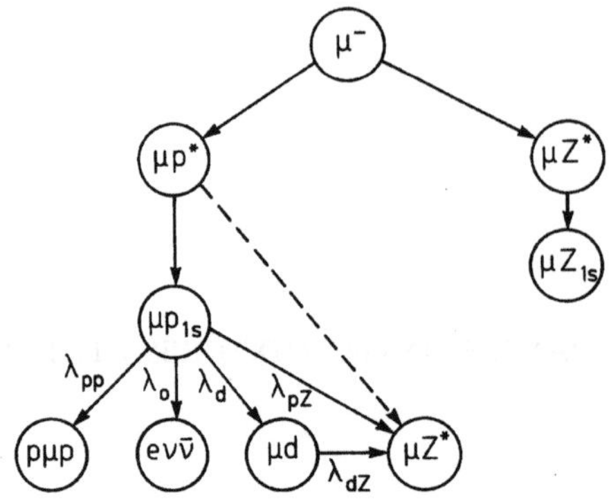

Fig. 1. Scheme of muon capture and transfer in natural hydrogen with a small admixture of an element Z. Only the main processes are shown.

In a hydrogen-argon gas mixture with a small relative argon concentration, the muon capture and transfer processes can therefore be sketched like in Fig. 1. Negatively charged muons are directly captured either by hydrogen or by argon. The muonic argon emits promptly muonic x-rays of characteristic energies and intensity patterns. The μp atom disappears through different channels. The time structure of the muonic argon x-rays resulting from muon transfer, for times greater than the experimental time resolution, is then a measure of the lifetime of the μp atom in the ground state.

2. DIRECT MUON CAPTURE IN A H_2 + Z GAS MIXTURE

The intensity patterns of the muonic x-rays resulting from direct capture in pure argon can be reproduced by a cascade calculation assuming a statistical initial angular momentum distribution in the muonic n=14 level.[12] The muonic cascade is pressure dependent because of the pressure dependence of the electron hole refilling rate.[13] For pressures between 1 and 170 bar, the radiative intensity of the Ar(2p-1s) transition is about 90% of the whole Lyman series intensities.[12]

The binding energy of the electrons in hydrogen is about the same as that of the outer electron shell in argon, such that one can assume that, in a H_2 + Ar gas mixture, the electron refilling rate in muonic argon is not essentially different from what it is in pure argon at comparable pressures. It seems therefore reasonable to assume that, in a H_2 + Ar gas mixture at 10-15 bar, 90% of the muons captured directly by argon transit through the 2p state to reach the ground state.

Of all muons transferred from the ground state of the μp atom to argon, only about 40% transit through the 2p state independently of pressure and gas mixture.[10] The corresponding Lyman series intensity pattern is essentially different from that measured in direct capture. This difference in the intensity structure of the muonic x-rays and the difference in the time distribution of the muonic events lead to two methods for the measurement of hydrogen-to-argon Coulomb capture ratios $A(H_2/Z)$: the time structure method and the intensity structure method.[2,14] The two methods used in the evaluation of muonic data in two H_2 + Ar gas mixtures (9.6 bar H_2 + 0.4% Ar, 14.9 bar H_2 + 2% Ar) gave comparable $A(H_2/Ar)$ capture ratios.[2] The mean of $A(H_2/Ar)$ = 0.096(7) is, however, only about half of the corresponding capture ratio for pions[15] and less than half of what is expected from the ratio of $A(H_2/He)$ and A(Ar/He) muon capture ratios (Table 1) and model predictions.[16]

Table I. Comparison of Coulomb capture ratios for muons and pions. The difference between $A(H_2/Ar)$ and $A(H_2/"Ar")$ capture ratios is explained in the text.

	Muons	Ref.	Pions	Ref.
A(He/Ar)	0.148(10)	[35]	0.161(10)	[15]
A(He/Ar)	0.208(39)	[36]	0.165(29)	[22]
$A(He/H_2)$	<0.8(2)	[37]	0.92(5)	[15]
$A(He/H_2)$	0.73-0.79	[38]	0.71-0.76	[38]
$A(H_2/Ar)$	0.096(7)	[2]	0.172(5)	[15]
$A(H_2/"Ar")$	≈0.05			

Through the analysis of the time spectra of the Lyman series of the muonic argon x-rays measured in the H_2 + Ar gas mixtures, one can determine the intensities of the np-1s transitions which do not proceed from muon transfer from the μp ground state (Fig. 2). These intensity patterns do not correspond to what is expected for direct capture in argon.[17] The 2p-1s intensity does not correspond to the expected 90%, but is much lower. The same is true for the prompt muonic x-ray intensities[18] in oxygen and sulphur (Fig. 3) measured in $H_2 + SO_2$. The intensity structure method can therefore not be used to determine $A(H_2/Z)$ capture ratios, and in the time structure method one cannot limit ourselves to the analysis of the time spectra of the K_α line.

By summing up the intensities of the muonic argon Lyman series of the prompt events on one side, i.e. all those events which do not proceed from muon transfer from the μp ground state (Fig. 2), and those resulting from normal transfer on the other side, and taking into account, as usually, the relative argon concentration and the muons lost through other channels than the transfer to argon, one obtains a new "per atom" $A(H_2/"Ar")$ capture ratio, which is much smaller than the previous one. In the H_2 + 2% Ar gas mixture, this ratio has been found approximately equal to 0.05. This is five times smaller than model predictions which were generally in reasonable agreement with measured capture ratios.[19]

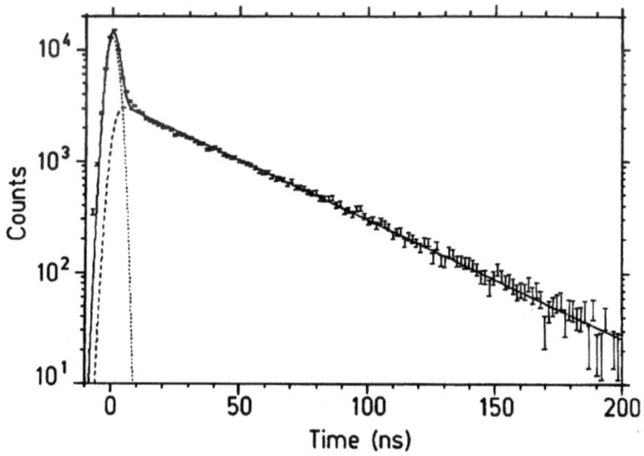

Fig. 2. Time spectrum (logarithmic scale) of the muonic 2p-1s transition in argon measured in a H_2 + 2% Ar gas mixture at 15 bar. The broken lines correspond to fits to the experimental data. The exponential function describes the events resulting from muon transfer from the ground state of μp atoms. The gaussian function corresponds to prompt events.

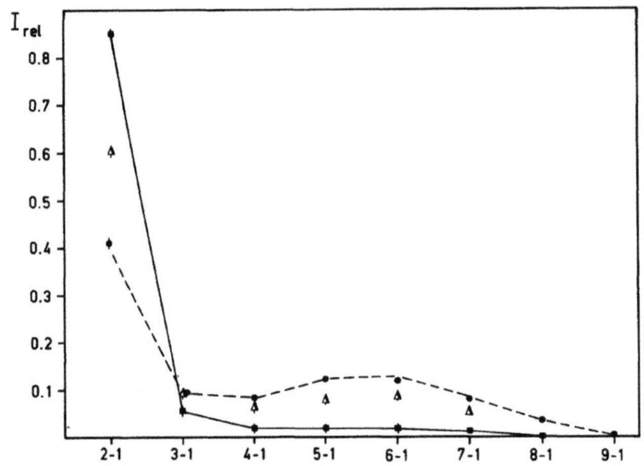

Fig. 3. Relative muonic x-ray intensities of the Lyman series measured in sulphur of SO_2. The squares correspond to the intensities measured in pure SO_2 at 2.4 bar. The intensities of the higher members of the Lyman series resulting from transfer from μp atoms in the ground state (open circles) are much larger and agree well with theoretical prdictions (broken line). The triangles represent the prompt intensities measured in a H_2 + 0.4% SO_2 gas mixture. The structure of these intensities does not agree with direct capture in SO_2.

In the H_2 + 0.4% SO_2 gas mixture, one can determine two such "per molecule" capture ratios A(H_2/"SO_2"). If one takes only the normal transfer component to oxygen,[18,20] one obtains 0.023, which is also about five times smaller than the model prediction. If one takes the total delayed oxygen intensity, the capture ratio increases to 0.030.

The densities of hydrogen and of the admixed gas are of different orders of magnitude. For this reason we assumed that transfer from excited μp states would give only small corrections. On the other side, it has been shown that in mixtures of noble gases, there is an important concentration dependence [21] of the per atom capture ratios of muons (up to 30%). In our case, a corresponding concentration dependence would give even smaller capture ratios. Surprisingly, pion capture in H_2 + Z gas mixtures does not show such a concentration dependence of per atom capture ratios.[15,22]

3. MUON TRANSFER FROM EXCITED STATES

The strong disagreement of these muon capture ratios A(H_2/Z) with corresponding pion capture ratios and with model predictions, and the fact that the intensity structure of the prompt muonic argon (or oxygen or sulphur) x-rays does not correspond to that of direct capture, leads us to envisage that the muon transfer to the admixed gas from excited μp states is important, even when the molar concentration is less than 1%.

Let us, as an example, take the A(H_2/"SO_2") capture ratio measured at 15 bar in a H_2 + SO_2 gas mixture[18,20] with a molar concentration of C(SO_2/H_2) = 0.4%. Similar considerations can be made for other A(H_2/"Z") capture ratios measured in H_2 + Z gas mixtures with comparable molar concentrations.

Our model predicts a capture ratio per molecule of A(H_2/SO_2) = 2/18. Let's assume that the prompt events measured in "SO_2" correspond to direct muon capture in SO_2 and to muons transferred from a unknown number x of μp atoms in excited states. Under the given experimental conditions of relative concentration, the model would then predict a capture ratio of :

$$A\left(\frac{(250\text{-}x)H_2}{SO_2 + xH_2}\right) = \frac{250 - x}{A(SO_2/H_2) + x} = \frac{250 - x}{9 + x}$$

If this is set equal to the corresponding experimental capture ratio of 0.03/0.4%, one finds for $x \approx 20$. This means that, to have agreement between the measured $A(H_2/"SO_2")$ capture ratio and the prediction of the model, about 8% of the muons captured by hydrogen would have to be transferred to SO_2 from an excited μp state.

In gaseous deuterium at 10 bar, a muon with an initial energy of 2 keV is captured in an atomic level and reaches the ground state in a time[1] of about 10^{-10} s. Starting from this time, one can estimate that a μp atom collides in mean about 10^2 times with H_2 molecules to reach the ground state. In our H_2 + 0.4% SO_2 gas mixture, the probability that, in an excited state, the μp atom would collide at least once with a SO_2 molecule is about 40%. If the discrepancy between the predicted $A(H_2/SO_2)$ and the measured $A(H_2/"SO_2")$ capture ratios should be explained through transfer from excited μp states (without 2s state), the mean reduced transfer rate from excited μp states had to be very high. This is not in contradiction with estimates.[23,24]

In the time distribution of the muonic x-rays in neon, measured in a H_2 + Ne gas mixture, one observes two time constants.[17] For the shorter one, an explanation might be given in terms of transfer from the 2s state of the μp atom via an intermediate muon-molecular state which should, however, be very long-lived.

According to Bracci and Fiorentini[24], the transfer cross-sections from excited μp states are strongly dependent on the chemical structure. In particular, the transfer to the positive ion should be favoured in a polar molecule. Thus one expects that in SO_2 transfer to sulphur should be favoured. The analysis of the transfer data, measured in $H_2 + SO_2$, showed that the capture ratio $A_T(S/O)$ through normal transfer from the μp ground state is compatible[20] with the estimates of the ratio of the reduced transfer rates $\Lambda_S/\Lambda_O \approx 1.4$ given by Fiorentini and Torelli.[23] The capture ratio A(S/O) measured in pure SO_2 is almost equal to the transfer ratio.[25] Thus, if the transfer from excited μp states really favours the positive ion in SO_2, one expects that the capture ratio A(S/O), deduced from the prompt events measured in $H_2 + SO_2$, is greater than 1.4. The analysis shows, however, that this ratio is clearly smaller.

However, as already known, the measured transfer data in $H_2 + SO_2$ presented some unexpected and yet unexplained effects,[20] such that one is not surprised to find new questions.

4. MUON TRANSFER FROM THE GROUND STATE

The muon transfer from the μd atom to helium is of particular interest in muon catalyzed fusion. Since the first calculation of the charge transfer rate assuming formation of an intermediate quasistationary state,[8] more refined calculations have been performed.[26-28] The measured and predicted transfer rates are in reasonable agreement with each other (Table II). The transfer rate to ^{3}He measured at room temperature[29] is only slightly lower than the theoretical predictions. The transfer rate to ^{4}He is, on the contrary, about 20% higher than the highest predicted transfer rate. The same is true for the transfer rate to ^{4}He measured at liquid hydrogen temperature.[30]

Muon transfer from μp to ^{4}He has been measured by using natural hydrogen and the triple gas mixture method.[10,31] Bystritskii et al.[31] extract from their data the Λ_{He} transfer rate independently of the transfer rate to the admixed gas (xenon). Jacot-Guillarmod et al[10] need, on the contrary, a precise value for the transfer rate to the admixed argon gas. Bystritskii et al. find a transfer rate to helium which agrees well with the theoretical predictions.[8,26,27] The three transfer rates measured by Jacot-Guillarmod et al., which yield a mean value of $\Lambda_{He} = 0.88(9)\cdot 10^8$ s^{-1}, agree with each other, but disagree by about a factor of two with the rate of Bystritskii et al. as well as with all theoretical predictions which range from 0.32 to $0.52\cdot 10^8$ s^{-1} (Table II).

Table II. Comparison between measured and predicted muon transfer rates from hydrogen to helium.

	Temp.	Exp.[$\cdot 10^8$ s^{-1}]	Ref.	Theory[$\cdot 10^8$ s^{-1}] Ref.[8],[26-28]
Λ_{p^4He}	RT	0.36(10)	[31]	
Λ_{p^4He}	RT	0.83(14)	[10]	
Λ_{p^4He}	RT	0.88(15)	[10]	0.32-0.52
Λ_{p^4He}	RT	0.92(14)	[10]	
Λ_{p^4He}	RT	0.03(1)	[32]	
Λ_{d^4He}	RT	3.68(18)	[29]	2.1-2.9
Λ_{d^4He}	20K	13.1(12)	[30]	3.4-10.7
Λ_{d^3He}	RT	1.27(11)	[29]	1.5-1.8

Table III. Comparisons of measured reduced muon transfer rates Λ_{Ar} from μp atoms in the ground state to argon.

Year	Ref.	pressure [bar]	$C(Ar/H_2)$ [$\cdot 10^{-4}$]	n(Ar) [$\cdot 10^{17} cm^{-3}$]	Λ_{Ar} [$\cdot 10^{11} s^{-1}$]	method
1965	[39]	45	1	1.1	1.20(19)	e
1969	[40]	10	12	2.9	1.46(14)	e
1988	[11]	100	6	15.0	1.42(16)	x
1988	[11]	140	20	68.0	1.46(5)	x
1988	[10]	9.6	41	9.6	1.47(8)	x
1988	[10]	13.3	200	65.0	1.41(5)	x
1967	[41]	26	33	21.0	3.48(60)	x
1971	[42]	900	20	440.0	3.28(78)	x
1982	[43]	2	10	0.5	3.61(220)	x
1982	[43]	3	20	1.5	3.55(112)	x
1982	[43]	3	10	0.7	3.77(136)	x
1982	[43]	4	10	1.0	3.81(167)	x
1980	[44]	600	2	29.0	9.8(15)	x

Very recently another muon transfer rate from μp atoms to ^4He has been published, which has been measured at low pressure using the muonic x-ray yield.[32] This last transfer rate is about one order of magnitude lower than the former two (Table II).

It is however not the first time that measured muon transfer rates do not agree with each other (Table III). One distinguishes presently, e.g. in argon, three classes of reduced muon transfer rates from the μp ground state to excited states of argon.[33] These transfer rates have been measured at room temperature, but under very different experimental conditions of pressure and concentration.[10] Up to now no correlation has been found between the reduced transfer rates and peculiarities of the experimental conditions.[11]

According to our present knowledge, the measured transfer rates are transfer rates from thermalized μp atoms in their ground state. One knows the dependence of the lifetime of these μp atoms from the pressure of the hydrogen gas and the density of the admixed gas. Our measurements, performed under very different experimental conditions, leading always to the same reduced transfer rates,[10,11] seemed to confirm that these dependences were well known.

One cannot brush aside measured transfer rates, which do not agree with ours, without convincing reasons. We are therefore constrained to distinguish three groups of experimental transfer rates to argon (and perhaps also to helium). These three classes of reduced transfer rates suggest then, that there were three kinds of thermalized μp atoms in their ground state, which under identical experimental conditions would have different lifetimes.

In a very recent experiment, performed to test the additivity of the transfer rates in a triple gas mixture, we anew obtained a transfer rate to argon in agreement with our former values.[17]

Another very recently performed measurement of muon transfer in a H_2 + Ne gas mixture shows that there is presently perhaps a similar problem with the reduced muon transfer rates to neon.[17]

5. MUON TRANSFER TO SO_2

In a letter published last autumn,[20] we presented an even more puzzling problem. In a measurement of muon transfer - to our knowledge from the ground state of the μp atom - to oxygen and sulphur of sulphur dioxide, we observed that the time spectra of the muonic sulphur x-rays had a time constant of about 100 ns (Fig. 4), corresponding to what is expected on the basis of approximate predictions for the reduced transfer rates to oxygen and sulphur.[23] The time spectra of the muonic oxygen x-rays should have had the same time constant and not another one. To our great surprise, the time spectra of all four observed muonic oxygen x-ray transitions, i.e. O(2-1), O(3-1), O(4-1) and O(5-1), had two time constants (Fig. 5), a longer one compatible with the time constant of the sulphur x-rays, and another one, for all four muonic oxygen transitions the same, of about 50 ns, i.e. about half of the longer one.

The experiment has been repeated one and a half year later under nearly the same experimental conditions. It fully confirmed the former observations.[18]

By analyzing the time spectra of the muonic oxygen x-rays in more detail, one observed that the shorter exponential decay does not start at the time of the muon capture (Fig. 6). More precisely, if the normal or longer time constant is the lifetime of the μp atoms in the ground state, the shorter time constant corresponds to the lifetime of another muon containing system, which does not yet exist, when the μp atom is in the ground state, but which has first to be formed.

One observes that the time spectra of the delayed muonic sulphur x-rays can be described, as expected, by a single exponential

$$\frac{dN_{\mu S\gamma}}{dt} = \lambda A_S e^{-\lambda t}$$

where λ is the sum of the disappearance rates of the μp atom in the ground state ($\tau = 1/\lambda$ is its lifetime). The time spectra of the muonic oxygen x-rays can be fitted in their delayed part with

$$\frac{dN_{\mu O\gamma}}{dt} = \lambda A_O e^{-\lambda t} + \frac{\lambda'\lambda''}{\lambda'-\lambda''} B(e^{-\lambda'' t} - e^{-\lambda' t})$$

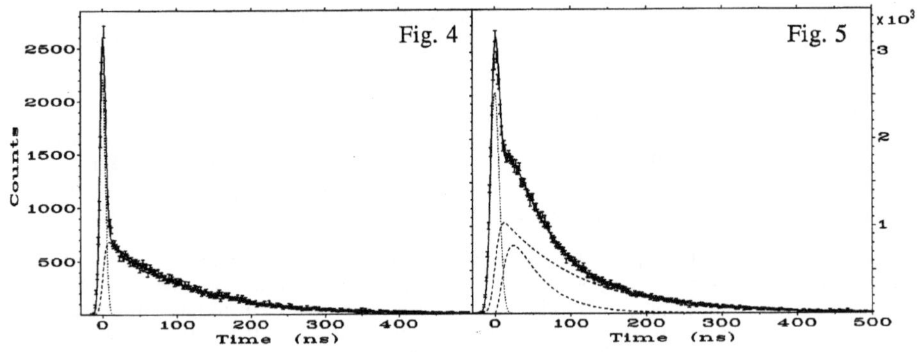

Fig. 4. Time distribution (linear scale) of muonic sulphur S(2p-1s) x-rays measured in a H_2 + 0.4% SO_2 gas mixture at 15 bar. The continuous line, which is a fit to the data, is a sum of a gaussian and an exponential folded with a gaussian. The time constant of the exponential is about 100 ns.

Fig. 5. Time distribution (linear scale) of muonic oxygen O(2p-1s) x-rays measured in a H_2 + 0.4% SO_2 gas mixture at 15 bar. The continuous line, which is a fit to the data, is the sum of the three functions represented with dotted and broken lines. The delayed part is fitted with the function given in the text, folded with a gaussian. The time constants of the three exponentials are 106 ns, 42 ns and 9.5 ns.

where λ' is the formation rate of the other muon containing system and λ'' the transfer rate to oxygen (other channels are neglected). Both rates, λ' and λ'', are greater than the disappearance rate λ of the μp atoms in their ground state. Consequently the other muon containing system does not originate from those μp atoms in the ground state, which transfer their muon to sulphur.

Because of the symmetry of the formula in λ' and λ'', one does not know which one is the formation rate.

In the H_2 + SO_2 gas mixtures, the shorter of the corresponding times were of the order of 5 to 10 ns. The precision of these times was too low to extract a pressure or concentration dependence. The longer of the two times shows a dependence on the SO_2 density, which is perhaps not linear.[18]

In a test run, the time spectra of the transfer x-rays in fluorine have been measured in a H_2 + SF_6 gas mixture. For all muonic fluorine x-rays one observes time spectra similar to those of oxygen in H_2 + SO_2. The shorter of the two times is there much longer and of about 40 ns. The longer one is only slightly shorter than the μp lifetime measured in the time spectra of the sulphur x-rays.

The relative muonic x-rays intensities in an element Z, resulting from muons transferred from μp atoms in their ground state, can be calculated.[9] In argon,[10] fluorine,[9,34] sulphur and oxygen,[20,33] the corresponding measured muonic x-ray intensities agree well with the predicted ones. It is important to note that the muonic x-ray intensities in oxygen of H_2 + SO_2, corresponding to transfer from the other muon containing system, have the same structure (Fig. 6) as if they would result from muon transfer from μp atoms in the ground state.[18] The same seems to prove right for the fluorine x-rays.[9,34]

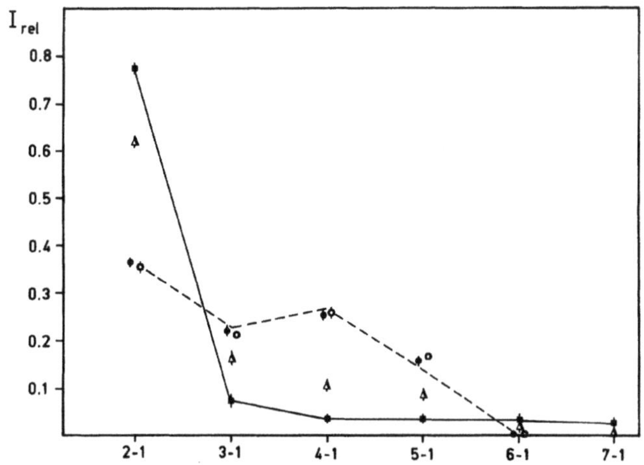

Fig. 6. Relative muonic x-ray intensities of the Lyman series in oxygen of SO_2. The triangles correspond to the intensities measured in pure SO_2 at 2.4 bar. The intensities of the higher members of the Lyman series measured in a H_2 + 0.4% SO_2 gas mixture at 15 bar, resulting from muon transfer from μp atoms in the ground state (open circles) are much larger and agree well with theoretical predictions (broken line). The intensities which proceed from muons transferred from the other muon containing system (black discs) are identical to those of muon transfer from the μp ground state. The squares represent the prompt intensities in a H_2 + 0.4% SO_2 gas mixture.

On the basis of these observations concerning the intensity structures of muonic oxygen x-rays, which appear delayed relative to a muon stop signal, we called the muon containing systems, which transfer normally their muon to oxygen and sulphur, "white" μp atoms, and those obscure objects, which transfer their muon in a different time "black" μp atoms.[33] Such "black" and "white" μp atoms (Fig. 7) have been found not only in the two first measurements of transfer in H_2 + SO_2 at 15 bar with a relative SO_2 concentration of 0.4%, but also in two recent experiments at SO_2 concentrations of 0.2% at 10 and 15 bar.[18] The normal or "white" μp atoms behave normally, i.e. their transfer rate is proportional to the density of the admixed gas. For the "black" μp atoms such a density dependence is not yet clearly established.

From the measured x-rays intensities, one can estimate the ratio of "black" μp atoms to "white" ones. In the H_2 + 0.4% SO_2 gas mixture, this ratio is about 1/4, i.e. of all muons which are transferred to oxygen about 30% come from "black" μp atoms.[18] In the H_2 + 0.2% SO_2 gas mixtures, these numbers are smaller by about 30%. A very preliminary analysis of the data taken in the H_2 + SF_6 gas mixture shows that there the numbers of "black" and "white" μp atoms are almost equal. One thinks immediately of a chemical effect. However, the H_2 + Ne data show also some anomalies, which cannot be due to chemical effects in the usual sense.[17]

CONCLUSION

After the experiment performed in 1982,[43] one could have imagined that the observed discrepancies between measured muon transfer rates from protium to argon were due to the two methods used. The new measurements[10,11] have not confirmed this assumption and called the older measurements in question. The muon transfer to sulphur dioxide[20] revealed a completely unexpected phenomenon, even if the measured data were received with great scepticism.[33] Since then, new measurements fully confirmed the first ones,[18] and other experiments showed that the phenomenon is not related to the dipole moment of sulphur dioxide. The recent data

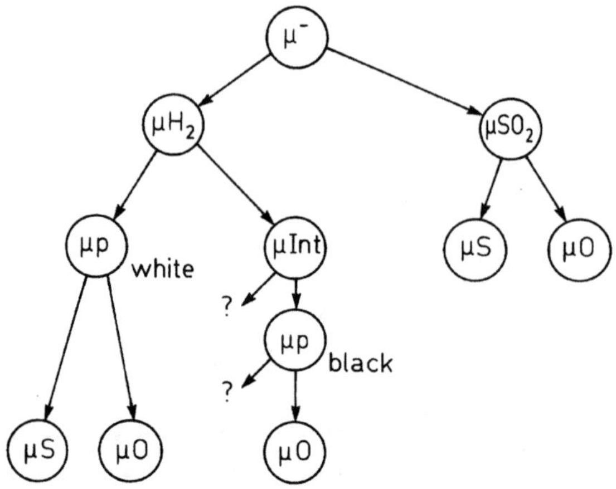

Fig. 7. Simplified scheme of muon capture and muon transfer in a H_2+ SO_2 gas mixture. Comments are given in the text.

measured in neon[17] might be a confirmation at the same time of what was observed in argon (Table III) and in sulphur dioxide.

I expected - and I am certainly not alone - that muon transfer from the ground state of muonic hydrogen atoms to elements with charge numbers $Z > 3$ would bring no particular problems with it. Till now, the new phenomenon remains, however, unexplained and the scheme given in Fig.7 is presently the only one I can propose.

REFERENCES

1. V.E.Markushin, Zh.Eksp.Theor.Fiz. 80:35(1981) [Sov.Phys.-JETP 53:16(1981)].
2. R.Jacot-Guillarmod, F.Bienz, M.Boschung, C.Piller, L.A.Schaller, L.Schellenberg, H. Schneuwly and W.Reichart, Phys.Rev. A 39:387(1989).
3. A.Bertin,A.Vacchi and A.Vitale, Lett. Nuovo Cimento 18:277(1977).
4. A.Bertin, M.Capponi, I.Massa, M.Piccinini, G.Vannini, M.Poli and A.Vitale, Nuovo Cimento A 72:225(1982).
5. L.Bracci, C.Chiccoli, P.Pasini, G.Fiorentini, V.S.Melezhik and J.Wosniak, Phys.Lett. A 134:435(1989).
6. P.Kammel, Lett.Nouvo Cimento 43:349(1985).
7. S.S.Gershtein, Zh.Eksp.Teor.Fiz 43:706(1962) [Sov.Phys.-JETP 16:501(1963)].
8. Yu.A.Aristov, A.V.Kravtsov, N.P.Popov, G.E.Solyakin, N.F.Truskova and M.P.Faifman, Yad.Fiz. 33:1066(1981) [Sov.J.Nucl.Phys. 33:564(1981)].
9. G.Holzwarth and H.-J.Pfeiffer, Z.Phys. A272:311(1975).
10. R.Jacot-Guillarmod. F.Bienz, M.Boschung, C.Piller, L.A.Schaller, L.Schellenberg, H. Schneuwly, W.Reichart and G.Torelli, Phys.Rev. A 38:6151(1988).
11. F.Bienz, P.Bergem, M.Boschung, R.Jacot-Guillarmod, G.Piller, W.Reichart, L.A. Schaller, L.Schellenberg, H.Schneuwly and G.Torelli, J.Phys.B:At.Mol.Opt.Phys. 21:2725(1988).
12. R.Jacot-Guillarmod, F.Bienz, M.Boschung, C.Piller, L.A.Schaller, L.Schellenberg, H. Schneuwly and D.Siradovic, Phys.Rev A 37:3795(1988).
13. R.Bacher, D.Gotta, L.M.Simons, J.Missimer and N.C.Mukhopadhyay, Phys.Rev.Lett. 54:2087(1985).
14. H.Schneuwly, Nucl.Instrum.Methods B 9:97(1985), 14:258(1986).
15. V.I.Petrukhin and V.M.Suvorov, Zh.Eksp.Teor.Fiz. 70:1145(1976) [Sov.Phys.-JETP 43:595(1976).

16. H.Schneuwly, V.I.Pokrovsky and L.I.Ponomarev, Nucl.Phys. 312A:419(1978).
17. R.Jacot-Guillarmod, F.Mulhauser, C.Piller, L.A.Schaller, L.Schellenberg and H. Schneuwly, contribution to this volume.
18. F.Mulhauser, R.Jacot-Guillarmod, C.Piller, L.A.Schaller, L.Schellenberg and H. Schneuwly, contribution to this volume.
19. H.Schneuwly, T.Dubler, K.Kaeser, B.Robert-Tissot, L.A.Schaller and L.Schellenberg, Phys.Lett. A 66:188(1978).
20. H.Schneuwly, R.Jacot-Guillarmod, F.Mulhauser, P.Oberson, C.Piller and L.Schellenberg, Phys. Lett. A 132:335(1988).
21. P.Ehrhart, F.J.Hartmann, E.Köhler and H.Daniel, Phys.Rev. A 27:575(1983).
22. A.V.Bannikov, B.Levay, V.I.Petrukhin, V.A.Vasilyev, L.M.Kochenda, A.A.Markov, V.I.Medvedev, G.L.Sokolov and D.Horvath, Nucl.Phys.A403:515(1983).
23. G.Fiorentini and G.Torelli, Nuovo Cimento A 36:317(1976).
24. L.Bracci and G.Fiorentini, Nuovo Cimento A 50:373(1979).
25. P.Oberson, diploma work, Institut de Physique de l'Université de Fribourg, June 1988, unpublished.
26. A.V.Kravtsov, A.I.Mikhailov and N.P.Popov, J.Phys.B:At.Mol.Opt.Phys.19:2579 (1986).
27. V.K.Ivanov,A.V.Kravtsov,A.I.Mikhailov, N.P.Popov and V.I.Fomichev, Zh.Eksp. Teor.Fiz.91:358(1986)[Sov.Phys.-JETP 64:210(1986)].
28. N.P.Popov, Muon Catalyzed Fusion 2:207(1988).
29. D.B.Balin, A.A.Vorobyov,An.A.Vorobyov, Yu.K.Zalite, A.A.Markov, V.I.Medvedev, E.M.Maev, G.G.Semenchuk and Yu.V.Smirenin, Pis'ma Zh.Eksp.Teor.Fiz. 42:236 (1985) [Sov.Phys.-JETP Lett. 42:293(1985)].
30. T.Matsuzaki, K.Ishida, K.Nagamine, Y.Hirata and R.Kadono, Muon Catalyzed Fusion 2:217(1988).
31. V.M.Bystritskii, V.P.Dzhelepov, V.I.Petrukhin, A.I.Rudenko, V.M.Suvorov, V.V.Filchenkov, N.N.Khovanskii and B.A.Khomenko, Zh.Eksp.Teor.Fiz. 84:1257(1983) [Sov.Phys.-JETP 57:728(1983)].
32. H.P.von Arb, F.Dittus, H.Hofer, F.Kottmann and R.Schaeren, Muon Catalyzed Fusion 4(1989) (in press).
33. H.Schneuwly, Muon Catalyzed Fusion 4(1989) (in press).
34. H.Daniel, H.-J.Pfeiffer and K.Springer, Phys.Lett. A 44:447(1973).
35. Y Budyashov, V.G.Zinov, A.D.Konin and A.I.Mukhin, Yad.Fiz. 5:830(1967) [Sov.J. Nucl.Phys. 5:589(1967)].
36. R.L.Hutson, J.D.Knight, M.Leon, M.E.Schillaci, H.B.Knowles and J.J.Reidy, Phys. Lett. A 76:225(1980).
37. F.Kottmann, In"Muons and Pions in Matter", JINR Report Nr. D14-87-799(Dubna 1987), p. 268.
38. J.S.Cohen, R.L.Martin and W.R.Wadt, Phys.Rev. A 37:1821(1983).
39. S.G.Basiladze, P.F.Ermolov and K.O.Oganesyan, Zh.Eksp.Teor.Fiz. 49:1042(1965) [Sov.Phys.-JETP 22:725(1966)].
40. A.Placci, E.Zavattini, A.Bertin and A.Vitale, Nuovo Cimento A 64:1053(1969).
41. A.Albergi Quaranta, A.Bertin, G.Matone, F.Palmonari, A.Placci, P.Dalpiaz, G.Torelli and E.Zavattini, Nouvo Cimento B 42:236(1967).
42. G.Backenstoss, H.Daniel, K.Jentzsch, H.Koch, H.P.Povel, F.Schmeissner, K.Springer and R.L.Stearns, Phys.Lett. B 36:422(1971), cited in Ref. 43.
43. E.Iacopini, G.Carboni, G.Torelli and V.Trobbiani, Nuovo Cimento A 67:201(1982).
44. H.Daniel, H.-J.Pfeiffer, P.Stoeckel, T.von Egidy and H.P.Povel, Nucl.Phys. A254:409 (1980).

DISCUSSION

D. HORVATH: Why do your time spectra have sometimes periodic "wiggles"? May they be due to second muons or muon precession? Could be a relation between the wiggles and the anomalous short component?

H.S.: An analysis of the time structure of these "wiggles" of all data taken during a running period of about two weeks revealed no regular periodicity. We therefore consider these "wiggles" as statistical fluctuations.

MUON TRANSFER IN $H_2 + SO_2$ GAS MIXTURES

F. Mulhauser, R. Jacot-Guillarmod, C. Piller, L. A. Schaller
L. Schellenberg, and H. Schneuwly

Institut de Physique de l'Université
CH-1700 Fribourg (Switzerland)

INTRODUCTION

The muon transfer process in binary and ternary gas mixtures with hydrogen looks like a well known process.[1] From the ground state of the μp atom, where the muon has a binding energy of about 2.6 keV, the muon is transferred to higher Z elements onto atomic orbits with much smaller radii than the outer electron shells of the atom Z, and where the muon has a comparable binding energy. The transfer process does therefore not depend on chemical structures and bonds in which the element, to which the muon is transferred, is involved.

According to our present knowledge, a negative muon stopped in an $H_2 + SO_2$ gas mixture is captured in an atomic orbit of either a hydrogen nucleus or one of the nuclei of the SO_2 molecule (Fig. 1). The μp atom in the ground state disappears by muon decay with a rate λ_0, by forming a pμp mesomolecule (λ_{pp}) or by transferring the muon to deuterium (λ_d), to oxygen (λ_O) or to sulphur (λ_S).

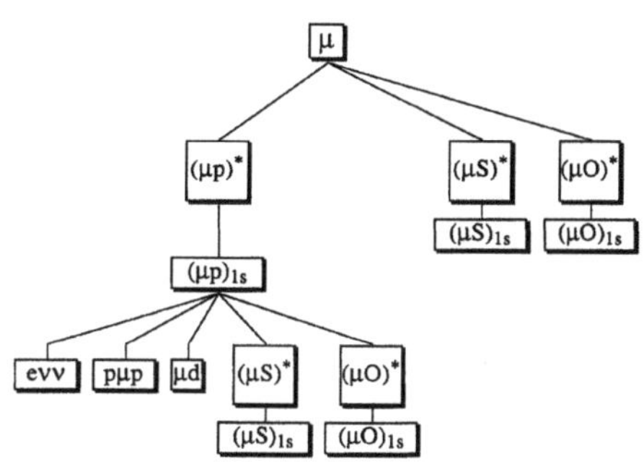

Fig. 1. Processes in a $H_2 + SO_2$ gas mixture.

Electromagnetic Cascade and Chemistry of Exotic Atoms
Edited by L. M. Simons *et al.*, Plenum Press, New York, 1990

The total disappearance rate λ of the μp atom can then be written :

$$\lambda = \lambda_0 + \lambda_{pp} + \lambda_d + \lambda_S + 2\lambda_O \qquad (1)$$

$1/\lambda = \tau$ is then the lifetime of the μp atom in the ground state under the corresponding experimental conditions of pressure, temperature and relative concentration.

TRANSFER IN H_2 + 0.4 % SO_2 GAS MIXTURE

Time and energy spectra of the muonic oxygen and sulphur x-rays have been measured in a hydrogen gas mixture containing 0.4 % SO_2 at 15 bar. The experiment has been performed at the μE4 channel of the Paul Scherrer Institute (PSI) in Villigen. The experimental setup and the data analysis were comparable to preceding measurements.[1,2] The energy spectra of the muonic oxygen and sulphur x-rays, which appear delayed with respect to the muon stop signal, are shown in Fig. 2.

Figs. 3a and 3b show the net time distributions of the muonic sulphur 2p-1s and 5-2 x-ray transitions. The latter is energetically situated between the oxygen 2p-1s and 3p-1s x-ray transitions. All time spectra of the sulphur x-rays have been fitted with the same function :

$$\frac{dN_{\gamma S}}{dt} = A\, e^{-(t/\sigma)^2} + B\, e^{-\lambda t} \int_{-\infty}^{t} e^{-(t'/\sigma)^2}\, e^{\lambda t'}\, dt' \qquad (2)$$

The fitted time constants $\lambda = 1/\tau$ were all compatible yielding a mean lifetime for the $(\mu p)_{1s}$ atom of τ = 107 (3) ns, called τ_S.

Figs. 3c and 3d show the net time distributions of the muonic oxygen 2p-1s and 3p-1s transitions. These time spectra and the time spectra of the other oxygen x-rays have not the shape of a single exponential and have been fitted using :

$$\frac{dN_{\gamma O}}{dt} = A\, e^{-(t/\sigma)^2} + B\, e^{-\lambda t} + \frac{C}{\tau_{O1} - \tau_m}\left[e^{-t/\tau_{O1}} - e^{-t/\tau_m}\right] \qquad (3)$$

where the second and third terms have also been convoluted with a gaussian corresponding to the time resolution of the detection system. After some trials, the parameter τ_m, corresponding to a rise time, was fixed to τ_m = 9.5 ns. The values obtained for τ_{O1} and $\tau_{O2} = 1/\lambda$ are given in Table 1. By setting $\tau_{O2} = \tau_S$, one obtains for τ_{O1} a mean value of 41 (4) ns.

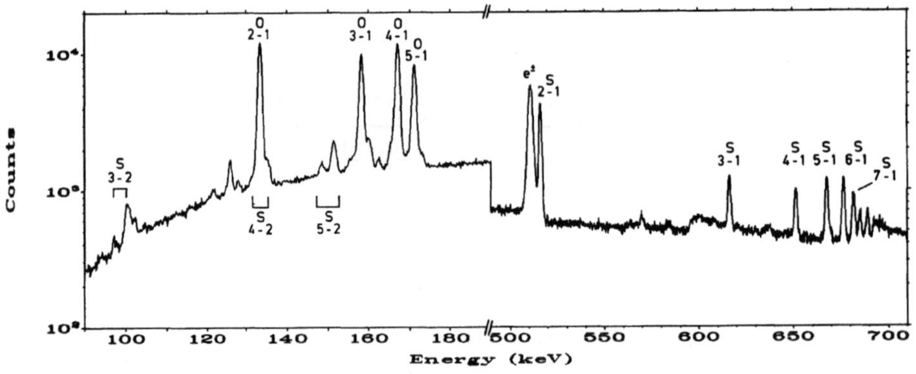

Fig. 2. Energy spectra of the muonic x-ray transitions in oxygen and sulphur, delayed with respect to a muon stop signal.

Table 1. Fitted time constants of the time spectra of the muonic oxygen x-ray transitions. a) both τ_{O1} and τ_{O2} free, b) τ_{O1} free and $\tau_{O2} = \tau_S$ fixed.

Transition	a) τ_{O1} [ns]	a) τ_{O2} [ns]	b) τ_{O1} [ns]	b) $\tau_{O2} = \tau_S$ [ns]
2-1	41.6 28	103.9 24	41.4 58	107
3-1	38.5 80	102.7 42	40.3 59	107
4-1	37.0 52	100.8 33	41.1 66	107
5-1	43.1 95	103.7 65	40.9 62	107

The relative muonic x-ray intensities resulting from muons transferred to oxygen and sulphur are given in Table 2. The intensity structure of the muonic Lyman series measured in sulphur agrees well with the structure predicted for muons transferred from thermalized μp atoms in the ground state.[1,3] The same proves true for the intensity patterns of the muonic oxygen x-rays, both for the longer and the shorter time components. The intensity of the shorter component is about 40% of the longer one. The ratio of the intensity of the Lyman series in sulphur to that in oxygen (per atom), corresponding to the time component $\tau_{O2} = \tau_S$, $A_T(S/O) = 1.24\ (9)$, should be equal to the ratio of the muon transfer rates from the μp atoms in the ground state.

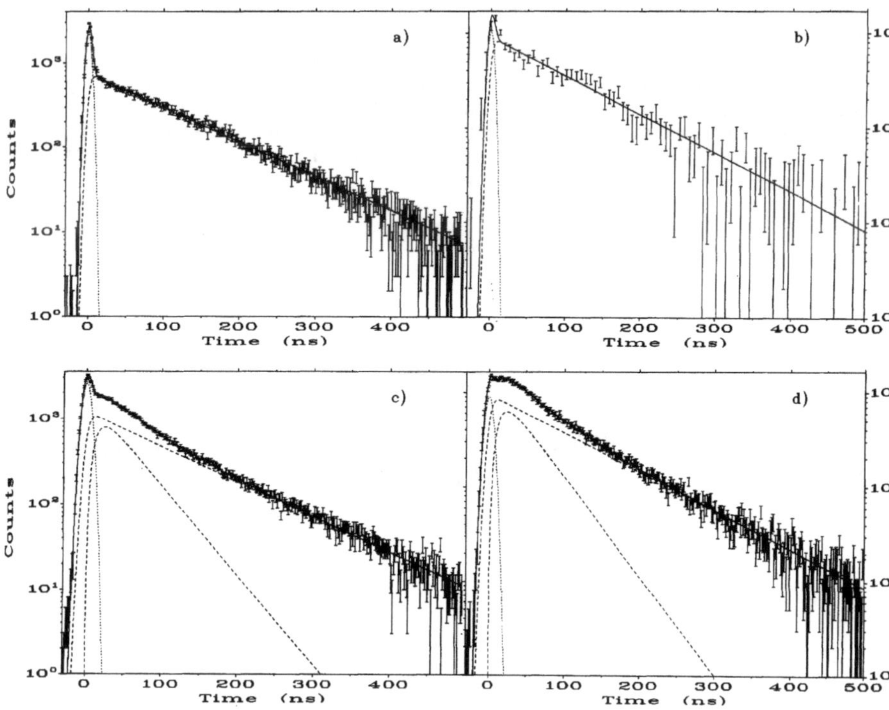

Fig. 3. Measured and fitted time spectra of the muonic a) 2p-1s and b) 5-2 transitions of sulphur, c) 2p-1s and d) 3p-1s transitions of oxygen.

Table 2. Relative intensities of the muonic Lyman series in sulphur and oxygen, delayed with respect to the muon stop signal.

Transition	Sulphur Experiment	Sulphur Theory	Oxygen Experiment τ_{O2}	Oxygen Experiment τ_{O1}	Oxygen Theory
2-1	1000	1000	1000	1000	1000
3-1	230	236	610	602	630
4-1	196	207	692	730	733
5-1	271	315	431	476	375
6-1	276	330	–	–	0
7-1	194	220	–	–	0

The intensity patterns measured in the prompt part of the time spectra do not agree with those measured in pure SO_2, and this neither for oxygen nor for sulphur. From the prompt and delayed (resulting from transfer) intensities, one can determine a per molecule capture ratio A(H_2/"SO_2"), which is approximately equal to 0.03, i.e. about a factor three less than what is expected from model predictions.[4]

TRANSFER IN H_2 + 0.2 % SO_2 GAS MIXTURE

After the unexpected effect observed in H_2 + 0.4 % SO_2 gas mixtures, we have performed further muon transfer experiments in H_2 + 0.2 % SO_2 gas mixtures at 10 and 15 bar. The measured time distributions of the muonic sulphur and oxygen x-rays fully confirm the observations made in the first measurements. Time spectra of the muonic oxygen 2p-1s transitions are shown in Fig. 4. The intensity patterns of the Lyman series in oxygen and sulphur are identical to those observed in the first measurements. From the time constants extracted from the sulphur time spectra, one can determine the muon transfer rate from the μp ground state to SO_2, reduced to liquid hydrogen density. These transfer rates turn out to be identical in the different experiments as expected[5] (Table 3).

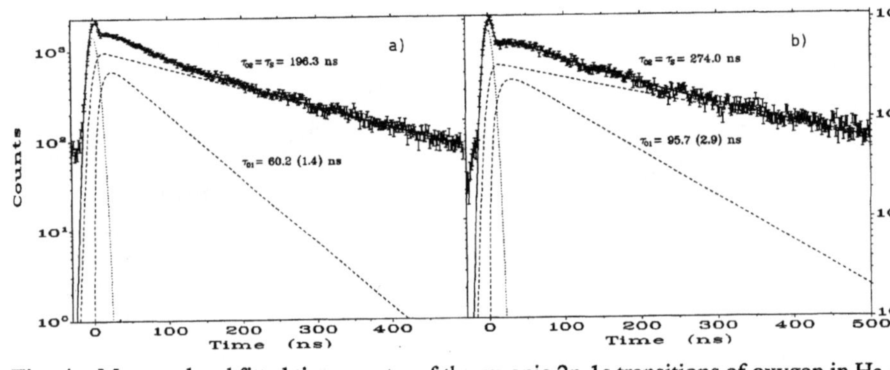

Fig. 4. Measured and fitted time spectra of the muonic 2p-1s transitions of oxygen in H_2 + 0.2 % SO_2 a) at 15 bar and b) at 10 bar.

Table 3. Time constants and reduced transfer rates Λ_{SO_2} for the three H_2 + SO_2 gas mixtures.

SO_2 concentration total pressure	0.4 % 15 bar	0.2 % 15 bar	0.2 % 10 bar
$\tau_S = \tau_{O2}$ (ns)	107 3	196 7	274 10
τ_{O1} (ns)	41 4	60 6	96 10
Λ_{SO2} ($\cdot 10^{11}$ s^{-1})	2.51 9	2.56 12	2.57 12

CONCLUSION

The surprising difference observed between the time spectra of muonic oxygen and sulphur x-rays in H_2 + 0.4 % SO_2 gas mixtures is reproduced in H_2 + 0.2 % SO_2 gas mixtures at 10 and 15 bar. The intensity of the unexpected component is large compared to the background. The intensity patterns of the muonic Lyman series in sulphur and in oxygen correspond both in their delayed part to what is expected for muons transferred from thermalized μp atoms in their ground state. At present, no explanation for the second oxygen time component has been found, yet. Such a second component has also been observed in fluorine of H_2 + SF_6.

REFERENCES

1. R. Jacot-Guillarmod, F. Bienz, M. Boschung, C. Piller, L. A. Schaller, L. Schellenberg, H. Schneuwly, W. Reichart, and G. Torelli, Phys. Rev. A 38:6151 (1988).
2. H. Schneuwly, R. Jacot-Guillarmod, F. Mulhauser, P. Oberson, C. Piller and L. Schellenberg, Phys. Lett. A 132:335 (1988).
3. G. Holzwarth and H.-J. Pfeiffer, Z. Phys. A 272:311 (1975).
4. H. Schneuwly, contribution to this volume.
5. S. S. Gershtein, Zh. Eksp. Teor. Fiz. 43:706 (1962); Sov. Phys.-JETP 16:501 (1963); G. Fiorentini and G. Torelli, Nuovo Cimento A 36:317 (1976).

DISCUSSION

D. HORVATH : Why does your background have a decay constant different from your signal ? Is it impossible that you introduce the anormalous, shorter lifetime by improper background subtraction ?

F. MULHAUSER : The problem of background subtraction has been studied very carefully. We can exclude that the shorter lifetime is introduced through improper background subtraction.

MUON TRANSFER TO LOW Z ELEMENTS

R. Jacot-Guillarmod, F. Mulhauser, C. Piller, L. A. Schaller
L. Schellenberg, and H. Schneuwly

Institut de Physique de l'Université
CH-1700 Fribourg (Switzerland)

INTRODUCTION

A negative muon, captured in a mixture of natural hydrogen with a small amount of another gas of atomic number Z, forms a muonic hydrogen or a muonic Z atom in an excited state. In the deexcitation process, muonic X rays are emitted. Because the cascade time is very short (of the order of 10^{-12} s at 10 bar), they appear promptly with regard to the incoming muon. After deexcitation of the $(\mu p)^*$ system (in about 10^{-10} s), which occurs essentially by collisions,[1] the $(\mu p)_{1s}$ atom can then disappear by muon decay with an associated rate λ_0, by formation of a $p\mu p$ mesomolecule (λ_{pp}), or by transferring the muon to deuterium (λ_d) or to the Z element (λ_{pZ}). The total disappearance rate of the μp atom, λ, is then :

$$\lambda = \lambda_0 + \lambda_{pp} + \lambda_d + \lambda_{pZ} \tag{1}$$

With a very low concentration of the Z element, one can assume that the transfer of the muon to the Z element proceeds mainly from the ground state. Otherwise, muonic X rays following a transfer from $(\mu p)^*$ would appear promptly. The transfer of the muon from the μp atom occurs to much lower n levels compared to direct capture in Z, and the associated angular momentum distribution is also very different. For argon, a statistical muonic angular momentum distribution (calculated at n=14) has been found by direct capture, while low angular momenta are favoured by transfer.[2,3] The intensity structure of the muonic Lyman series is also characteristic of the way of capture. The 2p-1s intensity is about 0.90 (depending of pressure) for direct capture, and 0.40 (quasi-independent of pressure) for transfer to argon. The muonic X rays emitted after muon transfer are delayed relative to the muon stop in the gas, and their time distribution is an exponential function of time :

$$\frac{dN_{\gamma Z}}{dt}(t) \propto e^{-\lambda t} \tag{2}$$

By fitting the delayed part of the time spectrum of these transitions, one deduces λ. The other rates λ_0, λ_{pp}, and λ_d being known, one can calculate, using eq. (1), the experimental transfer rate from protium to Z, λ_{pZ}. However, this rate depends on the atomic density ρ_Z of the Z element in the mixture. To compare with other experimental values measured at different concentrations, one normalizes them to the atomic density of liquid hydrogen ρ_0 :

$$\Lambda_{pZ} = \frac{\rho_0}{\rho_Z}\lambda_{pZ} = \frac{\rho_0}{\rho_Z}\left(\lambda - \lambda_0 - \lambda_{pp} - \lambda_d\right) \tag{3}$$

By using natural hydrogen, transfer to the Z element also occurs from μd. By calling λ' the total disappearance rate of muonic deuterium, eq. (2) becomes :

$$\frac{dN_{\gamma Z}}{dt}(t) \approx e^{-\lambda t} + \left[\frac{\Lambda_{dZ}}{\Lambda_{pZ}} \frac{\lambda_d}{\lambda - \lambda'}\left(e^{-\lambda' t} - e^{-\lambda t}\right)\right] \quad (4)$$

The second term of eq. (4) is small, because of the low values of the concentration of deuterium (150 ppm) and of the normalized transfer rate to deuterium[4] compared to other elements ($\Lambda_d = 0.168\ (26) \cdot 10^{11}\ s^{-1}$); it can therefore be treated as a correction.

Two experimental techniques are used to obtain the transfer rate to a given element Z, namely detection of the decay electrons or of the muonic X rays. Fig. 1 shows experimental transfer rates.[3,5-9] Some discrepancies, as in the case of argon, have been observed between experiments. However, they are not correlated to the type of method used and not yet explained. If one takes the most recent values[6-8] on Λ_{pXe} as a reference, and assumes a proportionality[10,11] between Λ_{pZ} and Z , or $Z^{2/3}$, Fig. 1 shows that the measured rates, specially for Z<20, are not in agreement with such Z-dependences.

The transfer rates to oxygen and sulfur have been deduced from our measurement of transfer to SO_2 molecules (see ref. 12 and contribution of F. Mulhauser *et al.* in this issue). Because of the unexpected results in SO_2, we decided to measure transfer to low Z elements and specially to atoms and simple molecules as neon and nitrogen. The experiments have been performed at the Paul Scherrer Institute (PSI) in Villigen. The experimental setup and the data analysis were comparable to preceding measurements,[3] all mixtures have been measured at 15 bar.

TRANSFER TO NITROGEN

A mixture of hydrogen with 4337 ppm N_2 has been chosen. The delayed energy spectrum of nitrogen is reported in Fig. 2a. The deduced intensity structure of the Lyman series (Table 1) is in agreement with theoretical predictions.[13] Fig. 3a shows the time spectrum of the muonic N(2-1) transition, it is composed of gaussian and exponential functions. The first one corresponds to X rays following direct muon capture, the second to delayed X rays emitted after transfer from the μp system. The delayed time spectra of all four transitions are pure

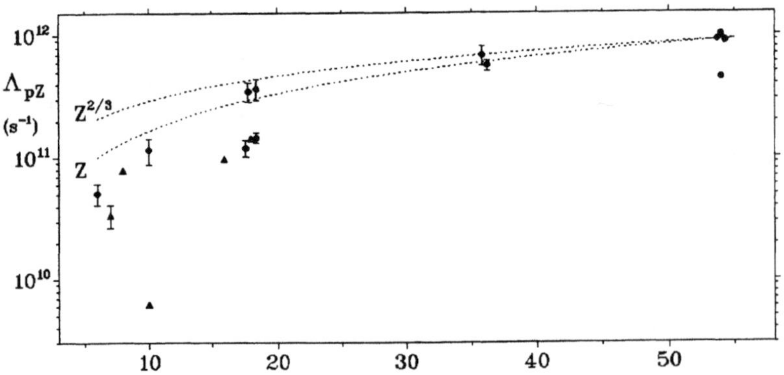

Fig. 1. Experimental transfer rates measured at room temperature, and in a pressure range between 1 and 50 bar. Points represent data from ref. 3-9, and triangles correspond to measurements made by our group.

Table 1. Relative intensities of the muonic Lyman series of nitrogen and neon after muon transfer. The sum of intensities is normalized to unity.

Transition	Nitrogen Experiment	Nitrogen Theory[13]	Neon Experiment	Neon Theory[13]
2-1	0.368 14	0.381	0.396 10	0.353
3-1	0.254 9	0.271	0.146 4	0.154
4-1	0.258 10	0.252	0.201 5	0.223
5-1	0.120 5	0.096	0.167 6	0.189
6-1	______	_____	0.090 4	0.081

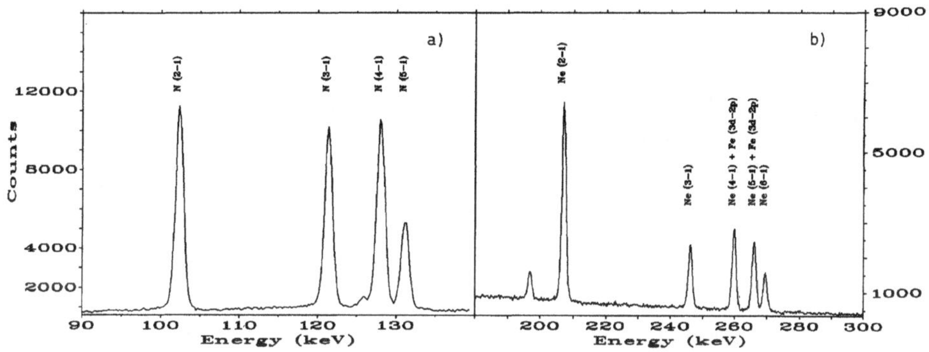

Fig. 2. Delayed energy spectra of the muonic Lyman series of a) nitrogen, and b) neon.

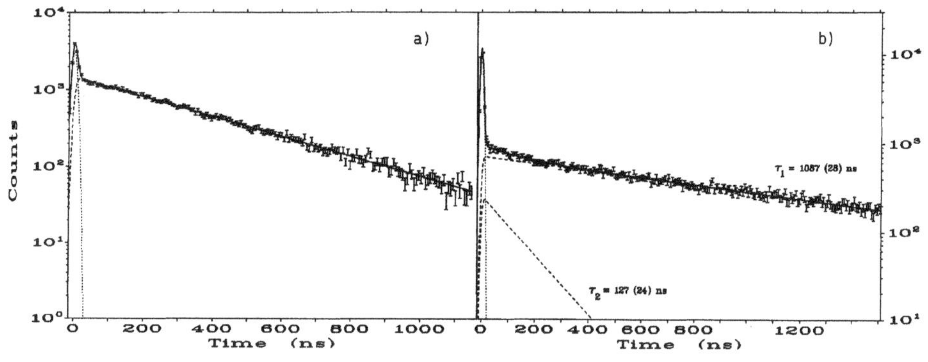

Fig. 3. Measured and fitted time spectra of the muonic K_α transitions of a) nitrogen, and b) neon.

exponentials with the same mean lifetime τ, the average value being $\tau=1/\lambda=334(3)$ ns. The deduced normalized transfer rate to nitrogen becomes then $\Lambda_{pN}=0.34(7)\cdot 10^{11}\ s^{-1}$. The transfer rate to nitrogen from deuterium ($\Lambda_{dN} = 1.00(30)\cdot 10^{11}\ s^{-1}$ from ref. 14) is greater than the one from protium. This has been observed only for helium, where recent calculations[15] predict about ten times higher transfer rates from deuterium than from protium at room temperature.

TRANSFER TO NEON

We have measured time and energy spectra of muonic X rays in a mixture of hydrogen with a concentration of 6930 ppm neon. The intensity structure of the Lyman series in the delayed spectrum (Fig. 2b) is reported on Table 1. As for nitrogen, the experimental and theoretical values are in good agreement with each other. The presence of muonic iron X rays is not due to transfer, but to X rays in random coincidence with an incoming muon. Their contribution to the total intensity of Ne(4-1) and Ne(5-1) has been taken into account. The time distributions of muonic Ne(2-1), Ne(3-1), and Ne(4-1) up to Ne(6-1) have been analyzed. Surprisingly, one needs two exponentials to obtain a good fit of the time spectra. If we compare the time spectra of the muonic K_α transition of nitrogen and neon (Figs. 3a and 3b), we observe that the prompt peak in nitrogen is much less important than in neon, and that the form of the spectrum just after the prompt peak is different. All three time spectra showed the same characteristics with compatible mean time constants. The average lifetimes of the flat and steep slopes were 1089(24) and 133(21) ns, respectively, and the intensity ratio between them about 4%.

Both exponential decays are not due to hidden delayed nuclear transitions, to background, or to an electronic malfunction of the detection system. These possibilities were tested by measuring a mixture of neon and argon without hydrogen. The gaussian time distribution of the muonic Ne(2-1) transition indicated clearly that the transition was prompt, as expected. Hence, the two decay components found in the H_2+Ne mixture are related to the presence of hydrogen.We assume that the flatter slope corresponds to the mean lifetime of the μp atom and the other to an unknown process, perhaps comparable to the one observed[12] in H_2+SO_2. However, it is worth while to notice that both transfer processes induce the same muonic Lyman intensity structure.

The normalized transfer rate to neon deduced from the flatter slope or longer time constant (Table 2) is about a factor 20 less than the rate measured by Alberigi Quaranta *et al.*.[6] Moreover, it is less by a factor of two than for deuterium.[4] The normalized rate deduced from the shorter mean time constant corresponds exactly to the value of Alberigi Quaranta *et al.*. In the same paper, and using the X-ray method in triple gas mixtures with xenon, Alberigi Quaranta *et al.* found for argon $\Lambda_{pAr} = 3.48\ (60)\cdot 10^{11}\ s^{-1}$, which is 2.4 times greater than our value.[3] Their technique of analysis was based on intensities and differed from ours, which uses the fit of the time distribution. To improve our measured transfer rate to neon, we have remeasured it by using a triple gas mixture, but with argon instead of xenon.

Table 2. Mean lifetimes and normalized transfer rates to nitrogen and neon.

	H_2 + 4337 ppm N_2	H_2 + 6930 ppm Ne Flat slope	 Steep slope
τ (ns)	334 3	1089 24	133 21
Λ_{pZ} ($\cdot 10^{11}\ s^{-1}$)	0.34 7	0.062 4	1.15 20

We have analyzed the time spectra of Ar(2-1), (3-2), (4-2), (5-2), (6-2) and Ne(2-1), measured in a triple gas mixture of H_2 + 1405 ppm Ne + 1390 ppm Ar. From the mean lifetime τ=428(4) ns, we deduced Λ_{pNe} = 0.056 (48)$\cdot 10^{11}$ s^{-1} (with Λ_{pAr} = 1.44 (4)$\cdot 10^{11}$ s^{-1}). One exponential sufficed to fit well all the time spectra. The statistics was, however, too poor to observe a 4% intensity effect in the time spectrum of Ne(2-1). The great uncertainty on Λ_{pNe} is due to the low contribution of only 3% of λ_{pNe} with regard to the total disappearance rate λ. By applying another method of evaluation, where we assumed that the ratio of the intensity of the Lyman series of neon to argon is equal to the ratio of their transfer rates, we obtained Λ_{pNe} = 0.062 (5)$\cdot 10^{11}$ s^{-1}, which is in good agreement with our preceding values. Using this method, the contribution of transfer from μd (Λ_{dNe} = 1.42(20)$\cdot 10^{11}$ s^{-1} by ref.14) is about 30% of the total transfer processes.

CONCLUSION

Our three values for the muon transfer rate from protium to neon, obtained from different measurements, agree well with each other, but disagree by nearly a factor of twenty with a preceding value.[6] The transfer rates to neon and nitrogen, which are expected to be greater from protium than from deuterium, have been found much smaller. They are three times less in nitrogen and twenty times less in neon.

By comparing the measured transfer rates (Fig. 1), e.g. to low Z elements like neon, oxygen and nitrogen, one observes that these transfer rates do not have a monotonous Z dependence.

REFERENCES

1. V. E. Markushin, Zh. Eksp. Teor. Fiz. 80:35 (1981); Sov. Phys.-JETP 53:16 (1981).
2. R. Jacot-Guillarmod, F. Bienz, M. Boschung, C. Piller, L. A. Schaller, L. Schellenberg, H. Schneuwly and D. Siradovic, Phys. Rev. A 37:3795 (1988).
3. R. Jacot-Guillarmod, F. Bienz, M. Boschung, C. Piller, L. A. Schaller, L. Schellenberg, H. Schneuwly, W. Reichart, and G. Torelli, Phys. Rev. A 38:6151 (1988).
4. A. Bertin, M. Bruno, A. Vitale, A. Placci, and E. Zavattini, Lett. Nuovo Cimento 4:449 (1981).
5. S. G. Basiladze, P. F. Ermolov, and K. O. Oganesyan, J. Exptl. Theoret. Phys. 49:1042 (1965); Sov. Phys.-JETP 22:725 (1966).
6. A. Alberigi Quaranta, A. Bertin, G. Matone, F. Palmonari, A. Placci, P. Dalpiaz, G. Torelli, and E. Zavattini, Nuovo Cimento B 47:92 (1967).
7. A. Placci, E. Zavattini, A. Bertin, and A. Vitale, Nuovo Cimento A 64:1053 (1969).
8. A. Bertin, M. Bruno, A. Vitale, A. Placci, and E. Zavattini, Phys. Rev. A 7:462 (1973).
9. E. Iacopini, G. Carboni, G. Torelli, and V. Trobbiani, Nuovo Cimento A 67:201 (1982).
10. S. S. Gershtein, J. Exptl. Theoret. Phys. 43:706 (1962); Sov. Phys.-JETP 16:501 (1961).
11. G. Fiorentini and G. Torelli, Nuovo Cimento A 36:317 (1976).
12. H. Schneuwly, R. Jacot-Guillarmod, F. Mulhauser, P. Oberson, C. Piller and L. Schellenberg, Phys. Lett. A 132:335 (1988).
13. G. Holzwarth and H.-J. Pfeiffer, Z. Phys. A 272:311 (1975), V. R. Akylas and P. Vogel, Comp. Phys. Commun. 15:291 (1978).
14. A. Placci, E. Zavattini, A. Bertin, and A. Vitale, Nuovo Cimento A 52:1274 (1967).
15. V. K. Ivanov, A. V. Kravtsov, A. I. Mikhailov, N. P. Popov and V. I. Fomichev, Zh. Eksp. Teor. Fiz. 91:358 (1986); Sov. Phys.-JETP 64:210 (1987).

V

MUON CATALYZED FUSION

"It seems very pretty, ...but it's *rather* hard to understand... Somehow it seems to fill my head with ideas - only I don't exactly know what they are!"

Lewis Carroll: Through the Looking-Glass

MUON CATALYZED FUSION

A SELECTION OF OLD AND NEW EXPERIMENTS AND OPEN PROBLEMS

C. Petitjean

Paul Scherrer Institute

CH-5232 Villigen PSI, Switzerland

1. Introduction

Muon Catalyzed Fusion (MCF) is a very active research field describing the phenomena that take place when a negative muon stops in a mixture of hydrogen isotopes. Figure 1 shows as an example the kinetic cycles induced in a deuterium-tritium mixture. This system received recently most of the attention, because it was shown that the dt cycle can be repeated hundreds of times.

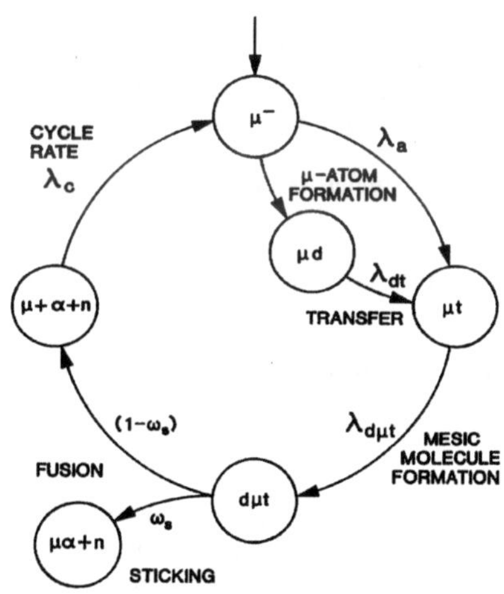

Figure 1. Kinetic cycle of dμt fusion.

In MCF many physics and chemistry processes play an important role, e.g.

- Atomic muon capture into an outer orbit, forming a muonic atom.
- Atomic cascade to the 1s ground state, by Auger process, electromagnetic radiation and other processes.

Table 1. Mesic Molecule Formations and Fusion Reactions of Hydrogen Isotopes

PROCESSES					Q VALUE	MMF RATE λ_{MMF}	FUSION RATE λ_f	STICKING ω
INITIAL		MESIC MOLECULE		FINAL	[MeV]	$[10^6 sec^{-1}]$	$[10^6 sec^{-1}]$	
$\mu d + p$	$\longrightarrow$	$p\mu d$	$\longrightarrow$	$^3He + \gamma + \mu$	5.5	3-6	0.3	0.85
$\mu d + d$	$\longrightarrow$	$d\mu d$	$\nearrow$	$^3He + n + \mu$	3.3	0.05-5	320	0.13
			$\searrow$	$t + p + \mu$	4.0			0
$\mu t + d$	$\longrightarrow$	$d\mu t$	$\longrightarrow$	$\alpha + n + \mu$	17.6	$\gtrsim 500$	10^6	~ 0.005
$\mu t + p$	$\longrightarrow$	$p\mu t$	$\longrightarrow$	$\alpha + \gamma + \mu$	19.8	7	0.06	0.95
$\mu t + t$	$\longrightarrow$	$t\mu t$	$\longrightarrow$	$\alpha + 2n + \mu$	11.3	1.8	15	0.14
$\mu p + p$	$\longrightarrow$	$p\mu p$	$\xrightarrow{(W.I.only)}$		-	2	0	-

- Acceleration mechanisms during the atomic cascade.
- Elastic scattering of the muonic atoms leading to thermalisation.
- Inelastic scattering processes inducing hyperfine transitions (spin flip).
- Transfer of the 1s muon to a heavier isotope or to impurities X ($\mu p \rightarrow \mu d \rightarrow \mu t \rightarrow \mu X$), with the release of the binding energy.
- Transfer of muons in excited orbits, e.g. $(\mu d)_n \rightarrow (\mu t)_n$. This fast process can reduce the initial μd-groundstate population significantly (factor q_{1s}).
- Formation of mesic molecules $p\mu d$, $d\mu d$, $d\mu t$, etc. in collisions of the tiny neutral muonic atoms with other hydrogen nuclei, see table 1. This process can be strongly enhanced due to a resonance mechanism, which is responsible for ultra high $d\mu t$ formation rates and for the characteristic temperature dependence of $d\mu d$ fusion.
- Nuclear fusion of the hydrogen nuclei, see table 1, with energy releases between 3 and 20 MeV.
- Sticking of the muon to the charged helium fusion product, which stops the continuation to the next fusion cycle.
- Reactivation of "sticking" muon during slow down of the μ-helium atoms.
- etc.

This list - although incomplete - shows that the phenomena within MCF are quite numerous and complicated. Therefore we are still far away from completing the understanding of MCF and have to deal with many problems that are open and still wait to be solved by future experiments and theories. This is a very fascinating situation!

Table 1 shows the list of possible fusion reactions that can be catalyzed with muons. Due to the complexity of MCF cycles, it is necessary to study them all in order to arrive eventually at a consistent description of MCF.

It is the goal of this lecture to elucidate some of the recent achievements and to present several open problems in MCF.

2. Resonant dμd Formation

The process of muon induced d-d (and p-d) fusion was discovered 33 years ago by Alvarez et al.[1] on photos taken with the Berkeley bubble chamber. Since the observed fusion yields were low, any hopes for large fusion output were discarded at that time.

In 1966, Dzhelepovs experimental group at Dubna found a dramatic temperature dependence of dμd formation ($\lambda_{d\mu d}$). This discovery lead to the well known temperature resonance curve of $\lambda_{d\mu d}$ [2], shown in figure 2, and to the theory by Gerstein, Ponomarev and co-workers of resonant mesomolecule formation[3].

The resonance mechanism - first proposed by Vesman[4] - is illustrated in fig. 3: The sum of collision energy ε_o and released binding energy ε_{11} of the dμd molecule is directly transferred into the vibrational excitation v of the electronic D_2 molecule.

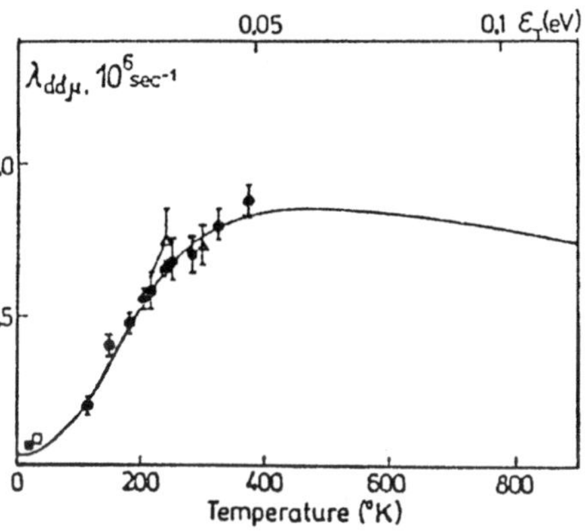

Figure 2. Resonant dμd formation as discovered by the Dubna experiments[2]. The curve is from the Dubna theory group[3].

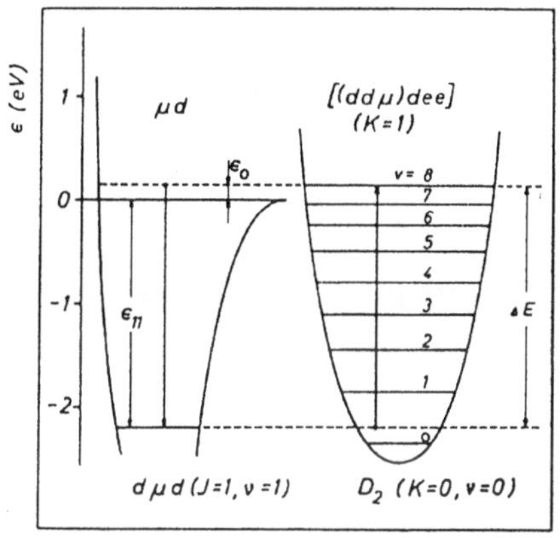

Figure 3. The resonance mechanism of dμd formation, as proposed by E.A. Vesman[4] and worked out by the Dubna theorists[3].

In experiments performed 1979/1980 at PSI (former SIN), strong hyperfine effects in the μd atom and dμd mesic molecule were discovered[5], see figure 4. Two very different temperature curves of dμd formation were observed (fig. 4a), depending on the μd spin configuration (F = 3/2 or 1/2), and a strong spin flip rate λ_d (fig. 4b) was found, which depopulates the upper μd hyperfine level F = 3/2.

The rates can now almost quantitatively be explained by the recent calculations of Menshikov et al.[6], see curves in figure 4.

The dμd resonance allowed the prediction of even much larger resonant formation rates of the dμt mesic molecule. Surprisingly, the behaviour of $\lambda_{d\mu t}$ is quite different from $\lambda_{d\mu d}$, - see chapter 4 – because the required kinetic energy of the μt atom for molecule formation is slightly negative ($\sim$ -10 meV) while for dμd the values are positive (+ 4/53 meV).

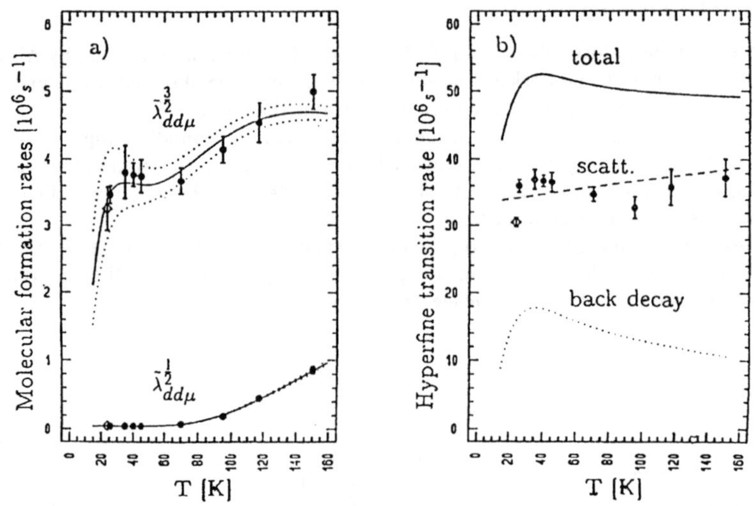

Figure 4. The complete picture of dμd resonant formation (a) and and μd (3/2)→ μd (1/2) spin flip (b), as measured at PSI[5].

3. The Wolfenstein-Gerstein effect in pμd fusion

The formation and fusion of pμd mesic molecules was studied extensively during the last 30 years. In 1960, Gerstein and Zeldovich[7] already made a detailed theory about the kinetics. On a suggestion of Wolfenstein, Gerstein[8] predicted an enhancement of the yield from pμd fusion if the deuterium concentration of the pd mixture is increased from "saturation" 1% to ~20%. This effect (in short W.G.-effect) is due to the μd spin transition

$$\mu d\ (F\ =\ 3/2)\ +\ d \xrightarrow{\lambda_d} \mu d\ (F\ =\ 1/2)\ +\ d' \qquad (1)$$

enhancing the lower spin μd population and giving a larger fusion rate. For illustration, see the kinetics scheme figure 5 from ref. 9.

In 1962/63 a detailed investigation of the pμd system was made by the experimental group of Ledermann[10] at Columbia. Indeed, the W.G.-effect was found as predicted (17% enhancement). Many years laters (1977) it was confirmed in experiments performed at PSI[9].

So far - so good! But with new calculations by Matveenko and Ponomarev[11] and with the direct observation of the μd spin flip rate λ_d (chapter 2, figure 4b), we know now that λ_d is much larger than estimated originally by Gerstein[8]. Therefore the W.G-effect should be considerably larger than observed by experiments[9,10], e.g. the enhancement should rather be 30-55% instead of 17% as observed in reality! How can this puzzle be solved? In ref. 9, it was postulated that spin flip on the proton,

$$\mu d(F = 3/2) + p \xrightarrow{\lambda_p} \mu d(F = 1/2) + p' \qquad (2)$$

with rate λ_p may be the explanation.

With a value $\lambda_p = 4.5{\cdot}10^6 s^{-1}$ the observed W.G.-effect could be explained satisfactorily[9]. However since no theoretical grounds were found until today for the λ_p-assumption, it is rather suggested, that in addition to the pμd fusion rate $\lambda_f^{1/2}$ from pd spin (1/2) overlap, there might also be a sizeable component $\lambda_f^{3/2}$ from pd spin (3/2) molecules[12]. Such a $\lambda_f^{3/2}$-contribution would also reduce the ratio of relative fusion yields at different c_d concentrations. So, this question is an open problem and has to be further investigated!

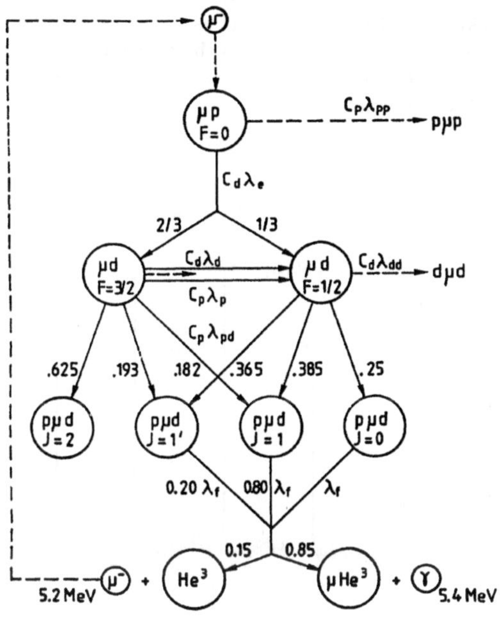

Figure 5. The kinetics scheme of the p-μ-d system. The yield of 5.5 MeV pμd fusion gammas depends on the μd spin flip strength (W.G.-effect).

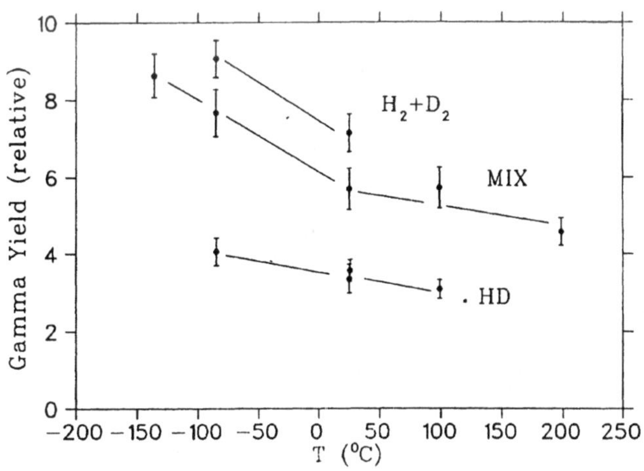

Figure 6. The pμd fusion yields at several temperatures and molecular compositions as measured by K. Aniol et al.[13].

More recently the group of Aniol et al.[13] has done a new series of pμd yield measurements at TRIUMF. As shown in figure 6, very surprising dependencies of pμd-γ-yields from temperature and from molecular species ($H_2 + D_2$ versus HD) were found. Such a behaviour might suggest the existence of a similar resonance mechanism for pμd formation as for dμd. Again this explanation is not too likely, because no weakly bound pμd-level exists as in the case of dμd and dμt. So, the TRIUMF results pose another new open problem to the understanding of the pμd system. It is connected with the puzzle of W.G.-effect-size? Or is the existence of odd parity states of the pμd mesic molecule[14] an explanation?

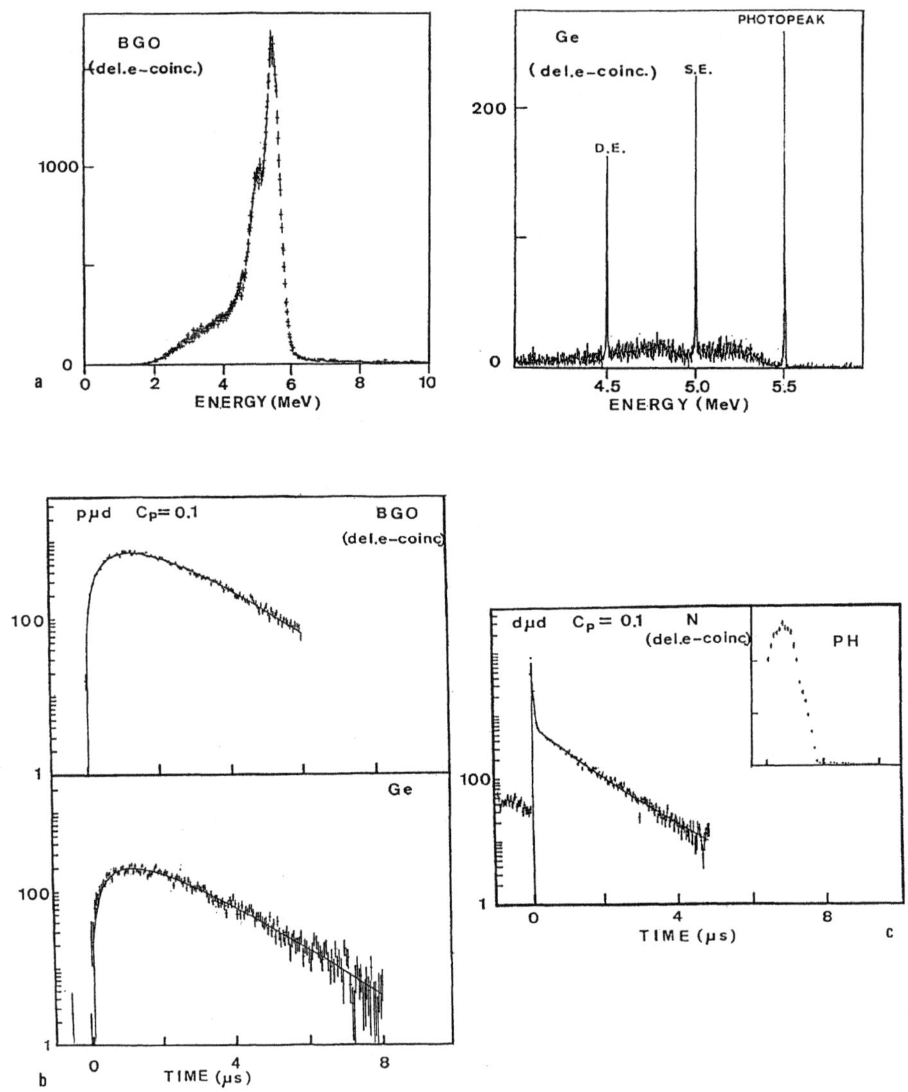

Figure 7. Spectra of new measurements in pμd fusion at PSI. (a) 5.5 MeV γ's measured with a BGO and a Ge detector, (b) the corresponding time distribution, (c) proton recoil spectra from 2.5 MeV dd fusion neutrons and time distribution.

In October 1988 a new effort was undertaken by the MCF collaboration at PSI[15] with the goal to get a deeper understanding of the pμd kinetics and to solve some of the listed problems. The measurements were done in liquid protium/deuterium mixtures at various concentrations and differing molecular compositions. The 5.5 MeV gammas from pμd fusion were measured with a high resolution Germanium diode and a high efficiency BGO detector.

Figure 7a illustrates some typical 5.5 MeV gamma energy spectra and fig. 7b the corresponding time distributions of fusion events. Simultaneously, the recoil spectra and time distribution of dμd fusion neutrons were measured (fig. 7c) with 2 NE213 liquid scintillation detectors using n-γ pulse shape discrimination. This method of combined measurement of p-d and d-d channels will allow us to determine simultaneously the development of μd population, μd spin transitions, pμd formation and pμd fusion. The analysis is in progress[15].

4. dμt fusion

The prediction of ultralarge d-t fusion rates[3] due to much stronger dμt resonant formation if compared with dμd, soon triggered major experimental programs in several laboratories (Dubna, LAMPF, PSI) to observe dμt fusion. A big problem of such experiments is the handling of the highly radioactive tritium isotope, which is needed in considerable quantities (grams or ten thousands of Curies) and requires special high vacuum equipment and safety precautions.

The first pioneering experiment was done 1979 by Dzhelepov's group[16] at low tritium concentrations c_t, reporting as a lower limit $\lambda_{d\mu t} > 10^8 s^{-1}$, confirming thus the main theory prediction. Surprisingly, in the range 100 K to 500 K no temperature dependence of the dt yield was found.

In 1982-84, large experimental series, using for the first time high densities and large tritium admixtures, were made at LAMPF and PSI.

At LAMPF, very broad emphasis was set on the study of the temperature behaviour[17]. A summary of results is given in figure 8, showing a very flat behaviour of $\lambda_{d\mu t}$ up to 300 K, and increasing rates at higher temperatures. The $\lambda_{d\mu t}$-rates consist of 2 components: dμt formation on the D_2 molecule ($\lambda_{d\mu t-d}$) and on the DT molecule ($\lambda_{d\mu t-t}$), respectively, both exhibiting a quite different resonance behaviour.

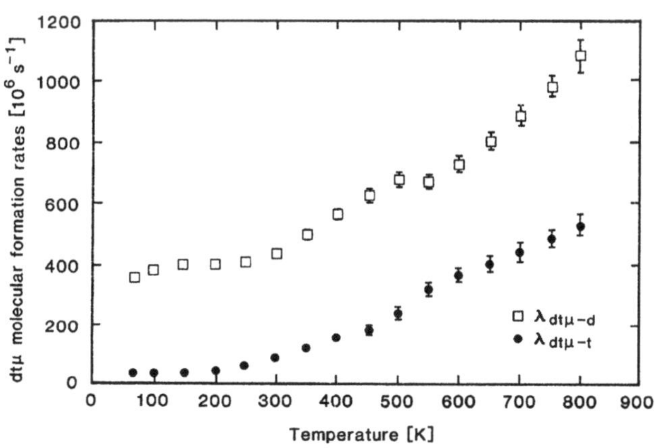

Figure 8. Temperature dependence of $\lambda_{d\mu t-d}$ (on D_2) and $\lambda_{d\mu t-t}$ (on DT) as measured by S.E. Jones et al.[17] at LAMPF.

At PSI, the main efforts were concentrated to the low temperature region (mostly 23 K to 35 K), trying to disentangle the complicated d-μ-t kinetics at constant temperature and at conditions where densities and molecular compositions could be widely varied[18]. In 1987, a second set of data was taken with improved experimental conditions. The range of measurements was extended to mixtures of frozen hydrogen isotopes and to nonequilibrated D_2-DT-T_2 compositions[19]. Some of the preliminary results are given in figures 9 and 10. Figure 9 shows the dt-cycle rates λ_c versus tritium concentrations c_t at various densities of gaseous, liquid and solid phases. The connection between the observable cycle rate and mesomolecule formation is given by the (simplified) formula

$$\frac{1}{\lambda_c} = \frac{c_d q_{1s}}{\lambda_{dt}} + \frac{1}{\lambda_{d\mu t}} \tag{3}$$

where $c_d q_{1s}$ is the probability of a free muon reaching the μd(1s) ground state and λ_{dt} is the d$\rightarrow$t transfer rate from that state. The solid line in figure 9 connects the results in liquid dt, corresponding to $\lambda_{d\mu t-d} \sim 4-5\cdot10^8 s^{-1}$ which is about 1000 times faster than the muon decay constant λ_o! All cycle rates are normalized to the density of liquid H_2 ($4.25\cdot10^{22}$ atoms/cm^3); thus very strong anomalous density effects of mesomolecule formation are visible. The points with a star demonstrate the effect of D_2 enrichment in nonequilibrated molecular compositions, since only $\lambda_{d\mu t-d}$ is resonant at low temperature.

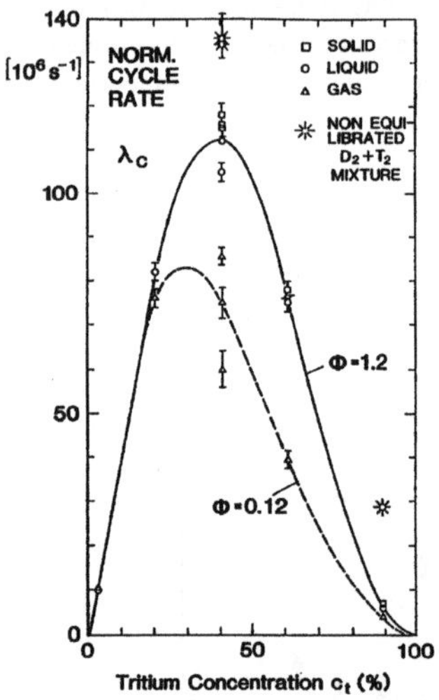

Figure 9. d-t cycle rates, normalized to the density of liquid H_2, measured 1987 by the collaboration at PSI[19] (preliminary data), plotted versus tritium concentration c_t.

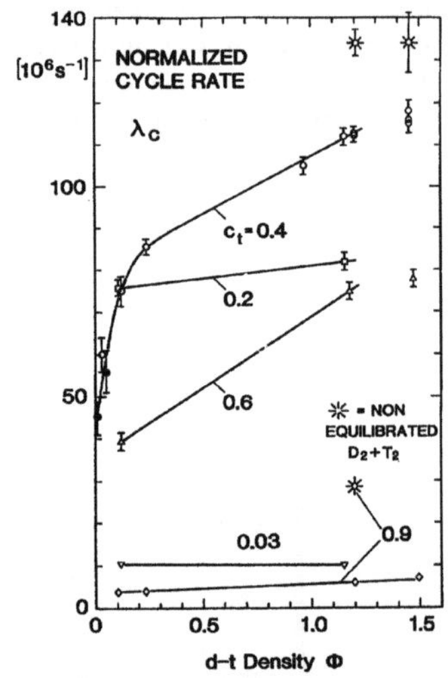

Figure 10. Normalized d-t cycle rates λ_c, measured 1987 by the collaboration at PSI[19], plotted versus density ϕ. A pronounced (non linear) density effect is visible at $c_t \gtrsim 0.4$.

The highest absolute cycle rate ever observed was measured (surprisingly!) at the lowest temperature 12 K in solid nonequilibrated d-t mixture, resulting in $\lambda_c = 2{\cdot}10^8 s^{-1}$. The fusion yield at this condition was determined as Y_f = 125±10 fusions, per muon.

Figure 10 plots the PSI-data (normalized cycle rates λ_c) versus the density. The non constant behaviour is expected from theory as being caused by triple collisions of μt atoms with the nuclei of the D_2 molecule. Resonance components become active with the main $[(d\mu t)_{11}$ dee)] energy level being slightly negative[20]. In triple collisions, the second molecule takes the excess energy over. This is also the explanation for the flat behaviour of $\lambda_{d\mu t-d}$ with T (see figure 8). Figure 10 clearly demonstrates that there is another more complicated density dependence of λ_c at $\phi \leq 0.2$. This effect has not been explained yet. It may be due to cascade processes and muon transfer in excited orbits. Moreover, the density effects disappear if the tritium concentration is lowered to $c_d \leq 0.2$. This is a very surprising phenomenon, because the theoretical calculations of q_{1s} transfer[21] suggest a rather strong density dependence at low c_t.

Thus, the dμt kinetics (see figure 1) poses still several open problems that we will have to study in future experimental and theoretical work. Here follows a list of questions (which is certainly not complete):

- what is the high temperature behaviour of dμt formation on D_2 and on DT. Are theoretical predictions of very large rates in the $\varepsilon \sim 1$ eV region[20] correct?
- why do we not see the effects of the hyperfine structure of the μt atom and dμt mesic molecule, as in the case of pure deuterium?
- why do experimental determinations of q_{1s} transfer disagree with theory predictions[21], see text?
- several experiments have revealed that energy acceleration mechanisms exist during the cascades and transfer of μd and μt atoms. What is their size and their influence on thermalisation and dμt kinetics?

5. Sticking ω_s

Soon after the experimental verification of extremely large d-t cycle rates, it became obvious that sticking may be the ultimate limit to dμt fusion output. Sticking ω_s is the probability that the muon remains trapped on the α emerging from the dt reaction:

$$\begin{aligned} d\mu t \quad & \xrightarrow{(1-\omega_s)} \alpha + \mu + n + 17.6\ MeV \\ & \xrightarrow{\omega_s} \mu\alpha + n + 17.6\ MeV \end{aligned} \tag{4}$$

We distinguish initial sticking ω_s^o, immediately after the fusion reaction, and final sticking ω_s which is smaller, because the initial $\mu\alpha$ system has 3.5 MeV recoil energy and can shake off the muons during slow down:

$$\omega_s = \omega_s^o\,(1-R) \tag{5}$$

R is called the reactivation probability and is about 0.3.

The fusion yield in the fast cycling d-t system is given by

$$\begin{aligned} Y_f &= \int \phi\lambda_c \cdot \exp - (\lambda_o + w\phi\lambda_c)t\ dt \\ &= \phi\lambda_c/(\lambda_o + w\phi\lambda_c) \\ Y_f^{-1} &= \frac{\lambda_o}{\phi\lambda_c} + \omega \end{aligned} \tag{6}$$

for $\phi\lambda_c \gg \lambda_o$: $Y_f = w^{-1}$, thus the maximum possible fusion yield is limited by sticking inverse. Pure dt sticking ω_s is obtained from w by correcting for sticking of other fusion channels than d-t[17–19].

If the cycle rate $\phi\lambda_c$ is known, sticking can be determined from the disappearance of fusion neutrons ($\lambda_n = \lambda_o + w\phi\lambda_c$). This method of sticking measurement was used by both collaborations at LAMPF and at PSI, but is indirect. It has however led to a rather controversial situation, see results plotted versus density in figure 11. While Jones et al.[17] claim a strong density effect of final sticking ω_s, the PSI data[18,19] suggest none or only a weak dependence. At high d-t density all experimental results are considerably lower than theory (ω_s (exp.) ~0.4% versus ω_s (theor.) ~0.6%), see figure 11.

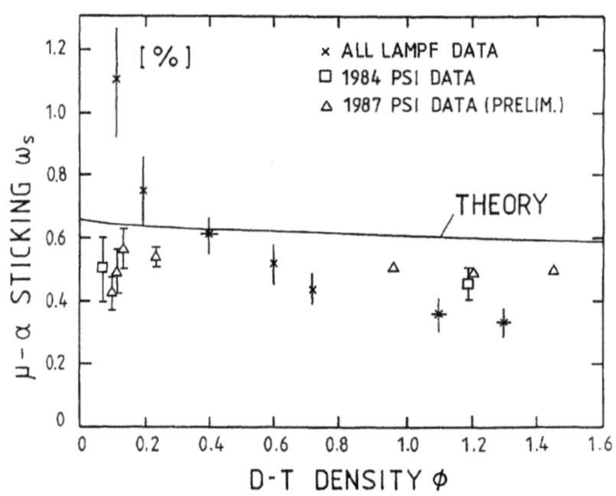

Figure 11. Final sticking ω_s plotted versus d-t density ϕ. LAMPF and PSI results from neutron disappearance rates. The solid curve is the expectation from canonical theory.

To clear up this situation, several experiments aiming at a direct observation are now underway at LAMPF, Rutherford Laboratory (RAL) and at PSI. A precision measurement is not easy, because the quantity ω_s to be determined is small ($\lesssim$0.5%). A promising method is now applied at PSI in collaboration with the Leningrad Nuclear Physics Institute (LNPI)[22]. The charges from d-t fusion products α's and $\mu\alpha$'s are collected in an ionisation chamber, developed at LNPI, and are analyzed with flash ADC's. Distinction of $\mu\alpha$'s from α's is made due to different recombination of charges which is caused by different ionisation density (singly versus doubly charged particles!). Figure 12 shows preliminary results of a test run made in April 1989. In the near future it is hoped, that the issue of sticking will be solved and a clearer assessment of possible applications of MCF can be made.

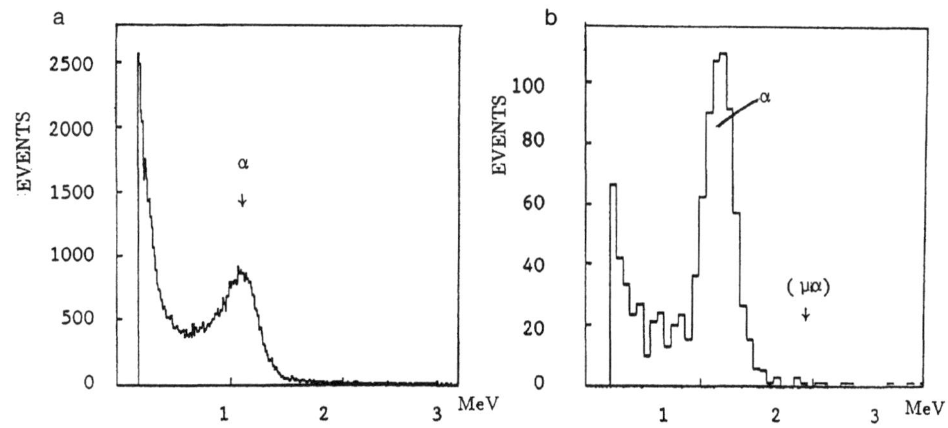

Figure 12. First preliminary test results of the new experiment at PSI, aiming at a direct observation of sticking. The energies of alphas were determined from the charge collection and are plotted (a) as raw data, (b) requiring neutron coincidences. The arrows indicate where the events of sticking $\mu\alpha$'s are expected.

6. "Supercold" Fusion

It was really a big jolt to the scientific community, when on March 23, 1989, the two electrochemists M. Fleischmann and S. Pons announced in the unusual way of a press conference, that they had solved the worlds energy problems in catalyzing deuterium fusion by running a heavy water electrolysis on palladium cathodes. Even better, in their paper[23] which appeared 2 weeks later, they report that the several watts of heat produced in their electrolytic cells, were accompanied by several 1000 neutrons and gammas per second, and that a similar amount of tritium was produced in the cells. This meant, the main fusion process was essentially **aneutronic** to the level 10^{-9}! To the author's question, where the nuclear energy goes, Fleischmann answered: "probably ^{4}He is formed and the fusion energy (24 MeV per fusion!) is transferred to the Palladium lattice". This was indeed hard to believe for a nuclear physicist, especially since Fleischmann had absolutely no proof for it except the heat!

Let us now first discuss the theoretical aspects of the whole issue. Figure 13 shows schematically in the frame of energy potentials the known cases of "hot" fusion and "cold" (muon induced) fusion. The basic problem for 2 approaching nuclei is to overcome the repulsive Coulomb forces until the much stronger nuclear forces take over. Fortunately the quantum-mechanical tunnel effect exists. Therefore it is not necessary that the Coulomb wall must be completely surmounted. In the case of hot fusion (e.g. plasma heating) it is sufficient to reach a temperature somewhere above 100 million degrees (or 10 keV mean kinetic energy of the nuclei) to trigger enough fusion processes to sustain a chain reaction.

In cold fusion, the heating is not necessary, because a negative particle provides the shielding from the positive nuclear charges. The muon is the ideal probe, since due to its mass, 205 times larger than the electron mass, it binds the hydrogen nuclei 205 times closer to each other, see figure 13. Fusion occurs in 10^{-12} sec for the $d\mu t$ and in 10^{-9} sec for the $d\mu d$ system.

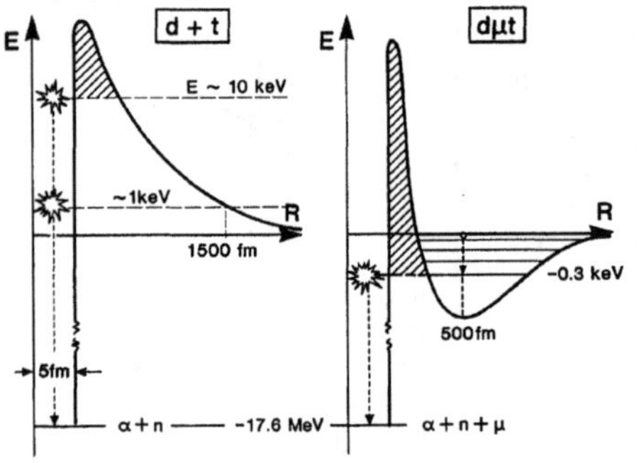

Figure 13. The potential schemes of "hot" (left side) and "cold" fusion (right side).

"Hot" fusion should be considered as the **dynamic case**, which is met either in high temperature plasmas or by acceleration of nuclei, while "cold" fusion applies to the **static case**, where the hydrogen nuclei remain at a certain equilibrium distance. If the observations reported by Pons & Fleischmann should be attributed to real cold fusion, then a mechanism would have to exist which provides a greatly increased shielding by electrons. E.g. a negative particle with 5 electron masses, would provide a charge density, 100 fold larger than a D_2 electron cloud. This would reduce the nuclear distances 5 fold and give rise to observable dd fusion processes. However, such a particle does not seem to exist and new studies[24] have not uncovered yet any theoretical grounds for an increased shielding effect (possibly thought to be a solid state phenomenon).

Anyway the interest was too great and the topic too hot, not to be pursued - so a program was initiated at PSI with the aim to conduct a search study as complete as possible of possible cold fusion phenomena in palladium. Our experimental investigations comprised:

1) Operation of 2 electrolysis cells, one with D_2O/LiOD based electrolyte, the other identical with H_2O/LiOH as a reference, both run with currents from 120 to 545 mA onto electrodes of 1mm and 2mm diameter Pd wires. A continuous search for excess heat was made by runing an absolute, constant cooling - constant temperature calorimetry with $\pm$ 0.1 watt sensitivity (performed by the electrochemistry group of Dr. O. Haas, PSI).

2) Operation of a NE213 neutron counter with n-γ pulse shaping discrimination and of a BGO γ detector, both placed close to the D_2O cell and embedded in elaborate bor paraffine shielding against neutron background from sources originating outside the cell (performed by the PSI-μCF collaboration).

3) Analysis of the Pd after the runs for the occurance of ^{3}He or ^{4}He isotopes (this part was made by Prof. P. Signer's group at the Institute of Crystallo- and Petrography at ETH Zürich).

4) Analysis of the D_2O, prior and after the runs for contents of the tritium isotope (made at the PSI laboratories).

It should be mentioned that the PSI efforts on "supercold" fusion were not directly aimed at low level neutron counting, as was reported first by S.E. Jones et al.[25], and later by several other groups[26]. These neutron levels were all reported to bee below 10^{-12} W and are therefore not very closely related to the the Fleischmann & Pons report of several watts of heat!

A detailed description of our experiments and results can be found in PSI preprints PR-89-17 and in the forthcoming paper accepted by Chimia[27]. Here only a short summary of our findings is given:

1) Throughout operation with a sensitivity of $\pm$ 0.1 W, no evidence of any excess heat was observed. Based on reference 23, a fusion signal of 1 W was expected.

2) No evidence for fusion neutrons nor γ's was observed. Our continuously measured upper limits for n/γ at the source are

$$I_n < 0.14 \text{ n/s} \quad (2.5 \text{ MeV-n})$$
$$I_\gamma < 0.1 \ \gamma\text{/s} \quad (5.5 \text{ MeV-}\gamma)$$

expressed in heat, this is $I_{n,\gamma} < 10^{-13}$ W.

3) No excess ^{3}He nor ^{4}He that could be attributed to fusion was found in the Pd after the runs. The ^{3}He level corresponded to the expected decay of tritium impurities contained in the D_2O already prior to the runs. In terms of heat (assuming d-d fusion via the regular channels p+t and n+^{3}He) this result sets an upper limit of 10^{-9} W (^{3}He).

From the ^{4}He measurements we can set an upper limit to fusion heat (assuming the d-d$\rightarrow^4$He channel to be predominant) of 10^{-6} W.

4) The tritium traces found in the electrolyte after the runs correspond to the original contamination of 10^{-9} moles/L, enriched due to an isotope separation factor of about 2, or several 10^{14} triton atoms. This represents an upper limit of fusion heat via the regular dd fusion channels of 10^{-4} W, assuming the produced tritium would go into the solution.

In conclusion, we can clearly state, that there was NO COLD FUSION AT PSI of the Fleischmann & Pons type. This finding does not exclude the existence of neutrons or neutron bursts at the level below 10^{-13} watt. Such events might however rather be attributed to acceleration mechanisms and belong to "hot" fusion!

References

1. L.W. Alvarez et al., Phys. Rev. 105 (1957) 1127.
2. V.M. Bystritsky et al., JETP 49 (1979) 232.
3. S.S. Gerstein and L.I. Ponomarev, Phys. Lett. 72B (1977) 80.
   S.I. Vinitsky et al., JETP 47 (1978) 444.
4. E.A. Vesman, JETP Lett. 5 (1967) 91.
5. P. Kammel et al., Phys. Lett. 112B (1982) 319 and Phys. Rev. A28 (1983) 2611.
   H. Zmeskal et al., Muon Cat. Fusion 1 (1987) 109 and contr. to the XII Int. Conf. on Few Body Problems (Vancouver, 1989).
6. L.I. Men'shikov et al., Sov. Phys. JETP 65 (1987) 656.
7. Ya.B. Zeldovich and S.S. Gerstein, Sov. Phys. Uspekhi 3 (1961) 593.

8. S.S. Gerstein, JETP 13 (1961) 488.

9. W. Bertl et al., Atomkernenergie 43 (1983) 184.

10. E.J. Bleser et al., Phys. Rev. 132 (1963) 2679.

11. A.V. Matveenko and L.I. Ponomarev, JETP 32 (1971) 871.

12. L.N. Bogdanova, V. Markushin and L.I. Ponomarev, private communication (1989).

13. K.A. Aniol et al., Muon Cat. Fusion 2 (1988) 63.

14. T. Ishihara et al., Dubna preprint E4-88-849 (1988).

15. C. Petitjean et al., "New results in muon catalyzed pd fusion", contr. to the forthcoming Oxford μCF-1989 workshop (September 1989).

16. V.M. Bystritsky et al., JETP 53 (1981) 877.

17. S.E. Jones et al., Phys. Rev. Lett. 56 (1986) 588.

18. W.H. Breunlich et al., Phys. Rev. Lett. 53 (1984) 1137 and 58 (1987) 329.

19. C. Petitjean et al., Muon Cat. Fusion 2 (1988) 37.

20. M.P. Faifman et al., Muon Cat. Fusion 4 (1989) 1.

21. A.V. Kravtsov et al., Muon Cat. Fusion 2 (1988) 183 and Phys. Lett. 132A (1988) 124.

22. C. Petitjean et al., "First direct measurement of final sticking ω_s in muon cat. dt fusion", contr. to the forthcoming Oxford μCF 1989 workshop (September 1989).

23. M. Fleischmann, S. Pons and M. Hawkins, J. Electroan. Chem. 261 (1989) 301 and erratum 263 (1989) 187.

24. B. Delley, "Effect of Electron Screening on Cold Nuclear Fusion Rates", subm. to Euro Phys. Lett.

25. S.E. Jones et al., Nature 338 (1989) 737.

26. H.O. Menlove et al., paper subm. to Nature.
A. Bertin et al., paper subm. to Nuovo Cimento.
A. De Ninno et al., paper subm. to Euro Phys. Lett.

27. J.P. Blaser et al., PSI Preprint PR-89-17, to appear in Chimia (Basel, Sept. 1989).

PRELIMINARY RESULTS ON MUON–CATALYZED pt FUSION

P. Baumann, H. Daniel, T. von Egidy, S. Grunewald
F. J. Hartmann, R. Lipowsky, E. Moser, W. Schott
Technical University of Munich
Garching, Federal Republic of Germany

P. Ackerbauer, W. H. Breunlich, M. Fuchs, M. Jeitler, P. Kammel
J. Marton, N. Nägele, J. Werner, J. Zmeskal
Austrian Academy of Science, Institute for Medium Energy Physics
Vienna, Austria

C. Petitjean, K. Lou
Paul Scherrer Institute
Villigen, Switzerland

H. Bossy, K. Crowe
Lawrence Berkeley Laboratory
Berkeley, CA, USA

R. H. Sherman
Los Alamos National Laboratory
Los Alamos, NM, USA

1 Introduction

Muon–catalyzed pt fusion certainly is not interesting for a possible energy production as dt fusion might perhaps be. Nevertheless there have been good reasons to perform a second experiment on the pt fusion process, after it was carried out for the first time a few years ago [1]. Several fusion channels exist for pt fusion [cf. Table 1]. Besides γ emission muon conversion [Reaction (3)], i. e. the transfer of the fusion energy to the catalyzing muon, and internal pair formation [Reaction (4)] should be possible. From rates and branching ratios of these processes one may draw conclusions about the nuclear structure of the few body system ^{4}He [2] and on spin–flip processes in the μt atom. This will be described in more detail now.

Table 1. Possible pt–fusion reactions. E denotes the energy of the γ quantum or, if stated, the kinetic energy of the muon.

Reaction		Q/MeV	E/MeV
$p\mu t \to {}^4He+\mu+\gamma$	(1)	19.8138	19.7614
$p\mu t \to \mu{}^4He(1s)+\gamma$	(2a)	19.8248	19.7738
$p\mu t \to \mu{}^4He(>1s)+\gamma$	(2b)		
$p\mu t \to {}^4He+\mu_{conv}$	(3)	19.8138	19.2194(μ)
$p\mu t \to {}^4He+\mu+e^++e^-$	(4)	19.7918	

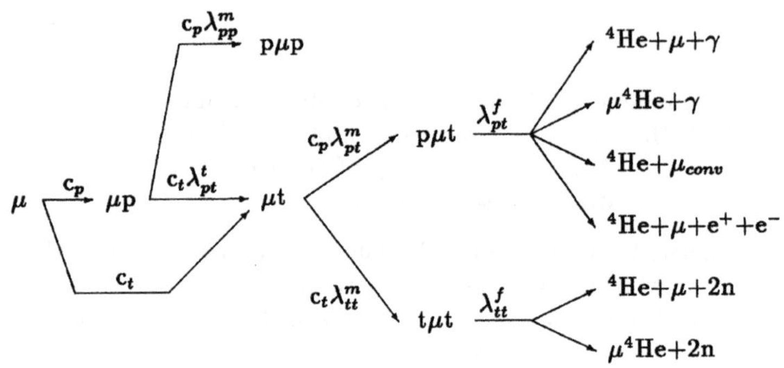

Figure 1: Reactions in a pure pt mixture. Muon decay as a possible side branch is not shown. Concentration of hydrogen isotope i: c_i ($i = p, d, t$ for proton, deuteron and triton, respectively), rates: λ^k_{ij} ($i, j = p, d, t$; $k = t, m, f$ for transfer, molecular formation and fusion, respectively).

2 pt fusion reactions

After formation of the μt atom – either by Coulomb capture or by μp formation and transfer – molecular formation takes place. Fusion starts from the rotational/vibrational state J=0/ν=0 of the $p\mu t$ molecule. The most probable reaction consists in sticking, i. e. the muon remains bound to helium. The side process tt fusion is not negligible but may serve to monitor the time evolution of the μt population.

If we go into more detail and take the hyperfine structure in the μt atom and $p\mu t$ molecule into account (see Figure 2) we see the following:

The 1s level of the muonic tritium atom is split into two levels depending on the spin orientation of the triton and the muon. The population of the three hyperfine levels of the formed muonic pt molecule depends on the state of the muonic tritium atom. Figure 2 shows the calculated transition probabilities according to Bakalov et al. [3]. The fusion process which takes place depends on the relative orientation of the spins of proton and triton: If these two spins are parallel – nuclear spin I=1 – we get a M1–transition under emission of a gamma ray. If the spins are antiparallel (I=0) fusion takes place via an E0–transition and gamma emission is forbidden. In this case a conversion muon or an electron–positron pair will be emitted.

If we change the hyperfine population of the μt(1s) state the yield Y^μ for conversion muon emission is much more affected than the yield Y^γ for gamma emmission.

3 Experiment

In the fall of 1988 our collaboration measured pt fusion at PSI. Our aim was to detect all types of emitted radiation. We looked in particular for converted muons, 19.8 MeV γ rays and muonic x rays from helium or from possible impurities. We registered also electrons from muon decay and neutrons from tt fusion. For the target cell pure iron was used to prevent fluorescent radiation from nickel, copper and zinc with their electronic K_α energies near to the K_α and K_β energies of muonic helium. The target cell had a volume of 40 cm^3. It was coated with a thin layer of silver to let muons hitting the wall disappear quickly. The set–up is shown in Figure 3.

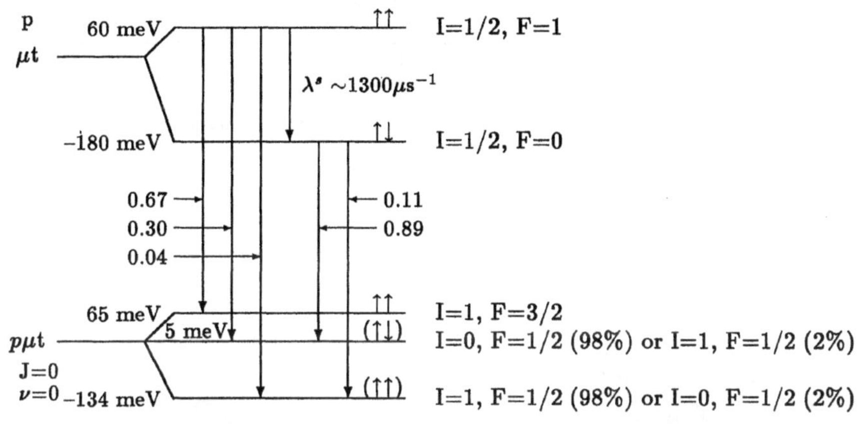

Figure 2. Hyperfine levels of the μt atom and pμt molecule as well as the transition probabilities according to ref. [3]. I denotes the nuclear spin, F the total spin. The two lower states of pμt are not pure. The distance between μt and pμt is drawn not to scale. The value of the spin flip rate λ^s in muonic tritium is taken from ref. [4]. The arrows represent the relative spin orientation of triton and muon in the μt atom and proton and triton in the pμt molecule, respectively.

We measured liquid mixtures with three different tritium concentrations between 0.045% and 0.8%. The target activity at maximum concentration was 800 Curies. The admixture of deuterium was not negligible.

To discover converted muons a plastic scintillation counter was used. The counter was thick enough to stop the muons. We were able to measure time and energy of the subsequent electron from muon decay as second signal with the help of a routing electronic. This gave us a clear signature for muons stopped in the counter.

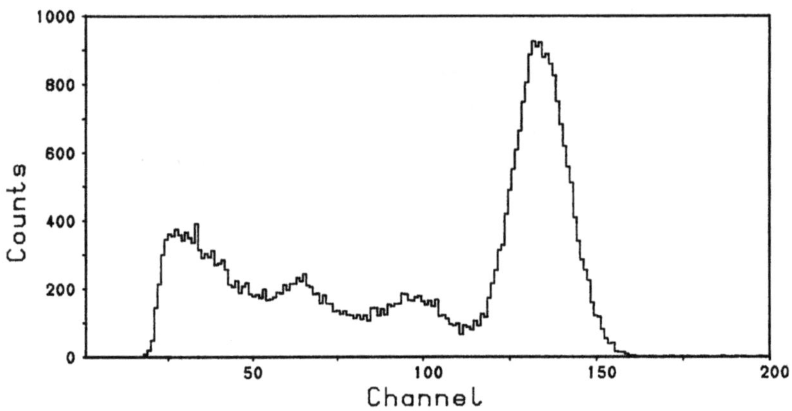

Figure 4. Energy spectrum taken by the counter for converted muons.

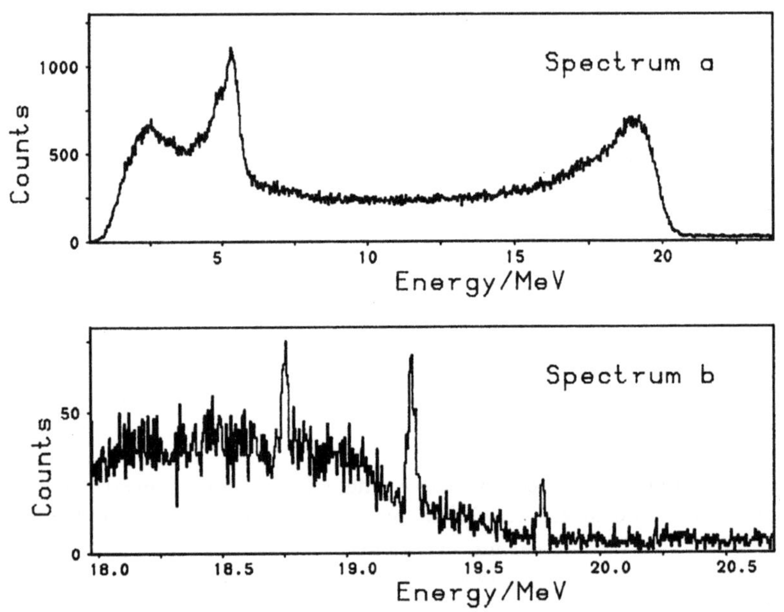

Figure 5. γ spectra. Spectrum a (above) was taken by the BGO detector, Spectrum b (below) was taken by the germanium detector (see text).

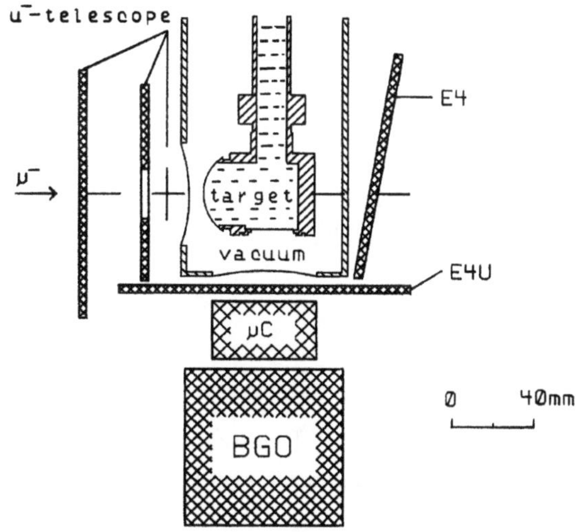

Figure 3. Cross section of the experimental set–up. E4 and E4U: Electron counters (NE102A), μC: muon counter, μ^-–telescope consisting of 3 scintillation counters (NE102A), BGO: BGO detector, used to detect gamma rays. The two Germanium detectors on the left and right sides of the target cell in beam direction as well as the neutron counters are not shown.

4 Preliminary Results

Figure 4 shows an energy spectrum taken by the muon counter described earlier. The conversion muon peak is clearly visible. The structure on the low–energy side has not been fully explained by now. The energy calibration of the counter was done with muons of known momentum coming directly from the μE4 channel at PSI.

Figure 5 presents energy spectra of high–energy γ rays observed by the BGO and by the germanium detector. In Figure 5a the pd and pt fusion peaks can be seen, in Figure 5b one can distinguish full energy, single escape and double escape peaks from the pt–fusion gamma. Background reduction has been achieved by demanding delayed coincidence with an electron from muon decay.

To obtain information on normalized fusion and normalized molecular formation rates we have to look at the time spectrum of the fusion γ rays (Fig. 6), detected with the BGO detector (dots). The solid line represents calculations based on reasonable assumptions for the relevant rates. They have been performed by solving the kinetic equations describing the pt fusion cycle. Obviously the measured time distribution is well reproduced by the calculations.

Table 2 shows the extracted values of disappearance and build–up rates. The disappearance rate corresponds roughly to the sum of the fusion rate and the muon decay rate. The build–up rate corresponds, in first approximation, to the pμt formation rate. The small build–up rate in the mixture with the lowest tritium concentration is not yet been accounted for. A possible explanation can be found in the relatively high deuterium concentration: Only 28% of the μ^- stopped in the liquid form pμt molecules.

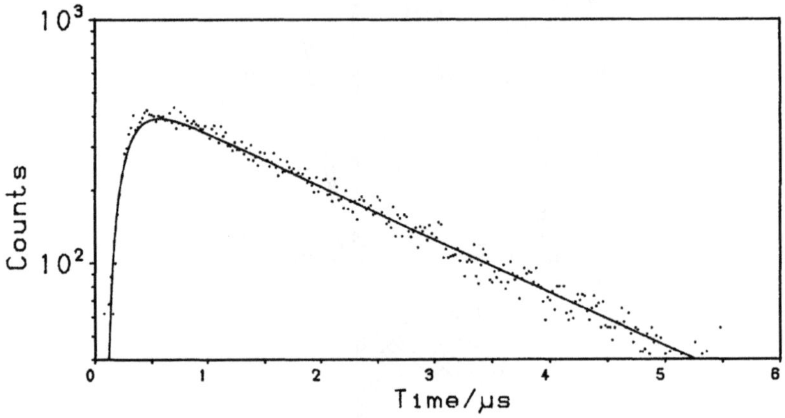

Figure 6. Time spectrum of the fusion gamma rays (dots) in comparison with calculations (solid line).

The lower section of Table 2 presents preliminary results on the ratio of the yields. This ratio is normalized to the value obtained from the 0.8% tritium mixture. In this "ratio of ratios" several uncertainties such as counter efficiency cancel. It is necessary to make many assumptions on various rates to get quantitative results from theory. Furthermore the delay of μt thermalization, which is caused by the deep Ramsauer–Townsend minimum in the elastic cross section, may assume importance. Nevertheless the experimental values for Y^{μ}/Y^{γ} show an increase with decreasing tritium concentration c_t expected in accordance with an increased population of the singlet state in $\mu t(1s)$ at higher c_t.

We wish to thank H. Angerer, H. Weiss and the mechanical workshop of PSI for their valuable help and assistance. Support by German Bundesministerium für Forschung und Technologie and Austrian Science Foundation is acknowledged.

Table 2. Preliminary results. Y^{μ} denotes the yield for muon emission, Y^{γ} the gamma–ray yield. The ratio Y^{μ}/Y^{γ} is normalized to the best value of $(Y^{\mu}/Y^{\gamma})_0$ observed at c_t*10^3=8.1 and c_d*10^3=0.76. For further explanations see text.

	mixture		
concentration			
c_t*10^3	8.1 ± 0.6	1.16 ± 0.06	0.45 ± 0.08
c_d*10^3	0.76 ± 0.05	0.402 ± 0.012	0.37 ± 0.06
build–up rate (μs^{-1})	7.18 ± 0.44	7.11 ± 0.36	5.51 ± 0.30
$p\mu t$ formation rate (μs^{-1}) [5]	6.5	6.5	6.5
disappearance rate (μs^{-1})	0.493 ± 0.004	0.518 ± 0.005	0.530 ± 0.006
fusion rate (γ) (μs^{-1}) [1]	0.06 ± 0.01	0.06 ± 0.01	0.06 ± 0.01
μ decay rate (μs^{-1})	0.455	0.455	0.455
$\frac{Y^{\mu}/Y^{\gamma}}{(Y^{\mu}/Y^{\gamma})_0}$	1.00 ± 0.02	1.20 ± 0.03	1.38 ± 0.04

References

[1] F. J. Hartmann, H. Bossy, H. Daniel, T. von Egidy, W. Neumann, H. Plendl, W. Schott, P. Weißgerber, P. Ackerbauer, W. H. Breunlich, M. Cargnelli, M. Jeitler, P. Kammel, J. Marton, N. Nägele, A. Scrinzi, J. Werner, J. Zmeskal, C. Petitjean, J. Bistirlich, K. Crowe, M. Justice and R. H. Sherman, Muon Catal. Fus. 2, 53 (1988).

[2] L. N. Bogdanova and V. E. Markushin, Muon Catal. Fus. 4, 103 (1989).

[3] D. D. Bakalov, S. I. Vinitskiĭ and V. S. Melezhik, Zh. Eksp. Teor. Fiz. 79, 1629 (1980) [Engl. transl.: Sov. Phys. JETP 52, 820 (1980)].

[4] L. Bracci, C. Chiccoli, P. Pasini, G. Fiorentini, V. S. Melezhik and J. Wozniak, Phys. Lett. A134, 435 (1989).

[5] L. I. Ponomarev and M. P. Faĭfman, Zh. Eksp. Teor. Fiz. 71, 1689 (1976) [Engl. transl.: Sov. Phys. JETP 44, 886 (1976)].

COLLISIONS OF MUONIC ATOMS

L. Bracci[(a)], C. Chiccoli[(b)], G. Fiorentini[(c)],
V.S. Melezhik[(d)], P. Pasini[(b)] and J. Wozniak[(e)]

a) Dipart. di Fisica dell'Univ. and Sez. I.N.F.N., Pisa, Italy
b) I.N.F.N., Sez. di Bologna and C.N.A.F., Bologna, Italy
c) Dipart. di Fisica dell'Univ. and Sez. I.N.F.N., Cagliari, Italy
d) Joint Institute for Nuclear Research, Dubna, USSR
e) Institute of Physics and Nuclear Techniques, Cracow, Poland

1. INTRODUCTION

The three-body problem is a classical topic in both classical and quantum mechanics. In this latter framework, several approaches have been developed (variational calculations, Faddeev method, adiabatic representation for instance). Here, we want to focus our attention to the very special case where the dominant force is the Coulomb interaction, and the continuous spectrum is investigated.

This problem, apart from its intrinsic interest, plays a very important role in the analysis of the processes connected with muon catalysed fusion. In fact, several elastic and inelastic processes may determine the yield of reactions per muon. We only mention two aspects where a detailed knowledge of the collision processes can turn out to play a critical role. First, the formation of the mesic molecule, for example in the case of the $dd\mu$ system, is dependent on the spin state of the muonic atom. Second, the formation rate is also energy dependent. Thus, in order to have an accurate estimate of the formation rate and of the rates for the ensuing processes, it is necessary to have a precise knowledge of the rates involving both elastic and inelastic collisions. Actually, even in the case of pure deuterium, the elastic processes

Electromagnetic Cascade and Chemistry of Exotic Atoms
Edited by L. M. Simons *et al.*, Plenum Press, New York, 1990

can play a relevant role in determining whether the muonic atoms taking part in the reactions are thermalized at the time of molecule formation. Furthermore, the molecular formation rate does depend on the fraction of F=1/2 and F=3/2 states which are populated. The situation is even more involved when a mixture of different isotopes of hydrogen is considered. The number of different processes which can occur is then rather large (muon transfer, spin flip, elastic collisions...) and only a detailed knowledge of the corresponding cross sections over a rather large range of energies can afford a reliable estimate of the rate of the processes of interest: mesomolecule formation, nuclear fusion, neutron yield.

Since many years, a systematic approach to the study of the three-body Coulomb problem has been developed by the Dubna group, on the ground of the adiabatic representation of the three-body Coulomb problem[1]. This method has been proved to give excellent results in the search of the spectrum of mesic molecules. The complete spectrum was indeed accurately calculated, and the existence of the weakly bound state in the $(dt\mu)$ molecule was brilliantly predicted.

Only more recently this method was extensively applied to the study of collision processes. There, the situation is not yet completely settled. Notable difficulties still exist for the collisions of non-symmetric systems, which are mainly connected with the high value of the threshold for muon transfer (several tens eV, depending on the particular process) in comparison with the collision energy.

First, we briefly review the general aspects of the adiabatic representation of the three-body Coulomb problem, then we discuss the method of solution of scattering problems, stressing some peculiar aspects of the technique which we use and finally we present some results.

2. THE ADIABATIC REPRESENTATION

We start with the Schroedinger equation for the three-body Coulomb system

$$(-\frac{1}{2M_t}\Delta_{\mathcal{R}} + H - E)\Psi = 0 \tag{1}$$

where $M_t = m_a + m_b + m_c$, m_a, m_b, m_c being the masses of the nuclei (a,b) and of the light particle (c), $\mathcal{R}$ is the center of mass coordinate and H can be written as $H = T_a + h_a + 1/R$, with

$$T_a = -\frac{1}{2M_0}((\nabla_{\mathbf{R}} + \frac{\kappa}{2}\nabla_r)^2 - \frac{(1+\kappa)}{2}\Delta_r) \tag{2}$$

$$h_a = -\frac{1}{2\mu_a}\Delta_r - \frac{1}{r_{ac}} - \frac{1}{r_{bc}} + \frac{1}{r_{ab}} \tag{3}$$

$$\frac{1}{\mu_a} = \frac{1}{m_c} + \frac{1}{m_a} \qquad \frac{1}{M_0} = \frac{1}{m_a} + \frac{1}{m_b} \tag{4}$$

$\mathbf{R} = \mathbf{r}_a - \mathbf{r}_b$, $\mathbf{r} = \mathbf{r}_c - (\mathbf{r}_a + \mathbf{r}_b)/2$, $\kappa = (m_a - m_b)/(m_a + m_b)$. After subtracting the CM motion, we get $(H - \varepsilon)\Psi(\mathbf{r}, \mathbf{R}) = 0$ with $\varepsilon = E - P_t^2/2M_t$, where P_t is the total momentum of the three-body sistem.

The main idea of the adiabatic representation consists in the expansion of the exact wave function over the basis of the two-centre Coulomb problem with nuclei at distance R from each other. This approach dates back to the thirties[2], and turns out to be powerful when the mass of the light particle is considerably smaller than the nuclear masses. Indeed, in that case the energy of the nuclear vibrational motion $E_{vib} \approx (m/M)^{1/2} E_{mol} \ll E_{mol}$ and still the Born-Oppenheimer approximation is a reasonable approach to the problem. In the case of muonic systems, which we will consider henceforth ($m_c = m_\mu$), this is not the case, and one needs an expansion over a sufficiently large basis of the two-body Coulomb problem in order to get a reliable result.

The properties of the adiabatic basis are extensively studied in the monograph [1]. Here we only recall that the eigenfunctions of the discrete spectrum can be labeled with parabolic quantum numbers k, n_2, m. For the case of nuclei with identical charge one has an additional quantum number (g,u) representing the parity of the wave function under inversion of the coordinates of the light particle.

The eigenfunctions $\Psi(\mathbf{r}, \mathbf{R})$ of the Hamiltonian H can be classified according to the eigenvalues of the mutually commuting conserved operators $\mathbf{J}^2$, J_z and parity P. We have

$$\Psi^{J\lambda}_{mm_J}(\mathbf{r}, \mathbf{R}) = \sum D^{J\lambda}_{mm_J}(\Phi, \Theta, \phi) F^{J\lambda}_m(\xi, \eta; R) \tag{5}$$

where Θ and Φ are angles describing the orientation of R and ξ, η, ϕ are the muon spheroidal coordinates. For the case of identical nuclei, the g,u quantum numbers have to be added.

When the above expansion is inserted into the Schroedinger equation and averaged over the angles, we get a system of equations

$$\sum (H^{J\lambda}_{mm'} - \epsilon\delta_{mm'}) F^{J\lambda}_m(\xi, \eta; R) = 0 \tag{6}$$

where $H^{J\lambda}_{mm'}$ are effective potentials arising from the average of the operators which appear in the Hamiltonian.

The above equation is solved by expanding the wave functions F in the adiabatic basis described above: $F_m^{J\lambda}(\xi,\eta;R) = \sum \frac{\chi(R)}{R}\varphi(\xi,\eta;R)$, where the functions φ are elements of the adiabatic basis and the coefficients χ(R) are the unknown coefficients of the expansion. When this final step is carried out and an average over the muon coordinates is taken, we get a system of (infinite) coupled equations

$$\{\hat{I}(\frac{d^2}{dR^2} + 2M\epsilon) - \hat{U}_{ii}\}\chi_i = \sum_{j\neq i}\hat{U}_{ij}\chi_j(R) + \sum\int_0^\infty dk\hat{U}_{ij}(k,R)\chi_j(k,R) = 0 \quad (7)$$

Here ϵ is the collision energy, reckoned from the ground state of the $(a\mu)$ atom, nucleus a being the heavier one, and $M = M_0/\mu_a$. The effective potentials $\hat{U}_{ij}$ are the average over the wave functions of the adiabatic basis of the formerly introduced potentials $H^{J\lambda}_{mm'}$. Each wave function component χ_i is actually a two-dimensional vector. The two components correspond to the u, g states in the cases of equal nuclei, or to states which dissociate into $(a\mu) + b$ or $(b\mu) + a$ systems for $R \longrightarrow \infty$. Accordingly, the matrix elements $\hat{U}_{ij}$ are 2×2 matrices. $\hat{I}$ is the identity matrix.

3. SCATTERING PROBLEMS

Application of system (7) to scattering problems so far has been extended only to the case when two channels are open. For the case of collisions with identical nuclei, the two channels correspond to the two possible states of the hyperfine structure of the muonic atom (F = 0,1 for $(p\mu)$ and $(t\mu)$, F = 1/2,3/2 for $(d\mu)$). The threshold for the excited state is .182 eV for the atom $(p\mu)$, .2373 eV for $(t\mu)$ and .0485 eV for $(d\mu)$. In the case of different nuclei, the two channels correspond to states where in the asymptotic region the muon is bound to the heavier or to the lighter nucleus in the ground state. The thresholds are 48.042 eV for the d-t collision, 134.709 eV for the p-d collision and 182.751 eV for the p-t case. We label σ_{11} and σ_{22} the elastic cross sections in the two channels, whereas σ_{12} and σ_{21} denote the cross sections for transitions from one channel to another. We also note that the inelastic processes, for both the symmetric and the non-symmetric cases, are muon transfer processes. Actually, with Coulomb interaction only, spin flip occurs only due to the exchange interaction connected with the symmetry of the total wave function.

In order to get the cross sections from system (7), we first truncate the system to a finite number (N) of equations, and look for c N-component wave functions (c=1 below threshold, c=2 above threshold) χ_i^c, i=1,2,...N, which obey definite boundary conditions. We have, for $R \longrightarrow \infty$,

$$\chi_i^c(R) = \delta_{ic} j_J(k_i R) - T_{ci} n_J(k_i R) \tag{8}$$

for $i \leq c$, where j_J and n_J are Bessel and Neumann functions for angular momentum J, whereas for $i > c$ the components are exponentially damped. The technique of solution of system (7), supplemented with boundary and normalization conditions, is based on an extension of Newton's method for the search of a root of a function f(x)

$$x_{k+1} = -f'(x_k)^{-1} f(x_k) \qquad ; \qquad f(0) = x_0 \tag{9}$$

Here, instead of the real variable x we have the set $\{\chi_j, \epsilon, \delta_\alpha\}$, where δ_α is the phase shift in channel α ($\delta_\alpha = \tan^{-1}(\mathrm{T}_{\alpha\alpha})$), ϵ is the energy and $\{\chi_i\}$ is the wave function describing the nuclear motion. The role of f(x) is played by a multilinear functional φ_i, such that the vanishing of any component corresponds to a condition which the solution of the scattering problem must fulfill (wave equation, boundary conditions, normalization condition)[3].

After truncation and discretization of momentum k, the system of equations has the following form:

$$\hat{I}\frac{d^2}{dR^2}\chi_i + \hat{I}(\bar{k}_i^2 - \frac{J(J+1) - 2m^2}{R^2})\chi_i - 2\sum_j \hat{Q}_{ij}\frac{d\chi_j}{dR} - \sum_j \hat{B}_{ij}\chi_j = 0 \tag{10}$$

where $\bar{k}_i^2 = 2M(\epsilon - E_i)$ are the momenta corresponding to the the collision energy in the i-th channel. $\hat{\mathrm{Q}}_{\mathrm{ij}}$ and $\hat{\mathrm{B}}_{\mathrm{ij}}$ are effective potentials. For any value of the pair ij both $\hat{\mathrm{Q}}_{\mathrm{ij}}$ and $\hat{B}_{ij}$ are 2×2 matrices, and $\bar{\mathrm{k}}_{\mathrm{i}}^2$ is a 2×2 diagonal matrix as well.

In order to get the desired accuracy, the final system must contain some hundred equations. In this respect, it is worth stressing how the correct asymptotic limit is recovered in the aforementioned approach. Taking into account the asymptotic form of the effective potential, in the asymptotic region the system has the following form, where the χ_1 component, whose asymptotic behaviour determines the δ_1 phase shift, is singled out:

$$\frac{d^2}{dR^2}\chi_1 + (\bar{k}_1^2 - \frac{J(J+1)}{R^2})\chi_1 - 2\sum_j q_{1j}\frac{d\chi_j}{dR} - \frac{\sqrt{(J-m)(J+m+1)}}{R}\sum_j b_{1j}\chi_j = 0$$

$$(11)$$

$$(\frac{d^2}{dR^2} + (\bar{k}_i^2 - \frac{J(J+1) - 2m^2}{R^2}))\chi_i - 2q_{i1}\frac{d\chi_1}{dR} - \frac{\sqrt{(J+m)(J-m+1)}}{R}b_{i1}\chi_1 = 0; \quad i > 1$$

q and b are matrices of the form

$$q = \begin{pmatrix} 0 & q_{12} & q_{13} & \dots \\ q_{21} & 0 & 0 & \dots \\ q_{31} & 0 & 0 & \dots \\ \dots & \dots & \dots & \dots \end{pmatrix} \quad b = \begin{pmatrix} 0 & b_{12} & b_{13} & \dots \\ b_{21} & 0 & 0 & \dots \\ b_{31} & 0 & 0 & \dots \\ \dots & \dots & \dots & \dots \end{pmatrix}$$

with $q_{i1} = -q_{1i}$, $b_{i1} = b_{1i}$. They are connected to the parts of the formerly introduced Q, B matrices which survive in the asymptotic region according to the equation $q = Q(1-\kappa)/2M$, $b = B(1-\kappa)/2M$. If we look for solutions which have an exponential dependence on R, for the components χ_i with $i > 1$, after substitution of $\chi_i = \sum_s \hat{C}_{is} A_s \exp(-p_s R)$ into eqs (11) and solution of the corresponding equations for p_s, we get the following expression[4]:

$$\chi_i = \frac{1}{2M(E_1 - E_i)}[2q_{i1}\frac{d}{dR} + \frac{(J-m)(J+m+1)}{R}b_{i1}]\chi_1 \tag{12}$$

Substituting the above expression into the the equation for χ_1, we find for χ_1 the following equation:

$$(1-\alpha(N))\frac{d^2}{dR^2}\chi_1 + (\bar{k}_1^2 - \frac{J(J+1)}{R^2}(1-\beta(N)))\chi_1 = 0 \tag{13}$$

where we have introduced the notation

$$\alpha(N) = \frac{(1-\kappa)^2}{2M} 4\sum_i^N \frac{|q_{1i}|^2}{E_i - E_1}, \quad \beta(N) = \frac{(1-\kappa)^2}{2M} \sum_i^N \frac{|b_{1i}|^2}{E_i - E_1}.$$

We denote by α_∞ and β_∞ the series extended to the complete adiabatic basis.

We stress again that in the above equation $\bar{k}^2$ is calculated using the mass M which is the reduced mass of the colliding nuclei (in units μ_a) whereas we expect that for example in channel 1 in the asymptotic limit the mass to be used is the reduced mass of the system $(a\mu) + b$. Now, this condition is actually fulfilled in the limit of $N \longrightarrow \infty$. This can be seen using the following sum rule

$$4\sum \frac{q_{1i}q_{i1}}{E_i - E_1} = 1/2 \tag{14}$$

which implies that $M/(1-\alpha_\infty) = \mathcal{M}$, where $\mathcal{M}$ is the corrected reduced mass.

The above sum rule is proved along standard lines,starting from the equality (which is true in the asymptotic limit)

$$4\sum \frac{q_{1i}q_{i1}}{E_i - E_1} = \sum \frac{<\psi_1|\frac{\partial}{\partial z}|\psi_i><\psi_i|\frac{\partial}{\partial z}|\psi_1>}{E_i - E_1} \tag{15}$$

and expressing $\frac{\partial}{\partial z}$ in terms of the commutator [z,H]. As a result, we get

$$4\sum \frac{q_{1i}q_{i1}}{E_i - E_1} = -i\sum < \psi_1|p_z|\psi_i >< \psi_i|p_z|\psi_1 >= 1/2 \tag{16}$$

In a similar way it can be proved that also the mass appearing in the centrifugal potential converges to the right value $\mathcal{M}$, due to the fact the the complete sum β_∞ over the whole basis turns out to equal α_∞. In conclusion, were it possible to include contributions from all the states, no problem with the asymptotic behaviour should arise for both the q and b terms.

When the system is truncated, on the other hand, we have to take into account the effect on the mass, and consequently on the momenta and the phase shifts, of this truncation. To this purpose, we compare the asymptotic form which is implied by the truncated system with the correct asymptotic limit. For the sake of simplicity, we only consider the case with J=0, for which the β term plays no role.

The correct asymptotic behaviour for χ_1 is governed by the equation

$$\frac{d^2}{dR^2}\chi_1 + k_1^2\chi_1 = 0 \tag{17}$$

where k_1^2 is the correct momentum, $k_1^2 = \frac{\bar{k}_1^2}{1-\alpha_\infty}$ The asymptotic behaviour derived from the truncated system is governed by the equation

$$\frac{d^2}{dR^2}\chi_1(1-\alpha) + \bar{k}_1^2\chi_1 = 0 \tag{18}$$

where $\alpha = \alpha(R)$ is conveniently represented as follows:

$$\alpha = \frac{\bar{\alpha}_D}{2M} + \frac{\bar{\alpha}_C}{2M} \tag{19}$$

with $\bar{\alpha}_D$ and $\bar{\alpha}_C$ representing the contribution of the discrete and the continuous spectrum respectively. For large R, the total contribution of the discrete spectrum is known to be

$$\bar{\alpha}(n_d) = 128\sum_i^{n_d} \frac{n^4(n_1-n_2)^2(n-1)^{2n}}{(n^2-1)^5(n+1)^{2n}} \rightarrow 0.28$$

This limit is saturated with a limited number of components of the discrete spectrum. In our calculations we have used 26 pairs of equations corresponding to the discrete spectrum and have verified that the limiting value is attained.

The situation is different for the contribution of the continuous spectrum. In that case, the range of R where saturation is attained grows with the number of states included into the system. We found that, in the conditions we chose for doing calculations (more than 200 components of the continuum spectrum) saturation occurs for $R < 20$. What is important is that R=20 is already in the asymptotic region, i. e. the wave function has an oscillating behaviour, although the wavelength is slightly uncorrect due to lack of saturation of the sum rule. When states with angular momentum $J \neq 0$ are considered, the centrifugal potential is multiplied by a factor $1-\beta$, (see eq. 13) where, for β, similar arguments to those previously presented for α apply. Saturation is attained only in the range $R < 20$. In figs. 1 and 2 the functions $\bar{\alpha}(R)$ (=$2M\alpha_C(R)$) and $\bar{\beta}(R)$ (=$2M\beta_C(R)$) are presented, together with the separate contribution of single shells (set of states $\varphi(k, R)$ for a lot of values of the discretized momentum k). We see that an increase of the number of states of the continuous spectrum extends the range of R where the saturation is attained.

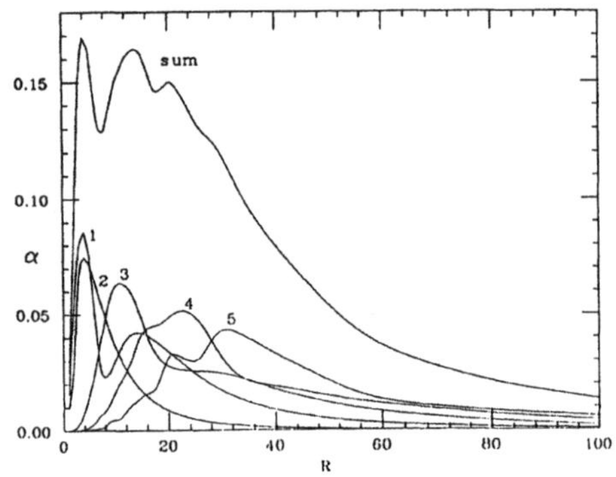

Figure 1

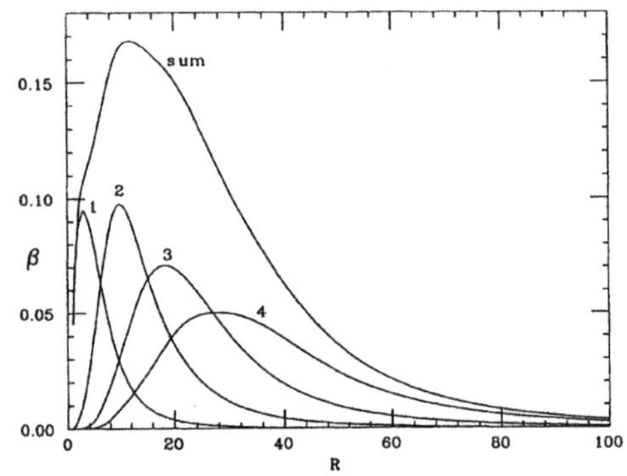

Figure 2

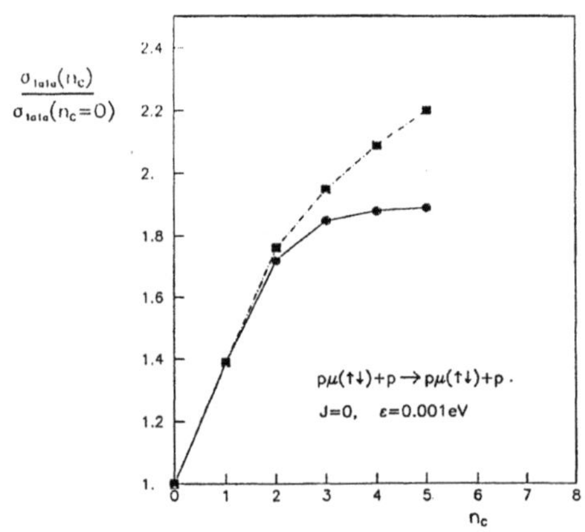

Figure 3

In order to account for the truncation of the system, we have used the following procedure[5]. In the system of equations we have used the momentum corrected for the contribution due to the discrete spectrum, i.e. we use the momentum $\frac{\bar{k}^2}{1-\alpha_D}$ instead of $\bar{k}^2$. In this case the wave equation for χ_1 differs from the correct one for a perturbative term ΔV such that

$$\Delta V\chi = -\alpha\chi'' + \bar{k}_1^2\left(\frac{1}{1-\alpha_D} - \frac{1}{1-\alpha_\infty}\right)\chi \tag{20}$$

This can be approximated as

$$\Delta V\chi = -\alpha\chi'' + \frac{\bar{k}_1^2\alpha_{C,\infty}}{(1-\alpha_D)(1-\alpha_\infty)} \approx -(\alpha(R)\frac{d^2}{dR^2} - \alpha_{C,\infty})\chi \tag{21}$$

The correction in β can be disposed of along the same lines. The total perturbation which accounts for the truncation of the system is

$$\Delta V\chi = (-\alpha(R)\frac{d^2}{dR^2} + \alpha_{C,\infty} + (\beta - \beta(R_{as}))J(J+1)/R^2)\chi \tag{22}$$

The effect of the above perturbative term on the phase shift can be obtained by standard perturbation theory. If χ and χ_N are the exact wave function and the one yielded by the truncated system with N equations, it is immediate to arrive at the following equation

$$\frac{d}{dR}[\chi_N\frac{d}{dR}\chi - \chi\frac{d}{dR}\chi_N] - \chi_N\Delta V\chi = 0 \tag{23}$$

which implies

$$\chi_N\frac{d}{dR}\chi - \chi\frac{d}{dR}\chi_N = \int_{R_{as}}^{R} dr\chi_N\Delta V\chi \tag{24}$$

where $R_{as} = 20$ (in mesic units) is the value of R where α and β begin to differ from the exact values α_∞ and β_∞ and consequently ΔV does not vanish.

By substituting in the LHS the asymptotic behaviour for the χ functions and using χ_N instead of χ in the R.H.S, we get for the correction to the phase shift the term

$$\Delta\delta = \frac{1}{\bar{k}_1}\int_{R_{as}}^{R_{max}} dr\chi_N\Delta V\chi_N \tag{25}$$

Note that the above correction is slightly dependent on R_{max} (the range of integration for the numerical solution of the system). This is correct, since it can be shown that the use of a truncated system, and consequently of incorrect boundary, conditions entails a dependence of the phase shift on the value of R_{max}, which is compensated by the R_{max}-dependent part of the above correction.

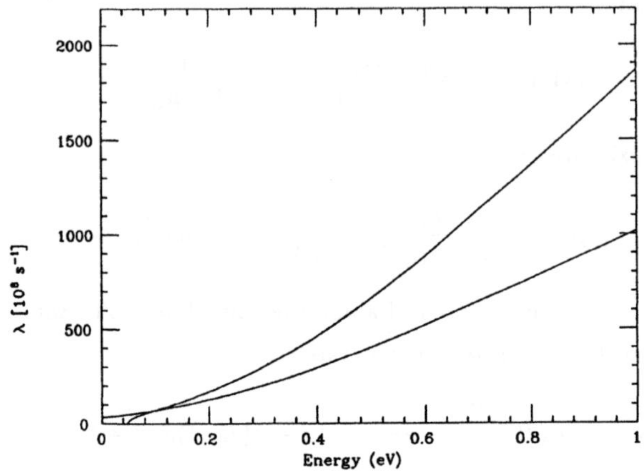

Figure 4

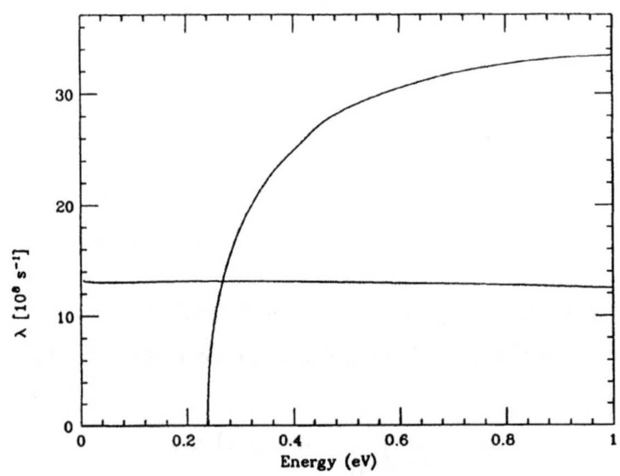

Figure 5

We remark also that in the above correction the dominant role is played by the α-dependent part. The β dependent part, which vanishes for J=0, gives only minor modifications to the phase shifts. On the other hand, the correction is important, since it ensures the convergence of the procedure when the number of equations is increased. This is true mainly at low energy. As an example, we show in fig. 3 the dependence of the elastic cross section for $(p\mu)+p$ collision at $\epsilon = 10^{-3}$ eV, for a value of the total spin S=1/2, on the number of shells of the continuous spectrum included into the system. Fig. 3 clearly shows that inclusion of the correction discussed above results in a convergent behaviour of the cross section with the increase of the number of states of the continuous spectrum included in the system.

4. SOME RESULTS

So far, we have used the above described method to estimate the spin flip rates in collisions with identical nuclei $(\mathrm{t}\mu + \mathrm{t}, \mathrm{p}\mu + \mathrm{p}, \mathrm{d}\mu + \mathrm{d})$, for energies below 1 eV, and to calculate the elastic and inelastic cross sections for collisions of $(\mathrm{t}\mu)$ atoms with tritium for energies up to some tens eV.

Since, with Coulomb interaction, the total spin S is conserved, we have calculated the transition cross sections for S=1/2 and 3/2 (for t-t and p-p) and for S=1/2, 3/2 and 5/2 for d-d. Taking into account the multiplicity of the states, the total rate is:

$$\lambda_{21} = \frac{1}{6}\lambda_{21}^{1/2} + \frac{1}{3}\lambda_{21}^{3/2}$$

$$\lambda_{12} = \frac{1}{3}\lambda_{12}^{1/2} + \frac{2}{3}\lambda_{12}^{3/2}$$

for the case of d-d, and

$$\lambda_{21} = \frac{1}{3}\lambda_{21}^{1/2} \quad , \quad \lambda_{12} = \lambda_{12}^{1/2}$$

for the case of p-p and t-t.

We took into account only channels with J=0,1. For J=0 48 states of the discrete spectrum and 190 states of the continuous spectrum were included, for J=1 only the discrete spectrum was considered. The results[7] are generally in fair agreement with the results of Bubak and Faifman[8], apart from the d-d case, where, however, the discrepancy does not exceed 10%. We report some results in figs. 4,5, where the spin flip rates for the case of ddμ (fig. 4) and ttμ (fig. 5) are shown.

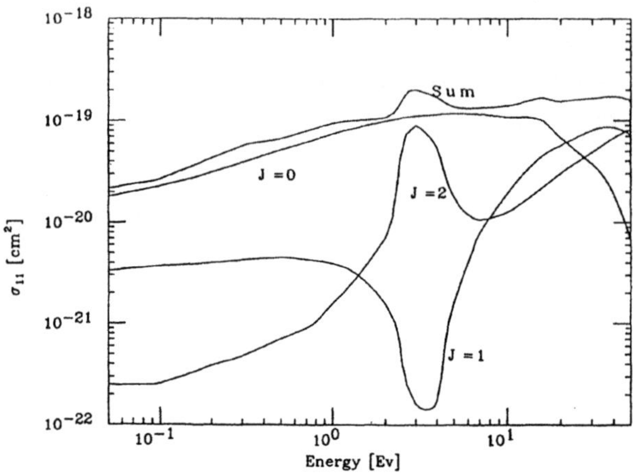

Figure 6

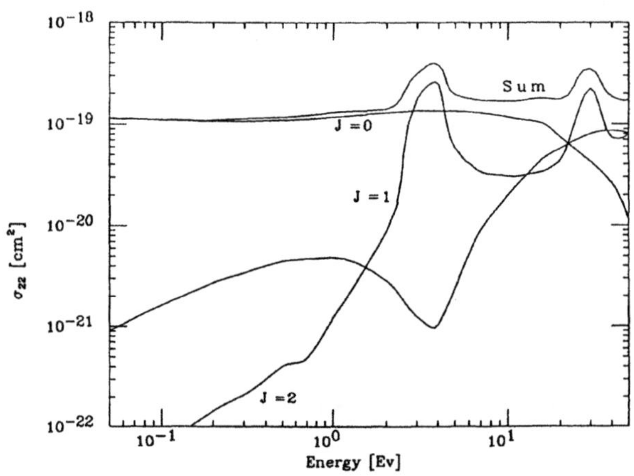

Figure 7

A comparison with experiment is possible for the d-d case, where accurate experimental results are available[9,10]. In the experiments an effective hyperfine transition rate is measured, which is affected by the back-decay process, i.e. the possibility that the (ddμ) molecule can be dissociated before fusion. The effective λ^{hf} is given by[11]

$$\lambda^{hf} = \lambda_{21} + \sum_M \lambda_{2M} \frac{\Gamma_{M1}}{\lambda_f + \Gamma_M} \tag{26}$$

where λ_{2M} is the formation rate, which is favoured in the excited state of the hyperfine structure,Γ_{MF} is the dissociation rate from the mesomolecular state M to d +$(d\mu)_F$, $\Gamma_M = \sum_F \Gamma_{MF}$ and λ_f is the effective fusion rate. Using the results of ref. 11 we find that at 23.8 K $\lambda^{hf} = (4 + 33.4)10^6 s^{-1}$, whereas at 34 K $\lambda^{hf} = (6 + 35.1)10^6 s^{-1}$. In both cases, the first figure is the contribution of the back-decay channel. The results are in reasonable agreement with those reported in refs. 9,10.

The method presented above has been applied also to the calculation of elastic and inelastic cross sections for collisions between (tμ) atoms and tritium in an energy range up to 100 eV. The angular momenta considered are J=0,1,2. Some results are presented in figs. 6,7, where the elastic cross sections σ_{11} and σ_{22} are shown.

An attempt has been made to estimate the effect of nuclear scattering on the energy distribution of $(t\mu)$ atoms in pure tritium. This can be interesting in connection with the measurement of hyperfine transition rate in d-t mixtures with high abundance of tritium. A very rough sketch of the energy distribution of $(t\mu)_{F=0}$ and $(t\mu)_{F=1}$ atoms can be obtained by assuming that fast (tμ) atoms are originated from Coulomb deexcitation from n=2 state. If λ_A, λ_C, λ_R are the Auger, Coulomb and radiative transition rates from the n=2 level, we estimate $P_f = \frac{1}{4}\frac{\lambda_C}{\lambda_A+\lambda_C} + \frac{3}{4}\frac{\lambda_C}{\lambda_A+\lambda_C+\lambda_R}$. At density $\rho = 10^{-2}\rho_{LH}$ we find $P_F \approx .44$. The kinetic energy of the (tμ) atoms after Coulomb deexcitation is approximately 1 keV, but collisions drastically reduce this energy. In order to have an estimate of the time dependence of the energy distribution, we simply study the distribution of the atoms in the two states of the hyperfine structure, assuming that the energy loss is completely determined by the elastic cross sections σ_{11} and σ_{22} starting from an initial energy of few hundred eV, and taking into account the transition rates λ_{12} and λ_{21}. This procedure is admittedly rough, but it can help in understanding whether after a time t $\approx 10^{-8}$ s thermalization occurs. We present in figs. 8,9 the energy distribution after this time, for the tμ atoms in the ground (n_0) and in the excited (n_1) state of the hyperfine structure. Both these distributions seem to be centered at energies higher than the thermal energy. A more reliable estimate is however necessary before concluding that thermalization requires

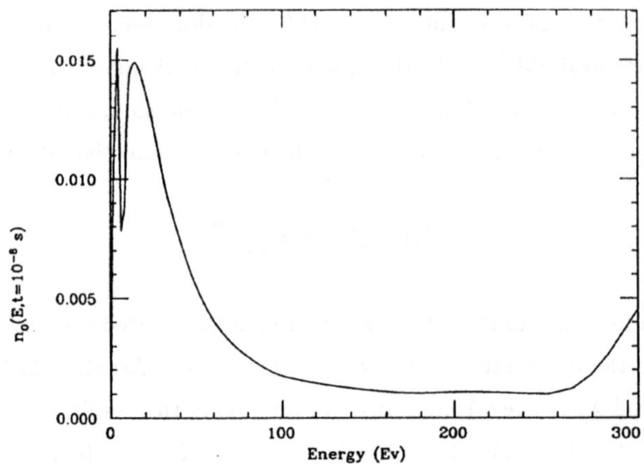

Figure 8

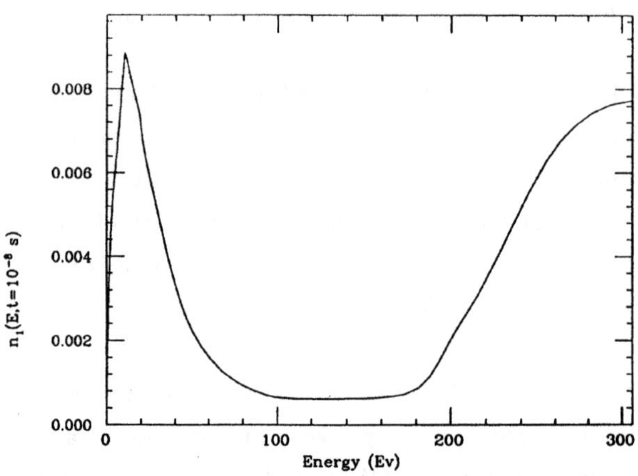

Figure 9

actually a longer time. In particular, the role played by molecular and screening effects is probably dominant and it is necessary to include these effects into the calculations.

REFERENCES

[1] S.I. Vinitskii and L.I. Ponomarev, Sov. J. Part. Nucl. **13**, 557 (1982)

[2] R. de L. Kronig, Band Spectra and Molecular Structure, Cambridge (1930)

[3] V.S. Melezhik, J. Comp. Phys., **65**, 1 (1986)

[4] L.I. Ponomarev, L.N. Somov and F.R. Vukajlovic, J. Phys. **B14**, 591 (1981)

[5] L. Bracci, C. Chiccoli et al., About the Boundary Conditions for the Three-Body Scattering Problem in the Adiabatic Representation, Pisa preprint (1989)

[6] V.S. Melezhik, L.I. Ponomarev and M.P. Faifman, Sov. Phys. JETP **58**, 254 (1983)

[7] L. Bracci, C. Chiccoli et al., Phys. Lett. **134A**, 435 (1989)

[8] M. Bubak and M.P. Faifman, JINR Comm. E4-87-464 (1987)

[9] W.H. Breunlich et al., Phys. Rev. Lett. **53**, 137 (1984)

[10] M. Nagele et al. Experimental Investigation of Muon-Induced Fusion in Liquid Deuterium, Nucl. Phys. (1988) in press

[11] L. Menshikov et al., Sov. Phys. JETP **65**, 656 (1987)

MORE ON THE COLD FUSION FAMILY

J. D. Davies

School of Physics and Space Research, Birmingham University, UK

J.S. Cohen

Theor. Div., Los Alamos National Laboratory, NM87545, USA

Alternative titles for this paper could be 'What the Newspapers Say' or 'Has Cold Fusion always been Hot - the Importance and Dangers of Tritium'. It is based on an April article in Nature[1] updated to July.

Sensational articles in the 4th estate in early Spring claimed 'cold fusion' as being responsible for the observations of A) massive heat during the electrolysis of $D_2O(+LiOD)$ with a Pd cathode after a long delay[2] B) 2.5 MeV neutrons during the electrolysis of D_2O(+ a witches brew[3]) with Pd or Ti cathodes[4]. For B) the rate was 0.41 neutrons $s^{-1}(\sim 3-4\sigma$ above background) and was assumed to come from $d + d \rightarrow^3 He + n$ fusion in the high density of deuterons (not deuterium) driven into the cathode by electrolysis. A) has been called aneutronic fusion since the alleged 10^4 neutrons s^{-1} therein is deficient by about $\times 10^9$ to explain the heat as the above $d - d$ fusion.

Before fusing the nuclei are held apart by their Coulomb repulsion. This barrier is penetrated by quantum-mechanical tunnelling rather than being surmounted. Note that the tunnelling is controlled more by the width of the barrier than by its height which is about $1MeV$. The mean life-time for a D_2 molecule to fuse is about 10^{64} sec. Despite some claims to the contrary N(d nuclei)/N(Pd atoms) probably does not exceed 1 within a deuterated cathode. So any model explaining the neutrons from 'cold' fusion (if real) must get the deuterons closer than the normal lattice sites inside the cathode. Replacing an e^- by a μ^- reduces the separation of the nuclei by about $m_\mu/m_e \sim 207$ and hence the mean fusion life-time[5] to 8.10^{-13}s for $d-t$ and 2.10^{-9}s for $d-d$ (the time difference is part explained by the resonant formation mechanism putting $d-d$ and $d-t$ into $J = 1$; the symmetric $d-d$ cannot avoid this centrifugal barrier by going to $J = 0$); the μ^- is released to catalyse further fusions during its 2.2μs life-time. Some channels of interest in muon-catalysed fusion (μCF) are given in table I.

Electromagnetic Cascade and Chemistry of Exotic Atoms
Edited by L. M. Simons *et al.*, Plenum Press, New York, 1990

Table 1. Some muon catalysed fusion channels.

$d+d+\mu \rightarrow$	$t(1.01MeV)+p(3.02MeV)+\mu$	(42%)	(1a)
	$^3He(0.80MeV)+n(2.45MeV)+\mu$	(58%)	(1b)
$d+t+\mu \rightarrow$	$^4He(3.5MeV)+n(14.1MeV)+\mu$		(2)
$p+d+\mu \rightarrow$	$^3He+\gamma(5.4MeV)+\mu$	(85%)	(3a)
	$^3He(0.2MeV)+\mu(5.2MeV)$	(15%)	(3b)

Consider fusion in nucleus-nucleus scattering. The rate is given by the product of barrier penetration and the strength of the nuclear interaction expressed as the 'astro-physical S-function', S.

$$\sigma = exp(-31.3\sqrt{\frac{\mu}{E}}).\frac{S}{E} \begin{cases} \mu = \text{ reduced mass in a.m.u.} \\ E = \text{C.O.M. energy in keV} \end{cases}$$

In the low energy limit, which applies for 'cold' fusion; $S \approx S_0$ and table II gives S_0 and μ for some of the hydrogen isotope combinations. S_0 is smallest for $p-d$ fusion as this is an electro-magnetic interaction and largest for $d-t$ since this goes via the resonant ^5He. However the reduced mass, and hence the Gamow penetration factor, favour *pd* fusion. Thus for bare nuclei the $d-t$ fusion rate exceeds that of $d-d$ for collision energies above 400eV and the $p-d$ rate, although very small, dominates below 200 eV.

Now consider 'cold fusion' where screening of the Coulomb repulsion by electrons or muons affects the barrier penetration. The characteristic separation in the isotopic hydrogen molecules is the bond length - normally somewhat greater than the distance R_0 at which repulsion between the nuclei becomes apparent. Allowing the nuclei easily to approach this close is roughly equivalent to colliding the bare nuclei together with a centre-of-mass energy $14/R_0$ eV where R_0 is measured in angstroms. (Such simple considerations have to do for now since an accurate calculation of the penetrability requires knowledge of the potential energy interaction od $d-d$ in the crystal, which is unknown for electrochemical cold fusion). For the ordinary hydrogen molecule, $R_0 = 0.7\AA$, so that the equivalent energy is about 20 eV, well below the value 400 eV at which $d-t$ fusion becomes faster than $d-d$. But with the ion's electron substituted by a muon, $R_0 = 4 \times 10^{-3}\AA$, so that the equivalent energy is about 3,500 eV and $d-t$ fusion is clearly favoured (by a factor of 36 in the zero- angular-momentum state[9]. Although muon-catalysed fusion presents the most dramatic evidence of the effect of screening on fusion rates, similar electron-screening effects have been seen in reactions such as (ref 10) $^3He+d \rightarrow ^4He+p$.

In discussing the new results any means of reducing R_0, vibrational excitation of the bond, a reduction in the bond length or even a flattening of the potential well could be effective. Explanation can range from a small increase in the barrier penetration for all the deuterons to a large increase for a small fraction.

Table 2. Parameters controlling fusion for hydrogen isotopes.

	$\mu(a.m.u.)$	S_0(keV barns)	reference
pd	0.67	$2.5.10^{-4}$	6
dd	1.00	$1.08.10^{2}$	7
dt	1.20	$1.15.10^{4}$	8

The fusion could be catalysed by semi-stable, charged quasi-particles. The earlier calculations of Van Siclen and Jones[11] suggested that reducing the hydrogen bond length by a factor of 2, achieved with m(quasi-particle) $\sim 2m_e$ (or by a density compression of the same amount), would give the neutron rate of Jones et al[4]. Large effective electronic muons are often associated with transition metal crystals like palladium; unfortunately the notion of heavy electrons is related to the non-local interactions of electrons with the crystal lattice; it is not easy to see how the local concept of tight binding can be squared with this. Correcting an error in Van Siclen and Jones' paper changes the required mass of the quasi-particle to $m_{qp} \sim 5-6m_e$. Both Jones' papers suggest the fusion of compressed pd to explain the excess heat from Jupiter and ^3He abundance in Hawaii volcanos; whereas a compression by $\times$ 2 is conceivable that by $\times$ 6 is most unlikely. For the shielding distances relevant to this paragraph, the $p-d$ fusion rate should exceed that of $d-d$. Several experiments have electrolysed an H_20/D_20 mixture and no 5.2 MeV γ have been observed[12].

Some[13] have suggested that 0.41 neutrons s^{-1} could be an artefact due to fusion catalysed by cosmic ray muons. This is most improbable. It is reasonable to take $0.1g^{-1}h^{-1}$ to be the upper limit for the rate of muons stopped in Utah basements[14]. The initial capture ratio is roughly proportional to nuclear charge[15] Z and $Z(Pd) = 46$. Moreover, any $d\mu$ would initially have about 1 eV of kinetic energy[16]; while thermalizing, most muons would be transferred to the high-Z atoms[15]. Lastly, the time for the formation of the muonic deuterium molecule is comparable to the muon life-time so that few fusions could be catalysed by each muon.

However assume that the above limitations can be removed, that channelling in the crystalline electrodes keeps muons from the high-Z atoms and that the deuteron density is so high that $d-d$ collision rates do not limit the process. Then the muon lifetime would be sufficient to catalyse 1,000 fusions. But this also assumes it is continuously available for fusion. One has $W_0 \sim 7\%$[17], the fraction of μ^- sticking to the ^3He at fusion, and $R_0 \sim 32\%$[18], the fraction of μ^- reactivated as the μ^3He thermalises in high density deuterium. Then $W_s = W_0(1-R) \sim 5\%$ and the maximum number of fusions per muon is $\frac{1}{W_s} \sim 20$. Stripping in high-Z elements is greater but usually results in capture[19]. Thus the above muon stopping rate and a $3g$ electrode gives a neutron production rate of $2 \times 10^{-3}s^{-1}$, well short of $0.41s^{-1}$. Several experiments have sought and not found 2.5 MeV neutrons from accelerator muons stopped in cathodes charged with deuterons; the most sensitive has the cosmic ray muon flux insufficient by $10^7 - 10^8$ to explain the observed fusion rate[20].

The 'aneutronic' fusion of Fleischman and Pons presents further difficulties. Charge symmetry gives equality of rates for reactions (1a) and (1b) in table I; this is slightly altered by Coulomb isospin mixing[21]. For bare nuclei (plasma fusion) the $p+t$ channel is slightly favoured but for μCF the $n+^3He$ channel is favoured by[22] $\times$ 1.4 - a consequence of fusion taking place from J=1. In any event, it would seem quite difficult to suppress the neutron channel or to hide the tritium from assay. Likewise the small branching fractions for reactions (3b) and for[23] $\mu pt \rightarrow^4 He\mu$ indicate that transfer of all the energy to a heavy electronic quasiparticle, even if it exists, is highly unlikely. Also the normally rare radiative channel $dd \rightarrow^4 He\gamma(24MeV)$ would have shown up in the γ-ray detector of Fleischman and Pons as pair production.

Conversely only heat would be detected if it were possible for the energy to go directly into the Pd crystal lattice - "Mossbauer fusion?" In a series of MIT preprints Hagelstein suggested coupling nuclear motion to lattice vibration. Then the elastic passage of high energy phonons could momentarily increase the vibration, and hence tunnelling probability, of many ions[24]. However such a process is expected to be strongly suppressed by the vastly different energy scales of the crystal lattice (eV) and fusion (MeV). Formation of excited ^4He with a subsequent cascade does not help as the first excited state is at some 20 Mev.

The above difficulties in interpreting the Fleischman and Pons experiment in terms of $d-d$ fusion lead one to consider the possibility of other nuclear reactions that might not produce neutrons or tritons. The electrolyte contained lithium so that $^6Li + d \rightarrow^4 He +^4 He$ fusion is possible at the electrode surface or with long lithium drift times into the palladium; the intermediate state ^8Be has a rich set of resonant energy levels to enhance the reaction, but it is difficult to see how the neutronic reactions $^6Li + d \rightarrow^7 Be + n$ and $^7Li + d \rightarrow n +^4 He +^4 He$ could be suppressed. Another difficulty is the smaller penetrability that comes from the greater nuclear charge, the higher reduced mass and the larger bond distance. For similar reasons, a cold fusion of d with Pd can be considered highly unlikely. Here again we note that there is the possibility of reaction with p as well as d, namely, $^7Li + p \rightarrow \alpha + \alpha$.

Now the Deryagin group in the USSR has made extensive observations of electrons, photons and neutrons emitted from surfaces freshly formed during fracture of solids. Somehow surfaces of cracks or isolated fragments become charged or there are micro-packets of plasma such that charged particles achieve energies of many keV. In particular 2.5 MeV neutrons were observed from sharply struck[25] LiD and were presumed to come from the fusion of deuterons having keV energies.

Deuteriding of palladium or titanium cathodes during electrolysis causes them to embrittle, swell and distort. Hence it was suggested that the neutrons of 'cold fusion' could come from deuterons accelerated in the electric field of cracks etc. This 'microscopically-hot' fusion would have deuterons with plasma-type energies. Neutrons occurring in bursts and production by other means of stressing Ti/Pd were predicted, in particular by rapidly cooling deuterated titanium[26]. Then the Scaramuzzi group observed 2.5 MeV neutrons occurring in bursts during the thermal shocking of Ti following pressure loading with D_2 gas[27].

That LiD is an insulator whereas deuterated Pd and Ti are not was a major objection to the model - it is possible that conduction electrons could rapidly quench any charge build up. However the Deryagin group now report the generation of neutrons during the fracture of transition metals, including titanium, in the presence of deuterided materials[28]. So either quenching is not a problem or some other model is necessary to explain Deryagin cold fusion.

If microscopically 'hot' fusion contributes to the neutrons of 'cold' fusion then considerably higher rates are predicted for a $D - T(D_20/T_20)$ mixture. But the dangers of such experiments are non-trivial. Hydrogen isotopes are very penetrating and easily exploded over a wide range of oxygen concentrations. Tritium is very radioactive - 10^4 curies $\cong 1gm$. However it is T_20 that is most worrying with a radiotoxicity more than $\times 10^4$ that of T_2 gas. Results may take a little longer.

Ignoring the experiments giving massive heat, which may have a chemical origin, there is little confirmation, in writing, for neutrons from 'cold' fusion - ref 29 for electrolytic neutrons and ref 20 for neutrons following thermal shocking. Menlove et al[31] report bursts containing 20-100 detected neutrons during thermal shocking experiments and for one particular, electrolytic measurement. They found a $Ti(6\%Va, 6\%Al, 2\%Sn)$ alloy gave reasonable reproducability: this alloy is very hard and known to become embrittled when hydrided. Any model must also account for a) the many null results[32], often at laboratories well versed in neutron detection and b) the verbatim reports, too frequent and too independent to be ignored, of a statistically significant result (or two) failing to be repeated.

For the 'microscopically-hot' fusion hypothesis, the probability of cracks depends very much on the type of Ti and/or Pd alloy, how it is treated and the attention to detail of the experimentalist. Then it could be that care taken by some group to avoid a too distorted cathode ensured either no fusion neutrons or a solitary appearance.

References

[1] J.S. Cohen and J.D. Davies, Nature 338, 705 (1989).

[2] M. Fleischmann and S. Pons, J. Electronalyt. Chem. 261, 301 (1989).

[3] W. Shakespeare, Macbeth Act 1, Scene 1.

[4] S.E. Jones et al., Nature 338, 737 (1989).

[5] L.N. Bogdanova, Muon Catal. Fusion 3, 359 (1988).

[6] G.M. Griffiths et al., Can. J. Phys. 41, 724 (1963).

[7] R.E. Brown & N. Jarmie, Phys. Rev. C. (in press).

[8] R.E. Brown et al., Phys. Rev. C35, 1999 (1987).

[9] C.-Y. Hu, private communication.

[10] S. Engstler et al., Phys. Lett. B202, 179 (1988).

[11] C. De W. Van Siclen and S.E. Jones, J. Phys. G12, 213 (1986).

[12] C. Van Eijk, private communication.

[13] inter alia, M. W. Guinan et al., UCRL-100881 preprint.

[14] J. Osborne, private communication.

[15] S.S. Gershtein and L.I. Ponomarev, Muon Phys. Vol 3, 141 (Academic, New York, 1975).

[16] W. Breunlich et al., PSI Newsletter (1989).

[17] L.N. Bogdanova et al., Phys. Lett. B161, 1 (1985).

[18] J. S. Cohen, Phys. Rev. Lett. 58, 1407 (1987).

[19] J.S. Cohen, Phys. Rev. A37, 2343 (1988).

[20] J.D. Davies et al., submitted to Nature.

[21] G.M. Hale, private communication.

[22] D.V. Balin et al., Phys. Lett. B141, 173 (1984).

[23] these proceedings.

[24] M.F. Russel, RAL-89-037 preprint.

[25] V.A. Klyvev et al., Sov. Tech. Phys. Lett. 12, 551 (1986).

[26] P. Flower, private communication.

[27] A. De Ninno et al., Europhysics Lett. 9, 221 (1989).

[28] A. G. Lipson et al., JETP Lett. 49, 588 (1989).

[29] A. Bertin et al., Nuovo Cimento (in press)
T. Mizuni et al., Electrochemistry, 57 p? (1989).

[30] N. Burgio et al., submitted to Nuovo Cimento.

[31] H.O. Menlove et al., submitted to Nature.

[32] eg. M. Gai et al., Nature 340, 29 (1989).
D. E. Williams et al., submitted to Nature.

VI

BEAMS AND METHODS

"Well, in *our* country," said Alice still panting a little, "you'd generally get to somewhere else - if you ran very fast for a long time, as we've been doing."
"A slow sort of country!", said the Queen. "Now, *here*, you see, it takes all the running *you* can do, to keep in the same place. If you want to get somewhere else, you must run at least twice as fast as that!"

Lewis Carroll: Through the Looking-Glass

THE ISIS FACILITY AT RAL THE PULSED MUON BEAM

J.D.Davies

Dept. Physics & Space Research, Birmingham Uni, UK

G.H.Eaton

ISIS Division, RAL, Chilton, Didcot, Oxon, OX11 0QX, UK

At the Rutherford Appleton Laboratory near Oxford a synchrotron provides 750 MeV protons to generate neutrons by spallation and fast fission at a uranium target. The protons are pulsed at 50 Hz so that neutron energies can be determined by 'Time-of-flight' along the many neutron beams. A π/μ test beam of reasonable duty cycle is obtained off a target, inside the accelerator, vibrating at 50 Hz into protons having incorrect orbits to be extracted. Pulsed neutrinos from the uranium target are investigated with a large liquid scintillator detector - KARMEN. Finally a thin transmission target in the extracted proton beam provides surface/cloud muons.

ADVANTAGES OF PULSED MUONS

The τ_μ = 2.2 μs muon life-time provides problems for continuous beams and also advantages, these are avoidable and exploitable respectively with pulsed muons.

1) Avoiding pile-up limits some experiments to having the possibility of one muon in the detector during many life-times, ie to beam intensities ~10 kHz. Beam rates can be increased by x 10^2 by injecting an intense bunch of muons in a time short compared with τ_μ and then following the behaviour of the whole ensemble.

2) The same and/or other experiments are limited by background from the accelerator, beam and/or continuous sources such as cosmic rays. The former can be avoided by starting the experiment after the beam pulse (there will be some, delayed, low energy neutral background) and the latter reduced by the low duty cycle (at ISIS, 50 Hz x 2.2 μs x 15 life-times ~1.6 10^{-3}).

Electromagnetic Cascade and Chemistry of Exotic Atoms
Edited by L. M. Simons *et al.*, Plenum Press, New York, 1990

3) μ^- can induce processes having life-times of several milli-secs, ie β-decay, when a pulsed beam enables the decay to be examined in the absence of beam and background.

4) Some experiments have intrinsically pulsed apparatus, eg micro-waves or lasers pulsed to get the required power densities - lasers have short, light storage times and typically 1-100 Hz rep rates.

5) Atomic lines widths can be reduced by monitoring over many life-times.

6) A well defined start can be provided.

As illustrations of the advantages of a pulsed beam for experiments consider the following well-established examples.

τ_μ^+ (τ^-)$_\mu$ Presently measurements to determine the Fermi constant G_μ and to test CPT are rate limited; avoiding pile-up means admitting and studying one muon every 20-30 μs. By starting many clocks simultaneously after a muon pulse the history of the whole ensemble can be followed. The increase in beam intensity can be fully exploited as use of the pulse beam avoids and reduces the otherwise limiting accelerator and cosmic ray background.

Ground State hyperfine splitting and Zeeman effects in μ^+e^- (in gas) test bound state QED and measure the fine structure constant and muon magnetic moment. Experiments are now limited by finite line widths; these can be narrowed by monitoring the time development of hyperfine transitions over many τ_μ. This is a common technique in atomic physics. A x 10 improvement is possible giving a 0.1 ppm value for m_μ/m_e.

Lamb Shift (2s-2p) measurements in μ^-p, μ^-d, $\mu^{-3}He$ test vacuum polarisation at the 50 ppm level (also μ-e universally) and measure the ^3He radius and the proton radius to 0.1% error. Kottmann and co-workers have obtained high μ^- stopping densities with a magnetic bottle. However, their laser to induce 2S→2p transitions is pulsed to get the required power density and appreciable event rates require a simultaneously pulsed muon beam.

There are Two Main Problems - the cost of providing pulsed protons and the very high event rate. For the latter one will learn and benefit from the high luminosity collider experiments.

ISIS - THE SPALLATION NEUTRON SOURCE

Buildings, apparatus and personnel from the former 7Gev NIMROD proton accelerator (with contribution from the old 4 Gev NINA electron synchrotron) have been used in constructing a 750 MeV 180 μA rapid cycling (50 Hz) proton synchrotron (fig 1). 70 MeV H^- from the linac are stripped for injection into the synchrotron. Here two proton pulses, on opposite sides of the machine are accelerated to 750 MeV and then single turn extracted with a kicker magnet ramping up between pulses. Fig 2 shows the time structure of the extracted proton beam which is transported some 150 metres to the well shielded, uranium target, the spallation neutron source. Table I lists the neutron beams available. The neutron energy spectrum is slightly hotter than that from a fission reactor. The uranium target has a lifetime of approximately one year of the current operating levels. The present operational emphasis is to obtain steady reliable beam - 31 days

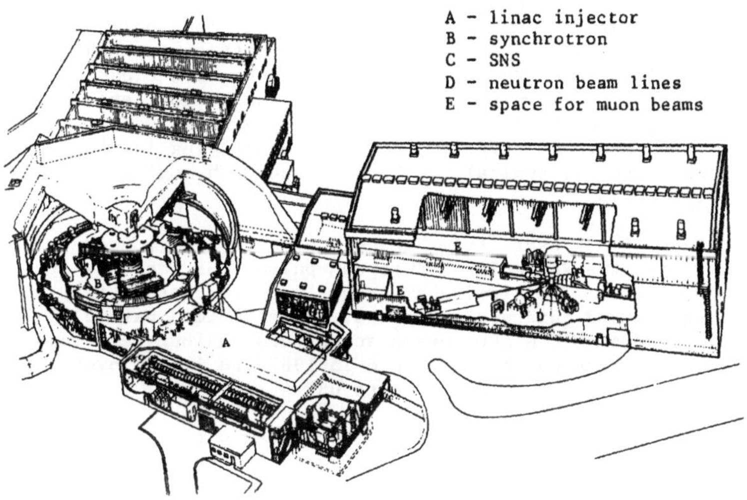

Fig. 1 The Rutherford Appleton Laboratory ISIS Facility

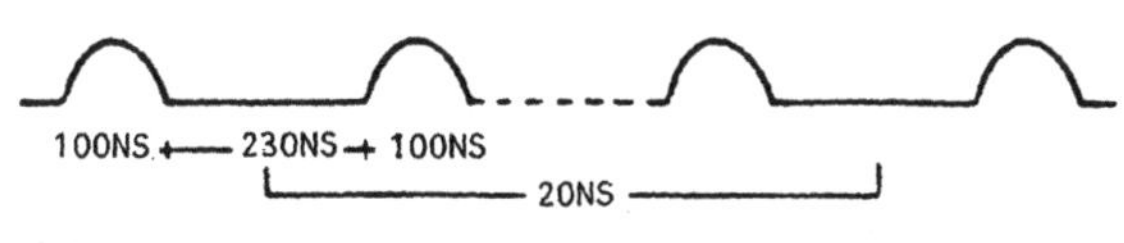

Fig. 2 Time structure of ISIS proton bursts

on, 11 days (including machine development) off for 9 months per year. The beam current is now 100 μA, 60 μA average.

The Pulsed Muon Beam

The initial suggestion of producing pulsed muon beams at ISIS from a thin transmission target in the extracted proton beam came from the Birmingham group. Low energy muon beams can be utilised from such a target in three ways:-

Surface muons: μ^+ in a narrow momentum band around 29 MeV/c from stopped π^+ decaying at the surface of the production target, 100% polarisation and high densities. Such beams are of increasing importance for μSR.
Cloud muons: $\mu^\pm$ up to 60 MeV/c from decays in the pion cloud around the target, no polarisation, significant e and π contamination, smaller spot size and momentum spread than obtainable from decay beams.
Decay muons: $\mu^\pm$ above 40 MeV/c coming from πs decaying inside a superconducting solenoid, typically 80% polarised with negligible e and π contamination.

A consortium of μSR users from France, Germany, Italy, Sweden, UK and Japan together with the RAL built a European pulsed μSR facility at ISIS aided by an EEC grant. The Birmingham group 'acquired' essential components, bending and quadrupole magnets, crossed-field particle separators - some from old NIMROD beams and others on long-term loan from DESY and CERN. The beam was commissioned in 1987 and is now over subscribed by x 3; 80% of the beam goes to μSR.

Figure 3 shows the 20m long surface/cloud beam with is conventional double bend. Significant are the 1m E x B particle separator, the considerable shielding and the lack of a beam switch yard. The muons have the time structure of Fig 2 broadened by $\tau_\pi \sim 26$ns. A 2.5 mm production target in 100 μA of protons drops the neutron fluxes by < 1% and provides $2.10^5\ S^{-1}\ \mu^+$ at 26.5 MeV/c and $1.7.\ 10^4\ \mu^-\ S^{-1}$ $(4.2\ 10^4\ \mu^-\ S^{-1})$ at 40(60) MeV/c. Spot size is 1.3 x 1.5 $(cm)^2$ and $\Delta p/p \sim 5\%$ all FWHM. The cloud beam has considerable numbers of electrons and pions, eg 4 e^- and 0.5 π per μ^- at 40 MeV/c.

The muon catalysed fusion programme (μCF) using low density gas targets is short of rate and sensitive to neutron backgrounds. The μ stopping density at 40 MeV/c is x2 greater than at 60 MeV/c and both are more than that from a comparable decay beam. There is a neutron background from the uranium target and from the neutral particle dump that degrades in energy with increasing time after the proton burst; roughly, neutrons have 2→1 MeV at 1→2 μs after the protons. Considerable $(CH_2)n$ and Pb shielding around the 2nd half of the beam pipe was necessary to shield the neutral particle dump. Another, potentially critical neutron background, had a 900 ns time constant and significant intensity above 3 MeV: this came from μ - Al interactions where the beam pipe is filled (vertically) inside the last quadrupole magnet. Fig 4 shows the effective life-time for μ^- stopping in elements of varying atomic number Z. Hence the beam pipe was sleeved with copper to give this background a much shorter time constant. The beam contamination provides a potentially overwhelming flash of particles during the burst, the e^+ by showering and the π^- through their 140 MeV mass. ± 60kV on the particle separator and conventional, double, circular Pb collimators immediately upstream of the experiment reduced the e^- contamination to tolerable levels but had little effect on the π^-. Also the proximity of the Pb collimation provided a diffuse source of background.

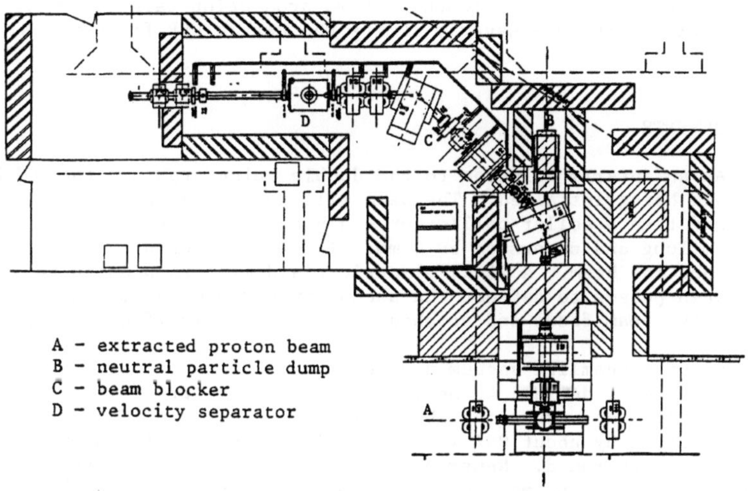

Fig. 3 Layout of the ISIS Pulsed Muon Facility

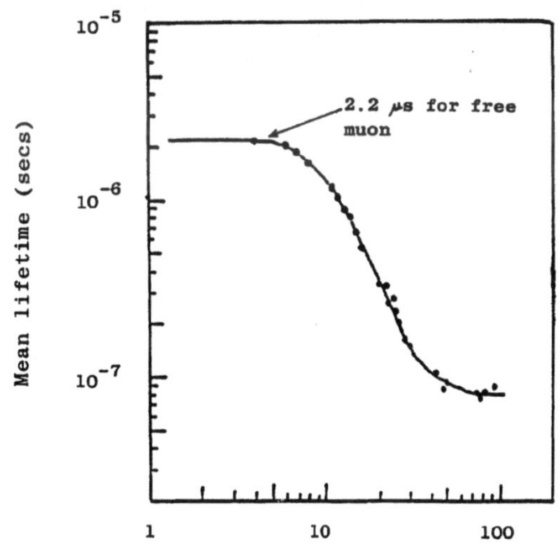

Fig. 4 Mean lifetimes of negative muons stopping in various materials versus the atomic number Z of the nucleus to which the muon is bound.

Figure 5 shows a μSR test spectrum obtained during beam commissioning; the higher beam intensities that can be used and the much lower background enable the μSR signal to be seen out to 20 μS compared with less than 10 μS at continuous beams. For μSR using digital recording the muon beam has to be collimated down with the thinnest 2.5 mm production target. Moreover, the double pulse structure severly limits the usable μSR frequencies.

BEAM DEVELOPMENTS

A) As an interim solution to the double pulse problem, the UPPSET device is now being installed. The beam will go through a parallel plate capacitor and then a soller collimator - a number of myler sheets 100 μ thick, 40 cm long and separated by 8 mm: plates and sheets are parallel to each other and the beam. A 15 kV electric field will ramp up in the approx 200 ns between pulses to deflect the second into the mylar sheets. These are 5m from the beam focus and so should not give a significant background.

The improved beamline vacuum now permits ± 180 kv on the E xB velocity separator which will deflect 40 MeV/c πs into the soller collimator sheets and electrons into a surrounding Pb collimator. This should reduce the beam contamination by about 99% with the collimation sufficiently distant not to provide background. Neutrons from μ^- - soller sheet interactions coming down the beam line may be a problem and so it could then be necessary to use a different material in the soller to give this background a short life time.

B) Fig 6 shows the beamline upgrade presented for EEC funding. The fields on the electrostatic septum will divide the 1st muon pulse to experimental areas A and C. The field will then ramp to zero allowing the 2nd pulse to reach area B. By altering the timing of the ramping relative to the muon bunch, either pulse can be time sliced to further improve the μSR frequency range. If approved areas A and C will be dedicated to μSR leaving area B for other users such as μCF. This facility will more than double the through put of μSR experiments and very much increase their frequency range and will provide floor preparation time for the other larger experiments. The second particle separator allows a $\pi/2$ rotation of the muon polarisation for μSR experiments.

C) If Japan joins the present μCF collaboration (Birmingham, Bologna, BYU, Culham, Delft, Idaho, LANL, RAL) then another beamline will be built off the same π/μ production target on the other side of the extracted proton beam. This would be a decay beam to complement the present surface/cloud line.

MUONIUM LAMB SHIFT

Heidelburg - Oxford - RAL - Southampton - Yale.
This is a laser experiment using Doppler-free, 2 photon spectroscopy to measure the 1S-2S energy difference in the purely leptonic (μ^+e^-); the data should give the 1S Lamb shift or the Rydberg constant. They have obtained 52 thermal muonium atoms in vacuo per muon double pulse, sufficient for the experiment and have found no unexpected background.

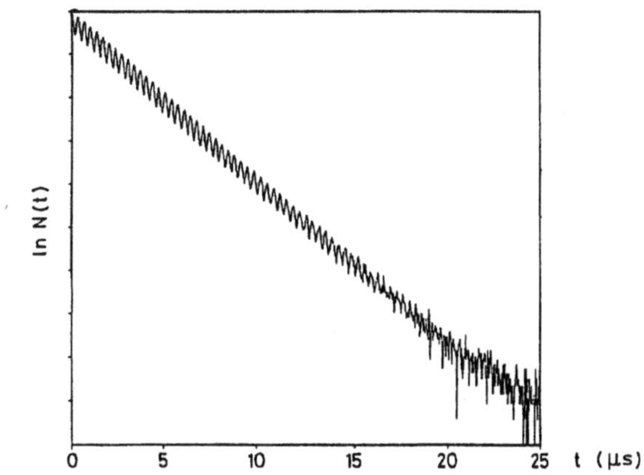

Figure 5. Test spectrum showing a 3 MHz precession signal in 22 mT transverse field. (Aluminium sample showing maximum (30 %) asymmetry in positron emission, ambient temperature, $8*10^6$ recorded events)

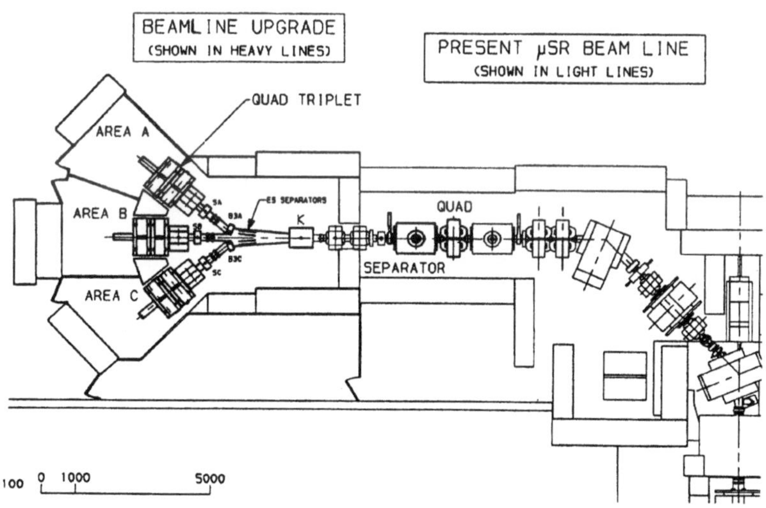

Figure 6. Proposed beam line upgrade.

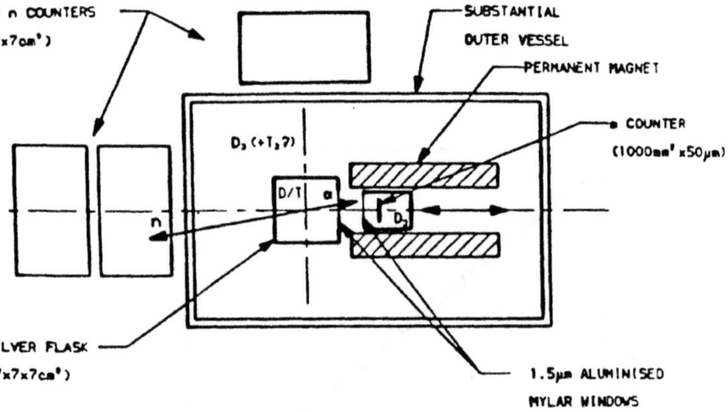

Figure 7. Layout of the experimental apparatus. The D/T mixture is contained in the silver inner vessel, surrounded by a substantial vessel. Alphas and neutrons are detected in a surface barrier detector and NE213 neutron detectors, respectively.

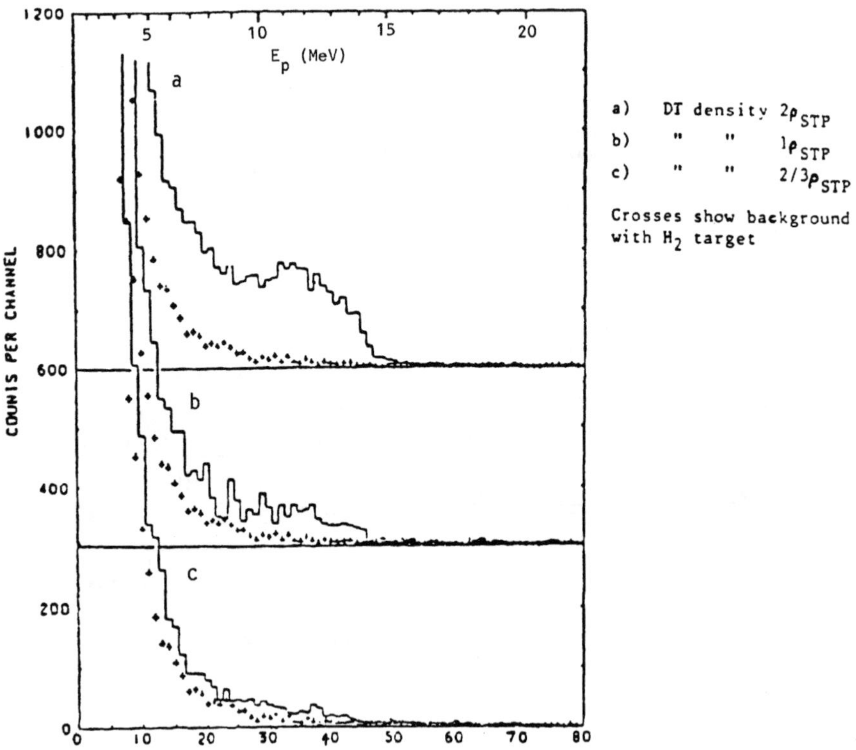

Figure 8. Fusion neutron → recoil proton energy spectrum.

MUON CATALYSED FUSION

The main experiment is a direct measure of W_s the μ-α sticking coefficient following muon catalysed $dt \rightarrow \alpha n$ fusion from the ratio of $N(\mu\alpha - n)/N(\alpha - n)$ co-linear coincidences. The short α range requires a small low density gas target. Fig 7 shows the apparatus which was developed at LAMPF where >99.9% of the μ stopped in the target flask walls giving a very large background. At ISIS 98% of the μ stopped in the silver flask walls giving a background with τ(effective) <100 ns (see fig 4). Then starting the experiment 1μs after the muon pulse avoids this background and reduces the energies of the neutrons from the neutral particle dump and from the uranium target to below a few MeV.

An event rate α (gas density $\rho)^2$ had been assumed factors ρ for stopping and for the number of collisions per muon. Also 10 ρ_{STP} had been the previous lowest density at which 14 MeV fusion neutron singles had been seen (detected as recoil proton with 0-14 MeV energy distribution). Fig 8 shows the neutron energy spectrum at 2, 1 and 2/3 ρ_{STP}- fusion neutrons can clearly be seen demonstrating the very low background at ISIS. Incidently, the (event rate/ρ^2) doubled in changing '$2 \rightarrow 1\rho_{STP}$', strong evidence for epithermal production.

<u>Other Proposals</u> in abeyance because of lack of beamtime. <u>Muonium →anti-muonium</u> (μ^+e^-) would be made from tungsten at 2000°C and the (μ^-e^+) detected via the e^+ in the very low background at ISIS.

<u>$\mu p \rightarrow n\nu$</u> The low backgrounds permit measuring the neutron time spectrum to many τ_μ; this would provide values for g_A and g_p.

TABLE 1. ISIS NEUTRON SCATTERING INSTRUMENTS

HET	-	High Energy Transfer Spectrometer
HRPD	-	High Resolution Powder diffractometer
IRIS	-	High Resolution Quasi Elastic Spectrometer
LAD	-	Liquid and Amorphous Materials Diffractometer
TFXA	-	Time Focussed Crystal Analyser Spectrometer
IOQ	-	Small Angle Scattering Instruments
CRISP	-	Critical Refraction Spectrometer
eVS	-	Electron Volt Spectrometer
POLARIS	-	Single Crystal Spectrometer
PRISMA	-	Coherent Collective Excitation Spectrometer (Italy)
SANDELS	-	Small Angle Neutron Diffractometer
MARI	-	Multi Angle Rotor Instrument (Japan)
SXD	-	Single Crystal Diffractometer
TEST	-	Neutron

FORMATION OF ATOMIC PIONIUM

H. Orth

GSI Darmstadt, Postfach 11 05 52, D-6100 Darmstadt

1. Introduction

Pionium (Pi = π^+e^-) is the bound state of a positive pion and an electron. It is therefore a new light hydrogenic system. Indeed the atomic energy levels of Pi resemble those of hydrogen and muonium very closely. However, the imprint of hyperfine interaction is removed due to the absence of spin angular momentum of the pion which reduces the multiplicity pattern of the gross structure terms. Besides we may argue if the theory of the Pi-atom starts with a Klein-Gordon equation of a pion in the Coulombic field of an electron or, conversely, with a Dirac equation of an electron in the Coulombic field of the pion.

Beyond this more academic look-out pionium offers new experimental perspectives[1]. The energy levels of bound state Pi are sensitive to the pion mass and the pion internal structure. A measurement of the 1S-2S splitting in Pi with an accuracy of 10 % of the Lyman α natural linewidth would result in a value for the pion mass better than 1 ppm, which compares to the best measurement of m_{π^+} = 139.5704(11) MeV/c ($\approx$ 8 ppm)[2]. If connected to precise Lamb shift results in hydrogen[3] and muonium[4] a possible measurement of the Pi Lamb shift ($2^2S_{1/2}$ - $2^2P_{1/2}$) accurate to 10^{-4} could result in an improved pion charge radius. On the basis of the present knowledge of the π^+ charge radius, the n=2 Lambshift of Pi has been estimated to be $\nu_{Lamb}(Pi)$ = 1050.4(3) MHz[5], which is to be compared with theoretical values of $\nu_{Lamb}(H)$ = 1057.867(11) MHz[6] and $\nu_{Lamb}(M)$ = 1047.5(3) MHz[7]. The errors in $\nu_{Lamb}(Pi)$ and $\nu_{Lamb}(M)$ are dominated by uncalculated higher terms of the recoil corrections.

This experiment served the objective to establish, that pionium atoms can be formed at the interface between a material surface and vacuum and that its presence can be uniquely monitored.

Electromagnetic Cascade and Chemistry of Exotic Atoms
Edited by L. M. Simons *et al.*, Plenum Press, New York, 1990

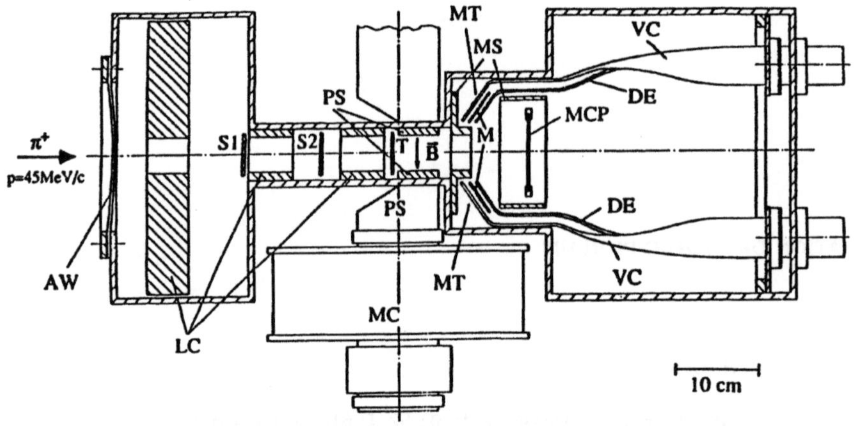

Fig. 1 Pionium apparatus: S1, S2 and T: beam scintillators; the neutralization foil is backed on the downstram surfach of T. B: magnetic sweeping field, MCP: microchannel plate to detect the slow π^+ and π^+e^-. MT: muon telescope. DE, M, VC: scintillators to measure dE/dx and total E of the muons as well as to veto beam positrons. VC is also used to detect the delayed decay positrons. MS: magnetic shielding for transverse fields at the MCP. AW: 86 μm thick aluminum vacuum window. LC: lead collimator. MC: magnet coils.

2. Experimental

The apparatus is shown in Fig. 1. It is designed to form Pi by a beam-foil technique similar to the first experiments with muonium in vacuum. The degraded pion beam at 45 MeV/c momentum was used as it was available at the πE3 channel of SIN. Pions entering through an 86 μm aluminium window, were collimated by 4 cm diameter lead collimators, and passed through three scintillators, S1, S2 and T. These scintillators served as active moderators giving a triple coincidence for incoming pions, π_{in} = S1 $\times$ S2 $\times$ T. A small fraction of the π^+ neutralizes immediately downstream of T in a 10 μm aluminum foil. 25 $\times$ 10^{-3} Tm magnetic bending power deflected most of the charged particles, thereby permitting the neutral ones to be collected and detected on a 75 mm diameter micro-channel plate (MCP) 15 cm downstream of the neutralization foil. The aim was to observe slow neutral particles in the time-of-flight spectrum and to monitor subsequently the decay of pions off the MCP surface $\pi^+ \rightarrow \mu^+ v_{\mu}$. The energy and time of this decay were measured in two muon telescopes laterally deplaced from the Pi beam. Each of these telesopes was made up of three scintillators with different thicknesses. In front was a dE/dx counter (DE, 0.2 mm thick) used to suppress the positron background, followed by a muon counter (M, 1.5 mm) in which the 4.1 MeV monoenergetic muons from pion decay stopped. The third scintillator (VC, 4 mm) was used for further rejection of background positrons passing through the whole telescope and also to record the delayed positrons from the decay of the stopped muons. A muon from π^+-decay was defined as the coincidence MU $= DE \times M \times VC$ which was required to happen in a 80 ns μ-gate started by a π_{in}. The positron from the μ-decay was accepted as a coincidence of either $M \times DE$ or $M \times VC$ in a 5 μs e-gate started by MU.

The trigger to initiate complete data read out (ADC and TDC) was the characteristic decay sequence of pion to muon to electron which was electronically set up by fast delayed coincidences. From the recorded 2.3×10^{10} π_{in} during 130 hours of data taking, the raw TOF spectrum of pions based on 8.5×10^9 π_{in} is shown in Fig. 2a. The prompt peak contains all the fast pion events not removed by the sweeping magnet. Only the tail includes the Pi atoms. The π_{in} rate was 5×10^4 Hz. A fraction of about 8 Hz of these pions resulted in a real good trigger mainly because of the small solid angle of the MCP and the muon telescopes but also due to the important magnet sweeping effect. A time cut at $t < 8$ ns corresponding to an energy cut $E > 300$ keV removes the prompt peak in Fig. 2a without rejecting Pi. A further 6 ns time cut on the π-lifetime spectrum reduces the beam-correlated background from scattered particles. Finally, requiring the remaining events to have a decay-μ^+ energy of 4.1 MeV in a window of $\pm$ 2 standard deviation of the energy resolution in the muon telescope leads to the spectra in Fig. 2b and 2c.

The data taken with the sweeping magnet switched "on" distinctly show two peaks in the time of flight of pions between the target foil and the channel plate. We interpret the events within a 19 ns to 70 ns time window as neutral pionium atoms. This is well supported by a detailed Monte Carlo simulation[1,5] in which the only empirical input are proton data from charge neutralizations experiments off foils[8]. This result, a total of 118 triggered neutral atoms, is used to extrapolate the number of Pi at the MCP. Including the solid angles and efficiencies of the detectors we obtain 3300 (400) pionium atoms on the 44 cm^2 area of the MCP. This corresponds to a yield of 4.0 (4) $\times 10^{-7}$ Pi/π_{in} or 1.1 (1) Pi/min. Under the assumption that the pionium atoms emerge isotropically from the neutralization foil, the Pi rate directly downstream of this foil is about $2s^{-1}$.

3. Conclusion and outlook

Atomic pionium has been formed in vacuum. These atoms are presently dispersed within a rather large flux of charged pions, decay products therefrom and scattered background. The measured rates and the energy distribution of the Pi atoms are in

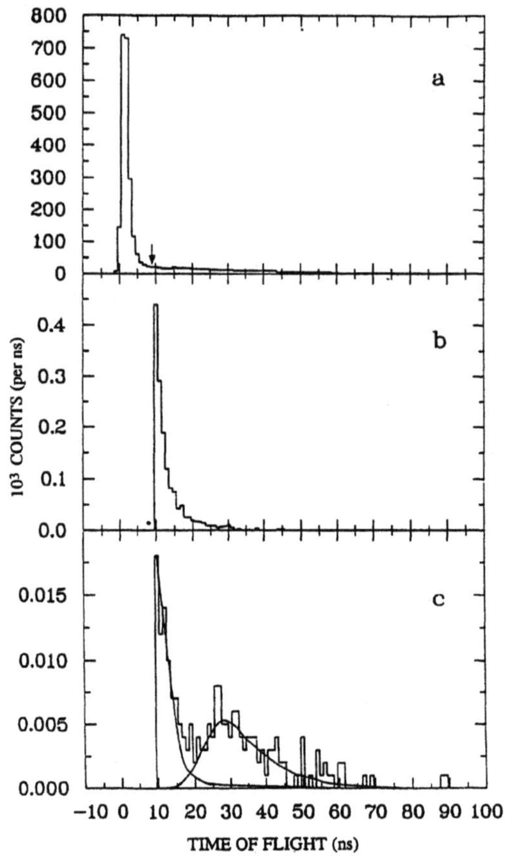

Fig. 2 Measured TOF spectra:

a) Raw TOF spectrum of the π^+ and Pi between T and MCP, the time cut at 8 ns (> 300 keV) is indicated by an arrow.

b) TOF of the slow π^+ and Pi with magnetic field off.

c) TOF with magnetic field. An exponentional fit to the background, and the result of the Monte Carlo calculation fitting only an amplitude is also shown.

good agreement with our expectations. For sensitive spectroscopic experiments such as those mentioned in the introduction, it will be necessary to improve the pionium yield. It seems that a compact low-momentum pion beam can be invented and realistically be built. With a pionium source of the second generation it may be possible to gradually cultivate a substantially higher pionium flux.

References

1. H.-J. Mundinger et al. Europhys. Lett. 8, 339 (1989)
2. R. Abela et al., Phys. Lett. 146B, 431 (1984)
3. S. R. Lundeen and F. M. Pipkin, Phys. Rev. Lett. 46, 232 (1981)
4. A. Bardertscher et al., Phys. Rev. Lett. 52, 914 (1984)
5. H.-J. Mundinger, Ph. D. Thesis University of Heidelberg, W. Germany
6. G. W. Ericson and H. Grotch, Phys. Rev. Lett. 60, 2611 (1988)
7. K. A. Wookle et al., submitted to Phys. Rev. A.
8. S. Kreussler, R. Sizmann, Phys. Rev. B 26, 520 (1982)

HIGH RESOLUTION SPECTROSCOPY OF X-RAYS FROM ANTIPROTONIC ATOMS USING A LOW-ENERGY CRYSTAL SPECTROMETER

G.L. Borchert, D.Gotta, O.W.B. Schult
Institut für Kernphysik, KFA Jülich, D-5170 Jülich

L.M. Simons
PSI, CH-5232 Villigen

K. Elsener
Institute of Physics, University of Aarhus, DK-8000 Aarhus C

K.Rashid
Center of Basic Sciences, University Grants Commission Islamabad, Pakistan

J.J. Reidy
Physics Department, University of Mississippi, University Miss.38677, USA

R.D. Deslattes, E.G. Kessler, T. Mooney
National Institute of Standards and Technology, Gaithersburg MD 20899, USA

The study of antiprotonic atoms in low-pressure gas targets became possible through the intense low-energy beams at LEAR. Strong interaction effects and the pressure dependence of the atomic cascade have been measured in the elementary systems $\bar{p}H$ and $\bar{p}He$ [1-8]. The deexcitation of very high atomic states and even molecular effects became visible in the spectra of antiprotonic noble gases [9] and C_nH_m compounds [10].

The results for strong interaction shifts ϵ and broadenings Γ of the atomic energy levels for $\bar{p}A$ ($A \leq 4$) are collected in table 1. The spin-isospin averaged $N\bar{N}$ scattering length and potential parameters derived agree in general with theoretical models [11-15]. The limited energy resolution of detectors (Li(Li), GSPC, XFC,...) does not allow the resolution of any fine (FS) or hyperfine structure (HFS), i.e. only spin averaged shifts and widths have been measured. A detailed investigation of the $\bar{N}N$ potential, however, suffers from partial cancellation of amplitudes in the averaging procedure. In particular, the spin dependence and the strong shifts are not accessible from indirect measurements, where the widths are determined from the intensity balance. In addition the very strong annihilation washes out any detailed structures of the real part of the potential. No strong interaction effects could be observed for antiprotonic deuterium. Shift and broadening of the 2p level are too small to be measured with the detectors mentioned above. On the other hand, the fraction of radiative decay of the 2p level is much smaller than for $\bar{p}H$ due to stronger annihilation, so that the yield of the $K\alpha$ transition is below the present detection limit [2,16]. No predictions have been made for the strong widths of $\bar{p}^3He$. The shift observed in experiment is not reproduced assuming π^0 exchange only [17]. For $\bar{p}^4He$ the agreement between theory and experiments is not convincing [4,6,13].

The measured X-ray intensities are reproduced by cascade models to a level of 10-20%. Neutral exotic atoms can be described well by the model of Leon, Bethe and Borie [18,19], where strength of the Stark mixing and velocity of the exotic hydrogen are parameters fitted to the observed intensities. The Stark effect is still dominating at 16 mbar leading to a complete mixture of angular momentum states for levels of principal quantum number $n \geq 8$. The atomic cascade has been measured to be similar for $\bar{p}H$ and $\bar{p}D$. For $\bar{p}He$ the cascade is successfully described, when in addition to Stark-mixing molecule-ion formation is considered [20].

With a crystal spectrometer energy resolution and energy calibration are superior by 2-3 orders of magnitude compared to previous detection methods [21]. In the cyclotron trap [22] 86% of the antiprotons can be stopped in less than a cm^3 at pressures below 30 mbar [8], thus providing a bright X-ray source. Low gas pressures are essential to saturate the X-ray intensities thus alleviating the rather low efficiency of a crystal spectrometer ($\sim 10^{-6}$). At pressures below $\sim$ 100 mbar the antiprotons annihilate predominantly from the 2p levels of $\bar{p}H$ and $\bar{p}He$ atoms. The acceptance of a Bragg spectrometer just matches the energy range of the $L\alpha$ transitions. Only at LEAR intensities of $5.10^5 \bar{p}/s$ and a small phase space of a beam are available which is essential to obtain sufficient counting rates.

Table 1. Strong shift ϵ and broadening Γ and X-ray yields at low pressures measured in $\bar{p}A$ ($A \leq 4$). The electromagnetic binding energies (E^*) include vacuum polarization [17].

The experimental results are from LEAR experiment PS175 [6,7,8], unless indicated otherwise: PS174 a:[2], c:[4], d:[1], e:[5]; PS171 b:[3].

	H	D	T	^3He	^3He
I,T	1/2, 1/2	1,0	1/2, 1/2	1/2, 1/2	0,0
$E(K\alpha)^*$/keV $E(L\alpha)^*$/keV	9.409 −1.737	(12.553) 2.317	(14.122) 2.605	(56.513) 10.440	(60.035) 11.129
$Y(K\alpha)$/% $Y(L\alpha)$/%	6.2± 40±5	<1(2α)a 33±4	- -	<0.5 26±4	<0.2 21±3
p/mbar	30	30	-	72	72
ϵ_{1s}/keV	-0.62±0.10 -0.73±0.05a -0.70±0.15b	- - -	- - -	- - -	- - -
Γ_{1s}/keV	1.13±0.17 1.13±0.09a 1.60±0.40b	- - -	- - -	- - -	- - -
ϵ_{2p}/eV	-	-	-	-17±4	-18±2 -7±5c
Γ_{2p}/eV	0.032±0.010 0.045±0.010d	-	-	25±8	46±5 35±15c
Γ_{3d}/meV	-	-	-	2.1±0.2	2.3±0.1 2.1±0.1e

Using a crystal diffractometer at least groups of the $\bar{p}H$ 2p hyperfine states can be resolved (fig.1). Following the theory of Richard and Sainio [11], shift and width of the $^3P_0$ level yield immediately the complex isoscalar scattering volume of the $\bar{p}p$ interaction and are sensitive to the contribution of $|n\bar{n}\rangle$ admixture. Up to now, no agreement has been obtained on the role of the spin effects [23,24]. Possibly, spin effects manifest themselves most clearly in the ratio of real to imaginary part of the scattering amplitudes. The strong shift is sensitive to the dynamical structures below threshold [25].

A direct measurement of the 2p shift and width is at present the only possibility to determine hadronic effects in $\bar{p}D$ atoms. Properties of the (isovector) $\bar{p}n$ interaction can be derived and compared to the charge-conjugate state $\bar{n}p$. Recent experiments yield different results for the isospin-dependent annihilation strength, derived from the extrapolation of $\bar{n}p$ scattering to threshold [26] and $\bar{p}D$, $\bar{p}^3He$, and $\bar{p}^4He$ annihilation at rest [27].

In antiprotonic helium, the strong interaction effects are determined to an accuracy of 10-35%. For the (I=0) nucleus ^4He the strong broadening is much larger than the FS splitting. For $\bar{p}^3He$ the HFS splitting is predicted to be in the range of -6 - +4 eV [17].

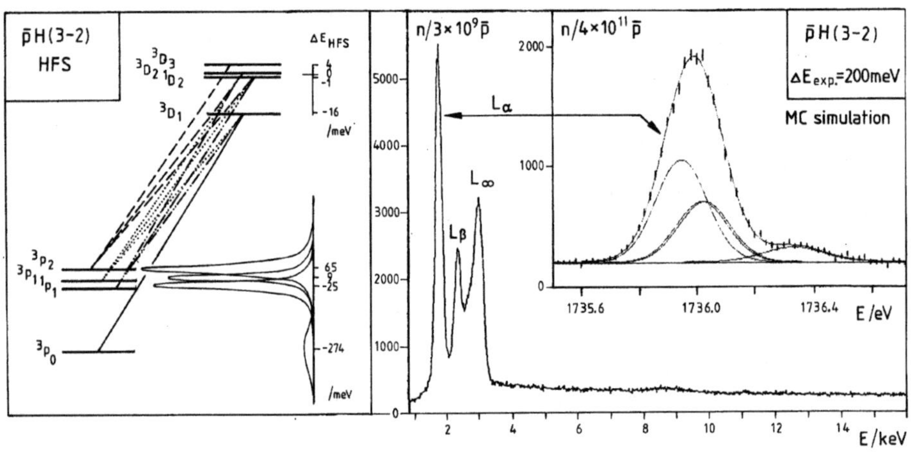

Fig.1. Low-lying HFS transition of $\bar{p}H$. The pattern of the $L\alpha$ transition with an energy resolution of $1.2 \cdot 10^{-4}$ (Monte-Carlo simulation) is compared to the X-ray spectrum measured with a Si(Li) at 30 mbar pressure (PS175) [8]. With an energy resolution of 200 meV it will be possible to determine the strong shift and broadening of the $^3P_0$ state and mean shift width of the group $^3P_2$, $^3P_1$, $^1P_1$ with an accuracy of 20% each. Transitions from the series limit of the atomic cascade can be resolved up to n=40.

Considering the experimental resolution of $\Delta E = 300$ eV in the measurement of the $\bar{p}^3He$ L X-rays, it might be, that the strong width is considerably smaller and evidence for a HFS pattern will appear in a measurement using a crystal diffractometer ($\Delta E \sim 5$ eV).

Besides the strong broadening which occurs only for the last transitions of the atomic cascade, a Doppler broadening is caused by the velocity of the antiprotonic atom. A kinetic energy of 1 eV yields a Doppler broadening of 54 meV for the $\bar{p}H$ L X-rays. A significant deviation from 1 eV kinetic energy has been found in pionic (liquid) hydrogen, where the kinetic energy of about one half of the $\pi^{-}p$ systems extend up to 70 eV due to Coulomb acceleration [28]. For transitions not affected by strong interaction (e.g. $\bar{p}D$ $(M\infty)$:E = 1.85 keV; $\bar{p}He$(5-4):E = 1.69 keV) the line broadening yields a direct measurement of the velocity (for the decaying state considered). It is expected, that the charged $\bar{p}He$ system is more thermalized than the neutral $\bar{p}H$ or $\bar{p}D$ system.

With decreasing pressure a drastic increase of the $L\alpha$ yields indicates dramatic changes of the population in the higher parts of the atomic cascade. The total intensity tends more and more to be shared between the circular transitions ($l_{i,f} = n_{i,f}-1$) and the transitions from the series limit ($L\infty$). The total intensity of the unresolved $L\infty$ transitions is about 10% and almost constant below a few 100 mbar. With an energy resolution of $\Delta E/E = 10^{-4}$ the population up to n=40 can be measured, yielding information directly from the part of the cascade which is governed only by the Stark-effect.

Electronic $K\alpha$ X-rays of e.g. the antiprotonic noble gases [9,10] yield the information on the state of the electron shell during the first step of the antiprotonic cascade. A detailed investigation is hindered by the limited energy resolution of the semiconductor detectors used.

The measurement of e.g. the $\bar{p}O$(7-6) transition (E=11.1 keV), where no electrons are present and nuclear size effects are still negligible, is a crucial test of the calculations providing the electromagnetic binding energies [17,29].

To summarize, with a low-energy crystal spectrometer a variety of open questions concerning the hadronic interaction, the capture process, and the atomic cascade can be attacked. The use of the high intensity beam at LEAR ($\geq 5\cdot 10^{5}\bar{p}/s$) in combination with stopping the antiprotons at low gas pressure in the cyclotron trap allows to achieve sufficient counting rates (~ 500/h).

References

1. C.A.Baker et al., Nucl.Phys. **A483**(1988)631
2. C.W.E. Eijk et al., Nucl.Phys. **A488**(1986)604
3. M.Ziegler et al., Phys.Lett. **206B**(1988)151
4. J.D.Davies et al., Phys.Lett. **145B**(1984)319
5. C.A.Baker et al., Nucl.Phys. **A494**(1989)507
6 R.Bacheretal., Proc. Antiproton 86 (Thessaloniki) eds. S.Charalambous, C.Papastefanou and P.Pavlopoulos, (World Scientific, Singapore 1987), p.223
7. R.Bacher et al., to be published in Zeit.f.Phys.A
8. K.Heitlinger et al., contribution to this workshop
9. R.Bacher et al., Phys.Rev. **A38**(1988)4395
10. J.J.Reidy et al., contribution to this workshop
11. J.M.Richard and M.E.Sainio, Phys.Lett. **110B**(1982)349
12. A.M.Green and S.Wycech, Nucl.Phys. **A377**(1982)441
13. O.Dumbrajs et al., Nucl.Phys. **A457**(1986)491
14. W.B.Kaufmann and H.Pilkuhn, Phys.Lett. **166B**(1986)279
15. A.M.Green and J.M.Niskanen, Progree in Part. and Nucl.Physics, ed. A.Faessler, vol.**18**(1987),p.93, and ref.therein
16. S.Wycech, A.M.Green and J.A.Niskanen, Phys.Lett. **152B**(1985)308
17. S.Barmo, H.Pilkuhn and H.G.Schlaile, Zeit.f.Phys. **A301**(1981)283
18. M.Leon and H.Bethe, Phys.Rev. **127**(1962)636
19. E.Borie and M.Leon, Phys.Rev. **A421**1980)1460
20. G.Reifenröther, E.Klempt, and R.Landua, Phys.Lett. **B191**(1987)15
21. R.Bacher et al., Proc. of the IV. LEAR workshop, Villars-sur-Ollon, 1987 (Harwood Academic, Chur 1988),p.745
22. L.M.Simons et al., in preparation
23. I.L.Grach,B.O.Kerbikov,andYu.A.Simonov, Sov.J.Nucl.Phys. **48**(1988)609 and ref. 4,29 and 30 therein
24. R.Bizarri,A.E.Kudryavtsev, and V.E.Markushin, Nuovo Cim. **100A**(1988)183
25. I.S.Shapiro, Phys.Rev. **35**(1978)129
26. G.S.Mutchler et al.,Phys.Rev. **D38**(1988)742
27. F.Balestra et al., Nucl.Phys. **A465**(1987)714
28. J.F.Crawford et al., Phys.Lett. **B213**(1988)391
29. E.Borie and B.Jödicke, Comput.Phys.,vol.**2**,no.6(1988)61

CONCLUSION

(Instead of)

"Would you tell me, please, which way I ought to walk from here?"
"It depends a good deal on where you want to get to," said the Cat.
"I don't much care where - " said Alice.
"Then it doesn't matter which way you walk," said the Cat.
"- so long as I get *somewhere*," Alice added as an explanation.
"Oh, you're sure to do that," said the Cat, "if you only walk long enough."

Lewis Carroll: Alice in Wonderland

PARTICIPANTS

Peter BAUMANN
Physics Department, Techn. Univ. München, James-Franck-Str.
D-8046 GARCHING, FRG

Günther BORCHERT
KFA Jülich, D-5170 JÜLICH, FRG

Luciano BRACCI
Physics Department, Univ. Pisa, Piazza Torricelli
I-56100 PISA, Italy

James S. COHEN
Theoretical Div., MS J569, Los Alamos National Lab.
LOS ALAMOS, NM 87545, USA

John D. DAVIES
Rutherford Appleton Lab., Chilton,
DIDCOT, OX 11 OQX, UK

Konrad ELSENER
CERN, EP Division,
CH-1211 GENEVA 23, Switzerland

Reinhart FROSCH
Paul-Scherrer-Institut
CH-5232 VILLIGEN, Switzerland

Semyon S. GERSHTEIN
Inst. High Energy Physics
SU-122284 PROTVINO, USSR

Detler GOTTA
KFA Jülich
D-5170 JÜLICH, FRG

F. Joachim HARTMANN
Physics Dept., E18, Techn. Univ. München, James-Franck-Str.
D-8046 GARCHING, FRG

Peter HAUSER
High Energy Physics, ETH Hoenggerberg
CH-8093 ZÜRICH, Switzerland

Ryugo S. HAYANO
Dept. of Physics, Univ. of Tokyo
7-3-1 Hongo, Bunkyo-ku, TOKYO 113, Japan

Klaus Heitlinger
Paul-Scherrer-Institut
CH-5232 VILLIGEN, Switzerland

Dezsö HORVATH
Central Research Institute for Physics
H-1525 BUDAPEST, P.O.Box 49, Hungary

Roland JACOT-GUILLARMOD
Institut de Physique de l'Université
CH-1700 FRIBOURG, Switzerland

J. KONIJN
NIKHEF-K
P.O.Box 4395, NL-1009 AJ AMSTERDAM, The Netherlands

Franz KOTTMANN
High Energy Physics, ETH Hoenggerberg
CH-8093 ZURICH, Switzerland

Kegang LOU
Paul-Scherrer-Institut
CH-5232 VILLIGEN, Switzerland

Valery E. MARKUSHIN
Dept. of Physics, Kurchatov Atomic Energy Institute
Pl. Kurchatova, SU-123182 MOSCOW, USSR

David F. MEASDAY
Department of Physics, Univ. of British Columbia
VANCOUVER, V6T 2A6 Canada

Vladimir S. MELEZHIK
JINR, Dubna
SU-10100 MOSCOW, P.O.Box 79, USSR

Francoise MULHAUSER
Institut de Physique de l'Université
CH-1700 FRIBOURG, Switzerland

Herbert ORTH
GSI
Pf. 11 05 52, D-6100 DARMSTADT, FRG

Paolo PASINI
INFN - Univ. di Bologna
Via Irnerio, 46, I-40136 BOLOGNA, Italy

Claude PETITJEAN
Paul-Scherrer-Institut
CH-5232 VILLIGEN, Switzerland

Rosa POGGIANI
INFN - Sezione di Pisa
Via Livornese, 582/A, I-56100 S. PIERO A GRADO, Italy

Leonid I. PONOMAREV
Dept. of Physics, Kurchatov Atomic Energy Institute
Pl. Kurchatova, SU-123182 MOSCOW, USSR

James J. REIDY
Dept. of Physics, Univ. of Mississippi
UNIVERSITY, MS 38677, USA

Günther REIFENRÖTHER
Institute for Physics of University of Mainz
D-6500 MAINZ, FRG

A. J. RUSI EL HASSANI
Inst. Intermed. Energy Phys., ETH Zürich - PSI
CH-5232 VILLIGEN, Switzerland

Wolfgang SCHOTT
Physics Dept., E18, Techn. Univ. München
James-Franck-Str., D-8046 GARCHING, FRG

Hubert SCHNEUWLY
Institut de Physique de l'Université
CH-1700 FRIBOURG, Switzerland

Marianne SIGNER
Paul-Scherrer-Institut
CH-5232 VILLIGEN, Switzerland

Leopold M. SIMONS
Paul-Scherrer-Institut
CH-5232 VILLIGEN, Switzerland

Gabriele TORELLI
INFN - Sezione di Pisa
Via Livornese, 582/A, I-56100 S. PIERO A GRADO, Italy

Johann ZMESKAL
Inst. Medium Energy Phys., Austrian Academy of Science
Boltzmanngasse 3, A-1090 VIENNA, Austria

INDEX

Adiabatic invariants, 197
Adiabatic ionization, 2-5
Adiabatic representation, 254
Antiproton - atom collisions, 47-51
Antiprotonic deuterium, 81
Antiprotonic hydrogen
  Balmer transitions, 120
  external Auger deexcitation, 100
  cascade, 81, 97-102
  chemical effects, 100
  collisions, 98
  formation
    in corotating beams, 21
    on hydrogen atom, 17-18
    on hydrogen molecule, 17-18
    on H^- ion, 19-22
  Lyman transitions, 121
  nuclear capture in, 100
  Stark mixing, 119
Antiprotonic X-rays, 295-300
Antiprotons
  in noble gases, 47, 134-137
  transfer of, 65
Atomic capture
  l-distribution after, 28, 34
  models of, 26-32
  n-distribution after, 34
  of antiprotons
    in hydrogen, 17-22
    in negative hydrogen ions, 19-22
  of negative muons
    in hydrogen, 1-16
    in hydrogen molecules, 18
  of negative pions, 56
  ratios, 34-37, 207
  semiclassical calculations of, 28-30
  semiempirical calculations of, 31-32
  solid state effects in, 36
Auger capture, 33
Auger deexcitation, 77, 89, 100, 128-133, 144

Barkas effect, 24, 48
Black and white muonic atoms, 213
Born approximation, 27

Cascade processes
  density effects in, 101, 105-109, 299
  in antiprotonic helium, 297
  in antiprotonic hydrogen, 97-102, 117-124, 297-299
  in hydrogen isotopes, 73-124, 97-103, 117-124
  in heavy atoms, 127-139
  in metals, 137
  in pionic hydrogen, 105-109
  l-distribution in, 28, 31, 34, 129, 201
  n-distribution in, 34, 128, 201
  standard model of, 74-82
Cascade time, 143, 147-149
Cascade trapping, 142-149
Chemical effects with
  pionic hydrogen, 58-63, 68-69
  antiprotonic hydrogen, 100
Classical-quantal coupling, 14-16
Classical trajectory Monte Carlo method, 10-12
Cold fusion, 269-277, 241-243
Collisional deexcitation, 69, 75-78
Collisions
  of antiprotons with atoms, 47-51,
  of light exotic atoms, 67-70
  of muonic atoms, 253-267
Condo trapping, 145
Coulomb capture *see* Atomic capture
Coulomb deexcitation, 89
Coulomb problem, 3-body, 254
Crystal spectrometer, 296
Cyclotron trap, 73, 106, 118

Day-Snow-Sucher mechanism, 80
Deeply-bound exotic atoms, 149-152, 167-176
Density effects, 87, 89, 101, 105-109, 136, 210
  in muon catalysed fusion, 238, 240
Diabatic states, 5-9
Doppler broadening of neutron lines, 85

Elastic scattering of muonic hydrogen, 256

Energy loss, 23-25, 48
for antiprotons, 50
for mesonic particles, 128
for muons, 25, 42-44

Fusion
see Cold fusion
see Muon catalysed fusion

Hadronic atoms, nuclear capture in, 80-82
Helium, mesons in, 142-149
muonic, metastable 2s state, 155-160, 163-166
Hyperfine levels, 247
Hypernuclei, 141

Internal transfer, 69
ISIS facility, 280

Kaon capture
in helium, 145-149
in scintillators, 64

Large mesic molecules, 54, 67

Mesic helium atoms, 142-149
Mesic molecules
and chemical effects, 54, 68-69
resonant formation, 233
Meson transfer
and mesic X-rays, 63-64
between hydrogen isotopes, 56-61
chemical structure and, 60-62
density effects in, 58
in liquids, 59-64
internal, 67
Metals, cascade in, 137
Muon bottle, 73
Muon catalysed fusion, 231-251
reaction channels, 232, 270
dd fusion, 233
density effects in, 238, 240
dt fusion, 237, 287
muon sticking in, 239
pd fusion, 234-237
pt fusion, 245-251
rates of, 232
temperature dependence of, 233-235, 237
Muonic atoms
in noble gases, 136-137
collisions of, 253-267
Muonic deuterium
diffusion, 82-84
Muonic helium
metastable 2s state, 155-160, 163-166
clusters, 164-166
Muonic hydrogen
acceleration of, 91
cascade, 78
and molecular effects, 91
deceleration of, 88
diffusion, 82-84, 92
Lamb shift, 280
metastable 2s state of, 84
scattering of, 256
Muonic molecules, 205
Muonium, 280, 287
Muon lifetime
in gas mixtures, 205-214, 217-220
with pulsed beam, 280
Muons in helium
residual polarization, 166
Muon spin resonance, 284
Muon sticking, 239
Muon transfer
and mesic X-rays, 207-214, 217-220
and muon lifetime, 207-214, 217-220
between hydrogen isotopes, 87
chemical structure and, 209
density effects, 210
from excited states, 208
from ground state, 209
to heavy atoms, 206
to helium, 210
to light atoms, 223-227
to neon, 226
to nitrogen, 225
to oxygen, 211, 218-220
to sulphur, 208, 211, 218-220

Neutral pion detection, 54, 86, 106-107, 111
Nuclear fusion
see Cold fusion
see Muon catalysed fusion

Optical potential, 150, 176, 179-195

Pion capture
and mesic X-rays, 63-64
chemical structure and, 60-62, 68-69
from excited states,
in alcohols, 60-62
in hydrogen, 54-55
in liquids, 59-64
in scintillators, 64
on nucleon clusters, 182
on single nucleons, 181
Pionic aluminium, 169
Pionic atoms
deeply bound, 167-176
energy levels, 184
strong interaction effects on, 189-19

Pionic hydrogen, 53-64, 67-69, 85-86
  cascade, density effects, 105-109
  kinetic energy of, 90, 113
  radiative capture in, 115
Pionic lead, 175
Pionic magnesium, 169
Pionic silicon, 172
Pionic X-rays, 62, 169 176
Pionium, 289-293
Pion transfer
  and mesic X-rays, 63-64
  between hydrogen isotopes, 56-61, 68
  chemical structure and, 60-62
  density effects in, 58
  in liquids, 59-64
Pressure effects *see* Density effects
Protonium *see* Antiprotonic hydrogen
Pulsed muon beam, 282-286

Runge-Lentz vector, 200

Sigma-minus atoms, 149
Silicon, muons in, 45
Spallation neutron source, 280
Spectral flux density, 26, 32, 41
Stark mixing, 77, 80, 89, 99, 119, 144
Stopping power, 23-25, 48
  for antiprotons, 50
  for muons, 25, 42-44
Surface analysis with muons, 44-45
Surface muons, 282

Time-dependent Hatree-Fock method, 12-14, 27

MIX
Papier aus verantwortungsvollen Quellen
Paper from responsible sources
FSC® C105338
www.fsc.org

If you have any concerns about our products,
you can contact us on
ProductSafety@springernature.com

In case Publisher is established outside the EU,
the EU authorized representative is:
Springer Nature Customer Service Center GmbH
Europaplatz 3, 69115 Heidelberg, Germany

Printed by Libri Plureos GmbH
in Hamburg, Germany